John Domijan

Y0-ALK-384

ELECTROCHEMISTRY
AT SOLID ELECTRODES

MONOGRAPHS IN ELECTROANALYTICAL CHEMISTRY AND ELECTROCHEMISTRY

CONSULTING EDITOR:
Allen J. Bard

ELECTROCHEMISTRY AT SOLID ELECTRODES
by Ralph N. Adams

OTHER VOLUMES IN PREPARATION

ELECTROCHEMISTRY AT SOLID ELECTRODES

Ralph N. Adams
DEPARTMENT OF CHEMISTRY
UNIVERSITY OF KANSAS
LAWRENCE, KANSAS

1969

MARCEL DEKKER, INC., New York

COPYRIGHT © 1969 by MARCEL DEKKER, INC.

ALL RIGHTS RESERVED

No part of this work may be reproduced or utilized in any form or by any means, electronic or mechanical, including photocopying, microfilm, and recording, or by any information storage and retrieval system, without permission in writing from the publisher.

MARCEL DEKKER, INC.
95 Madison Avenue, New York, New York 10016

LIBRARY OF CONGRESS CATALOG CARD NUMBER 68-28980

PRINTED IN THE UNITED STATES OF AMERICA

For Kotty,
and all my girls

PREFACE

The original outline for this monograph was jotted down during a research discussion with the late Professor N. Howell Furman in the spring of 1953. Professor Furman's encouragement of my enthusiasm for writing such a book can be attributed to his remarkable kindness and patience with his students. As one of the most eminent electroanalytical chemists of any time, he surely knew I was hardly prepared for the task. Sixteen years later, I hope the final product will prove to be somewhat worthy of his confidence.

This material comes into print at a time when it may be more useful than originally envisaged. The past few years have seen an upsurge of interest in solid electrodes. This was generated in part by fuel cell and battery systems studies and recent commercial successes in preparative electrochemistry. Also, quantitative techniques at solid electrodes have become fashionable and are even used for routine analytical work. We "solid electrode types" have come a long way in 15 years.

The specific aim of this monograph is to provide a thorough and modern discussion of the utility and applications of solid electrode systems. The approach is predominantly experimental. Indeed, the theory is examined in terms of how, and with what confidence, the pertinent equations can be realized at real solid electrode surfaces. This experimental approach reflects the style of work in the author's laboratory over the years.

It must be emphasized strongly that this experimental success has depended on the excellent theoretical progress which began in the 1950s. It is hoped that the relevant chapters adequately credit these developments but a few examples may be noted here. Perhaps the most important single impetus to this area was given by Delahay via his monograph in 1954. There followed a series of contributions from various European, U.S.A., and U.S.S.R. laboratories. Particularly pertinent to stationary electrode systems was the work of Reinmuth and Shain. The several contributions of Nicholson and Shain made single sweep and cyclic voltammetry commonplace in the study of complex organic electrode reactions. Similarly, Levich's book, followed by studies of the Frumkin school, Riddiford, and Albery and Bruckenstein paved the way for the present utility of rotated disk electrodes.

Two main topics lie largely outside the scope of this book. First, no chapter on inorganic electrode reactions has been included. Also, more important to the concept of the book, there is no organized treatment of adsorption phenomena. Critics may well ask, "How in the world can one write a book on *solid electrodes* and not adequately treat *adsorption*?" I can only answer in the modern vernacular by saying "adsorption is just not my thing and I've told the story like it is." Unlike some, I do not feel that adsorption is *the* single, most important part of *every* electrode reaction. I do not believe that very much of what appears in this book about overall electrode processes is incorrect because adsorption is largly bypassed. When adsorption processes are better understood, and particularly when their role in the overall electrode process can be assessed better, then these discussions can be properly modified. In the meantime, I believe the present level of understanding provides useful insights.

This book would not have been possible without the hard work of my students and co-workers, both graduate and undergraduate, and postdoctoral associates. Any success the book enjoys is due entirely to their efforts. I am pleased that each of them, in his own way, has contributed to the present status of electrochemistry at solid electrodes. I am indebted to them for their loyalty and perseverance, especially during difficult times in our laboratory. My old friend and co-worker, Dr. J. K. Lee, has played a particularly important role over the years by advising and directing the research activities of my laboratory.

I would also like to thank Professor Jacob Kleinberg, Chairman, Department of Chemistry, University of Kansas, and Professors Ray Q. Brewster, Calvin Vanderwerf, and Arthur Davidson, former chairmen, who provided many opportunities for me to work on the manuscript. I wish to thank Professor Allen Bard for helpful criticism of the manuscript and Professor Fred Anson, Dr. N. L. Weinberg, and Dr. Lennert Eberson for advice on some of the chapters.

Three delightful ladies helped through the preparation of this manuscript. I am particularly grateful to Alberta Rogers who prepared most of the early drafts and Judy Becker and Sharon Teamenson who completed the task.

Another lovely lady persevered in this endeavor. Without the help and encouragement of my wife, Gini, the writing of this book would not have been accomplished. Perhaps my greatest indebtedness is to my parents, whose unselfish efforts made possible the entire process.

R. N. A.

Lawrence, Kansas
August 1968

CONTENTS

Preface vii

PART 1 FUNDAMENTALS OF ELECTROCHEMICAL METHODS AT SOLID ELECTRODES 1

1. Introduction 3

1-1.	Historical Development of Voltammetry	3
1-2.	Current–Voltage Curves	5
1-3.	Electrode Systems	11
1-4.	Polarization and Polarized Electrodes	13
1-5.	Relation to Dropping Mercury Polarography	14
References		16

2. Scope and Limitations of Solid Electrodes 19

2-1.	Background Processes and Potential Limits	19
2-2.	Potential Ranges in Nonaqueous Media	29
2-3.	Residual Currents	36
2-4.	Cathodic Reactions	37
2-5.	Anodic Oxidations	38
2-6.	Sensitivity	38
2-7.	Half-Wave Potentials	40
References		41

3. Mass Transfer to Stationary Electrodes in Quiet Solutions 43

3-1.	Introduction	43
3-2.	Linear Diffusion to Plane Electrodes	45
3-3.	Cylindrical Diffusion	61
3-4.	Diffusion to Spherical Electrodes	62
3-5.	Convection in Unstirred Solution	63
References		64

xi

4. Mass Transfer by Forced Convection 67

4-1.	Nernst Diffusion Layer	67
4-2.	Hydrodynamics and Forced-Convection Electrodes	71
4-3.	Stationary Electrodes in Flowing Solution	76
4-4.	Rotated Disk Electrodes	80
4-5.	Applications of RDE to Electrode Kinetics and Mechanisms	92
4-6.	Mass Transport with Turbulent Flow	102
4-7.	Rotated Wire Electrodes	104
4-8.	Vibrating Wire Electrodes	107
References		107
Special Bibliography on Rotated Disk Electrodes		110

5. Current–Potential Curves 115

5-1.	Convective Mass Transport	115
5-2.	Quiet Solutions	118
5-3.	Rapid Voltage Sweep Methods at Stationary Electrodes	122
5-4.	Single-Sweep Peak Voltammetry	124
5-5.	Peak Polarograms of Systems with Coupled Chemical Reactions	139
5-6.	Electron Transfer with Follow-up Chemical Reactions	140
5-7.	Cyclic (Triangular Wave) Voltammetry	143
5-8.	Conclusions	159
References		160

6. Electrochemical Methods Employing Controlled Current 163

6-1.	Current Sweep Voltammetry	164
6-2.	Chronopotentiometry	165
6-3.	Chronopotentiometric Study of Electrode Processes: Methods Employing $i_0\tau^{1/2}$ Variation	172
6-4.	Chronopotentiometric Study of Electrode Processes: Application of Current Programs	177
6-5.	Practical Measurement of Transition Times	183
References		184

7. Electrode Surface Conditions 187

7-1.	Introduction	187
7-2.	Adsorbed Hydrogen Films on Platinum and Gold	189
7-3.	Oxidation of Platinum and Gold Electrodes	191
7-4.	Effect of Electrode History	205
7-5.	Operating Procedures for Platinum Electrodes	206
References		208

CONTENTS xiii

PART 2 EXPERIMENTAL AND APPLICATIONS 211

8. Investigation of Electrode Processes 213

8-1. Introduction 213
8-2. Evaluation of Diffusion Coefficients 214
8-3. Correlation of Electroanalytical Techniques: Electrode Reactions Without Chemical Complications 231
8-4. Determination of Heterogeneous Rate Constants 240
8-5. Electrode Processes with Coupled Homogeneous Chemical Reactions 244
8-6. Physicochemical Methods for Studying Electrode Reactions 255
References 262

9. Fabrication of Electrode Systems 267

9-1. Electrolytic Cells 267
9-2. Working Electrodes 270
9-3. Reference Half-Cells 288
9-4. Instrumentation 291
References 300

10. Applications to Organic Compounds 303

10-1. Basic Patterns for Anodic Oxidation of Aromatic Compounds 305
10-2. Aromatic Hydrocarbons 308
10-3. Primary Aromatic Amines 327
10-4. Secondary Aromatic Amines 345
10-5. Tertiary Aromatic Amines 351
10-6. Aromatic Diamines 356
10-7. Aromatic Hydroxy Compounds 363
10-8. Sulfur Compounds 369
10-9. Miscellaneous Aromatic and Heterocyclic Systems 370
10-10. Aliphatic Hydrocarbons 372
10-11. Aliphatic Acids 372
10-12. Aliphatic Alcohols and Aldehydes 375
10-13. Aliphatic Amines and Amides 375
10-14. Aliphatic Halides 377
10-15. Reduction Processes 377
References 378

Author Index 385

Subject Index 399

PART 1
FUNDAMENTALS OF ELECTROCHEMICAL METHODS AT SOLID ELECTRODES

1 INTRODUCTION

1-1. Historical Development of Voltammetry 3
1-2. Current-Voltage Curves 5
1-3. Electrode Systems 11
1-4. Polarization And Polarized Electrodes 13
1-5. Relation To Dropping Mercury Polarography 14
References 16

1-1. HISTORICAL DEVELOPMENT OF VOLTAMMETRY

Since electrolysis with the dropping mercury electrode was not initiated until the late 1920s, initial voltammetric studies began with solid electrodes. The origins may be traced to the classic research of LeBlanc (*1*). Following an investigation of the decomposition voltages of acid and base solutions, LeBlanc studied the electrolysis of metal ions. The results have little in common with modern measurements in this field but did serve as an impetus for further study. Working with Nernst at Göttingen in 1897, Salomen (*2*) obtained current–voltage curves for the electrolysis of silver ion and indicated the existence of limiting currents. Caspari (*3*) examined the anodic oxidation of bromide and iodide at platinum electrodes.

Technical interest in the electrolytic reduction of organic compounds was already widespread in Germany at this time. Haber (*4*) and Russ (*5*) examined the electrochemical properties of gold, platinum, and other less noble metal electrodes with respect to organic reductions and oxidations. The effects of electrode history and pretreatment techniques were noted qualitatively by these workers. It is interesting that some of the problems developed in these early studies are still under active investigation.

Theoretical studies of rotated electrodes and stirred solutions developed with the work of Nernst and Merriam (*6*) and Brunner (*7*), among others.

Fundamental contributions to diffusion processes were made by Weber (*8*), Sand (*9*), and Karaoglanoff (*10*).

More attention was given to limiting currents in the 1920–1930 period by Wilson and Youtz (*11*), Glasstone (*12*), and Glasstone and Reynolds (*13*). The first comprehensive studies of diffusion conditions at solid electrodes were made by Laitinen and Kolthoff (*14,15*). Zlotowski (*16*) and Rogers et al. (*17*) investigated automatic recording of solid electrode polarograms.

Considerable activity by Skobets and co-workers (*18,19*) in Russia developed in the late 1940s and has continued. Delimarskii's group has been especially active in fused salt voltammetry, with some 40 publications in this area since 1946 (*20,21*). The Frumkin school has predominated over a long period in the measurement of adsorption and capacity effects at solid electrodes (*22*).

The growth from 1950 on of publications dealing with solid electrode voltammetry parallels the rapid increase of the general polarographic literature as cited by Kolthoff and Lingane, (*23*), although the total number of papers is, of course, smaller. Some of the increased activity can be traced to an appreciation of modern electroanalytical techniques which employ solid electrodes. A rise in the applications involving limiting current measurements for routine analytical work is evident.

Solid electrode techniques have played an increasing part in the recent interest of the chemical industry in organic oxidation–reduction processes. Isolated but substantial applications of solid electrodes appear in studies of organic semiconductors, the rates and mechanisms of photochemical and radical ion reactions, electrochemiluminescence, and other areas. Measurements at solid electrodes have become increasingly reliable in the last few years. Significant advances have been made in the interpretation of complex electrode reactions. Information from solid electrode determinations has reached a level of acceptance and is, in fact, commonly used in theoretical studies. For example, correlations of potentials with molecular orbital calculations appear regularly.

The extensive fuel-cell programs which have developed in the last few years have a strong emphasis on catalytic electrode surfaces and the technology of the cell design. However, this area has contributed to solid electrode methodology, especially with regard to electrode surface interactions.

There is every reason to believe that the current interest in solid electrode voltammetry will continue and perhaps increase in the near future. This monograph was begun about 11 years ago and it now appears fortunate

1-2. CURRENT–VOLTAGE CURVES

A. Relation to Classical Potentiometry

Voltammetry may be defined generally as the measurement of current–voltage relationships at an electrode immersed in a solution containing electroactive species. More specifically, it is the determination of the potential of a single electrode during the course of a sustained electron-transfer reaction at the electrode surface, i.e., while a net current flows through the electrochemical cell. At first glance, this process appears considerably different from the more familiar classical potentiometry. Actually, for purposes of unity in definition, classical potentiometry may be considered a special case of the more general technique of voltammetry. Both methods may be examined with almost identical experimental equipment.

Consider a 1-cm² platinum-foil electrode immersed in a well-stirred solution which is ca. 0.1 M in Fe(II) and Fe(III) and 1 M with respect to sulfuric acid. A conventional saturated calomel half-cell (SCE) completes the electrode assembly. Figure 1-1 shows the circuit which may be used to measure the potential of the platinum electrode under classical potentiometric conditions. The voltage divider, comprised of the battery B and the variable resistance R, is adjusted until the galvanometer G shows no deflection upon momentarily closing tapping key K. The voltage between points 0 and X on the voltage divider is then equal to the cell voltage. The value is read on voltmeter V. The accuracy of the measurement is limited by the voltmeter. Provided the tapping key is not closed at off-balance conditions for too long an interval, the measurement is made at essentially zero current flow.

In the present case, the potential of the platinum electrode is found to be +0.44 V vs. the SCE.* This is the formal redox potential of the Fe(III)/Fe(II) system in 1 M sulfuric acid referred to SCE. The Nernst equation

---

* The convention adopted in this book is the analytical or European system. For a cogent discussion of the significance of potential conventions and their utility, the reader is referred to the treatments of P. Delahay, *New Instrumental Methods in Electrochemistry*, Wiley (Interscience), New York, 1954 (p. 28), and W. Rieman, J. D. Neuss, and B. Naimen, *Quantitative Analysis*, McGraw-Hill, New York, 3rd ed., 1951.

expresses the relation between the potential of the platinum electrode and the concentrations of Fe(III) and Fe(II) in the *bulk* of the solution:

$$E_{Pt} = E^{\circ\prime} + 0.059 \log \frac{[Fe(III)]}{[Fe(II)]} \tag{1-1}$$

where $E^{\circ\prime}$ is the formal potential of the Fe(III)/Fe(II) system in 1 M sulfuric acid. If, by any means, the bulk concentrations are altered—for

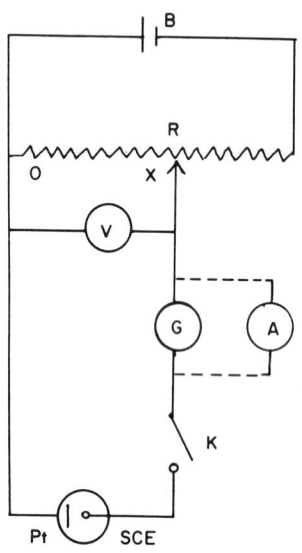

Fig. 1-1. Simple circuit for potentiometry and voltammetry.

example, by oxidative or reductive titration—then Eq. (1-1) indicates that E_{Pt} must vary. This is the basis of classical potentiometric titrations

A current–voltage curve of this system may be determined by simply locking the tapping key in the closed position. Since fairly large currents will flow, a milliammeter is now substituted for the galvanometer. If the voltage divider is adjusted in an incremental manner to values less positive than +0.44 V, considerable current will flow through the cell. The applied voltage can then be varied to values more positive than +0.44 V, and current flow in the opposite sense is obtained. Figure 1-2 is an experimental plot for the solution under discussion. These data were obtained using the apparatus of Fig. 1-1. A 6-V storage cell powdered the divider and the current was measured with a milliammeter. The platinum electrode was a 1-cm² foil, sealed in soft glass. The SCE was of conventional half-cell design.

Figure 1-2 is a respectable current–voltage curve. From it, one finds that the potential at zero current is +0.44 V vs. SCE, a fact established

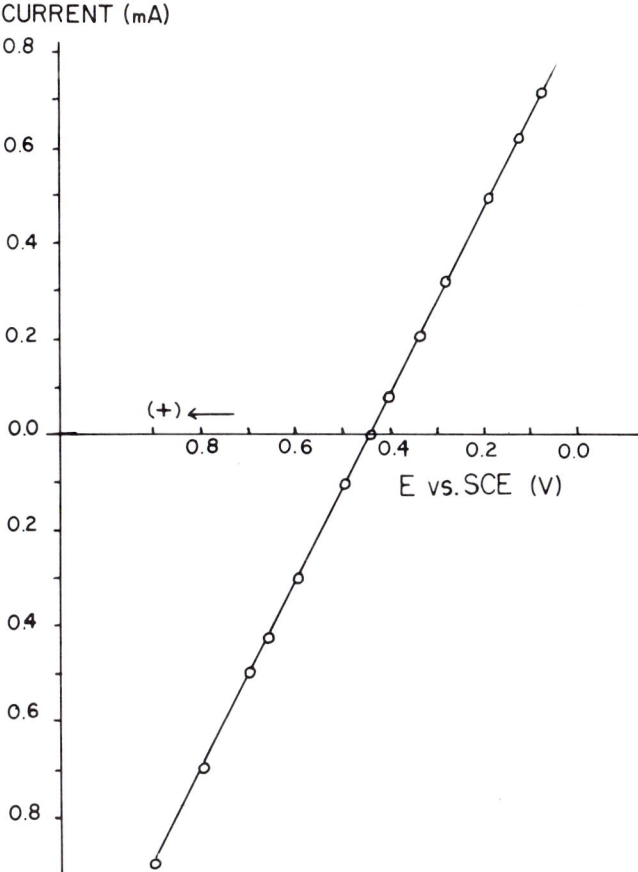

Fig. 1-2. Current–voltage curve for an Fe(III)/Fe(II) system.

by the previous potentiometric measurement. It is apparent that the current flow is an approximately linear function of the applied voltage. In other words, the electrolytic cell could be replaced by an ordinary resistor. The value of this resistance is easily calculated from the slope of the current–voltage curve to be 510 Ω. An independent measurement of the solution resistance (ac bridge) gives a resistance of 480 Ω. Voltammetric measurements would appear to give fairly accurate values of the solution resistance or conductance. This is not surprising since, in electrical measurements, a common means of measuring an unknown linear resistive element is to pass a known current through the resistor and determine the resulting voltage drop.

The merits of a linear array of experimental data is well documented in the chemical literature. Nevertheless, very little useful *electrochemical information* is available in Fig. 1-2. The situation is remedied by examining a more dilute solution of Fe(III)/Fe(II).

Let the solution now be ca. 10^{-4} M in both Fe(III) and Fe(II) in the same 1 M sulfuric acid. An 18-gage platinum wire, 2 cm long, rotated at 600 rpm, is substituted for the foil. (The solution itself is not stirred in this case.) Otherwise the equipment is identical with that of Fig. 1-1. This time an entirely different plot of current vs. applied voltage is seen in Fig. 1-3. The zero current potential is still about 0.44 V, and voltages slightly displaced from this formal potential result in rapid increases in current. However, the current is no longer linear with applied voltage; instead it soon reaches a limiting value which is independent of voltage over a considerable range.

The explanation of this curve is straightforward. At an applied potential several tenths of a volt less positive than +0.44 V, Eq. (1-1) is not satisfied unless the ratio [Fe(III)]/[Fe(II)] is considerably less than unity. The only conceivable way in which this ratio can be altered is for electrons to be transferred to the Fe(III) species. It is not necessary to change the entire bulk concentration. We may write Eq. (1-1) as

$$E_{Pt} = E^{\circ\prime} + 0.059 \log \frac{[\text{Fe(III)}]_{x=0}}{[\text{Fe(II)}]_{x=0}} \qquad (1\text{-}2)$$

where the subscript $x = 0$ specifies the concentrations* at the electrode surface as opposed to the bulk concentration defined at $x = \infty$. Electron transfer merely alters the concentration ratio *at the electrode surface* until Eq. (1-2) is satisfied for the particular value of E_{Pt}.

This same transfer of electrons constitutes a flow of current in the ammeter circuit. At any other E_{Pt} less than +0.44 V, current will continue to flow, the magnitude depending upon the ratio of [Fe(III)]/[Fe(II)] which must be maintained *at the electrode surface* to satisfy

---

* The precision of voltammetric measurements at solid electrodes rarely justifies the use of activities. In this and all subsequent discussions, concentrations of electroactive species will be used rather than activities.

CURRENT–VOLTAGE CURVES

Eq. (1-2). The platinum electrode is functioning as a *working cathode*. On the other hand, at values of E_{Pt} more positive than 0.44 V, current flows in the opposite sense. Now the ratio of ferric/ferrous at the electrode surface must be greater than unity and electrons are abstracted from the Fe(II) species. The platinum electrode is now an anode.

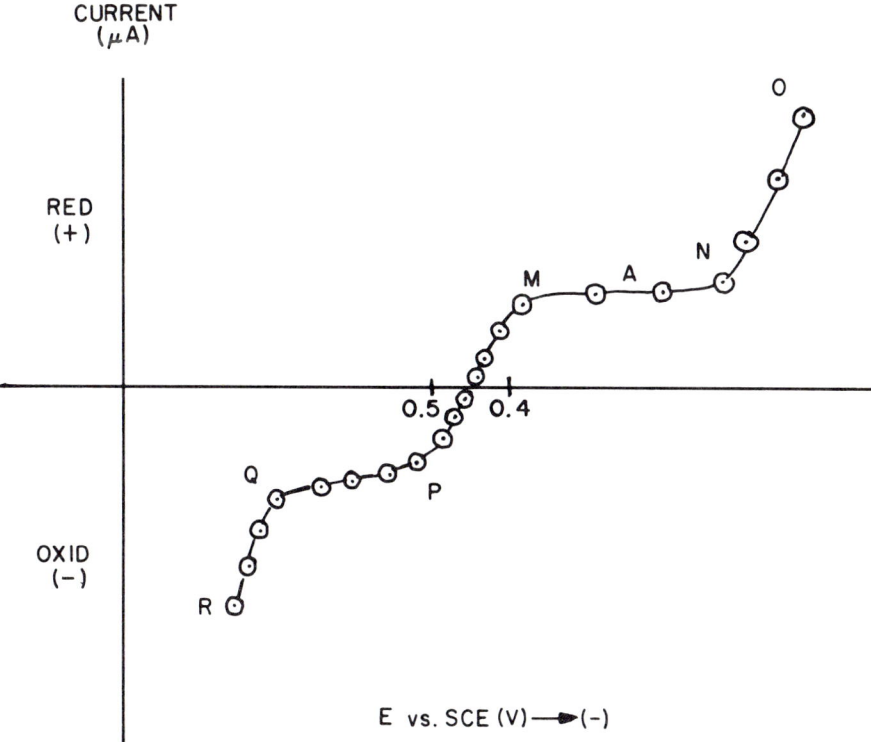

Fig. 1-3. Current–voltage curve of a dilute Fe(III)/Fe(II) system.

At voltages sufficiently removed in either direction from the $E^{o\prime}$ value, the current reaches a limiting value. Under these conditions ferric or ferrous ions are being reduced or oxidized, respectively, as rapidly as they reach the electrode, by whatever mass-transfer process is operative. More specifically, at point A in Fig. 1-3, where E_{Pt} is considerably more negative than $E^{o\prime}$, Eq. (1-2) indicates that

$$\frac{[\text{Fe(III)}]_{x=0}}{[\text{Fe(II)}]_{x=0}} \ll 1 \tag{1-3}$$

It is evident from Eq. (1-3) that the concentration of ferric at the electrode surface must decrease to a level negligibly small with respect to the bulk concentration. That is,

$$[Fe(III)]_{x=0} \ll [Fe(III)]_{x=\infty} \qquad (1\text{-}4)$$

or

$$[Fe(III)]_{x=0} \to 0 \qquad (1\text{-}5)$$

Thus the ferric concentration at the electrode surface is approximately zero. As a result of the concentration gradient established, ferric ions diffuse toward the electrode. If diffusion is the only mode of mass transfer, the current is limited by the rate of diffusion of ferric to the electrode surface. The current at the plateau regions (*MN* and *PQ*) in Fig. 1-3 is often called the *diffusion current*. Diffusion is seldom the only mass-transfer process at a practical solid electrode. Hence the term *limiting current*, i_L, is to be preferred.

There remain to be explained the extreme portions of Fig. 1-3 (*ON* and *QR*). At *O*, the E_{Pt} is such that the supply of electrons exceeds the rate at which ferric can reach the electrode for reduction (for this particular bulk concentration of ferric). Hence the next most readily reducible species undergoes reaction—in this case, hydrogen ions. The current is now split between two electrode processes. Since a generous supply of hydrogen ions is available in the 1 *M* sulfuric acid *background electrolyte*, this second electrode process has no observable limiting current. (The branch *QR* is due to oxygen evolution. Had the background electrolyte been 1 *M* hydrochloric acid, the oxidation of chloride ion would have been the second oxidation process.)

The curve of Fig. 1-3 is known as a *current–voltage curve*, a *current–potential curve*, a *polarogram* (polarographic wave), or a *polarization curve*. Current–voltage curve, polarogram, or polarographic wave are all acceptable. The first is to be preferred since it most aptly describes the experimental measurement. We will reserve current–potential curve to describe a more refined technique discussed later. Since the electrochemical system not only undergoes several stages of concentration polarization but also has associated electrochemical reaction polarization, the term polarization curve is a catch-all expression which has little to recommend its continued usage in polarography. It properly belongs to the area of electrode kinetics, and even there could be replaced by a more definitive term.

B. Conventions of Voltammetry

Figure 1-3 illustrates the preferred plotting of current–voltage curves inspired by standard polarographic practice. Applied voltages or potentials are plotted in an increasing positive sense from right to left. Cathodic current corresponding to a reduction process is plotted upward and is positive by definition. Anodic reactions are then always below the zero current line.

As in classical polarography, the potential at $i = 0.5i_L$ is the half-wave potential, $E_{1/2}$. A discussion of the methods of evaluating i_L and $E_{1/2}$ from experimental polarograms is reserved for a later section.

The SCE is almost universally used as a reference electrode, and voltages are plotted directly vs. SCE. Since most tabulations of half-wave potentials are referred to SCE, there is rarely any need to convert to the hydrogen electrode scale.

Currents are normally expressed in microamperes (μA) but in some cases may extend into the milliampere (mA) region. The common idea that limiting currents can be obtained only with microelectrodes or very low concentrations is not justified. In most cases, however, it is seldom necessary or even desirable to operate with the concentrated solutions which give rise to milliampere-level currents.

It should be noted that it is conventional practice to run polarograms by varying applied voltage and measuring the resulting current. Actually, early workers in solid electrode voltammetry controlled the current density and measured the resulting voltage changes at the working electrodes. This technique recently has been revived and is discussed more fully later.

1-3. ELECTRODE SYSTEMS

A. Working Electrodes

Strictly speaking, any electrode may be called a working electrode whenever a net current flows through the cell. Indeed, even potentiometric electrodes are "working" except in the rare instance when absolute null balance is obtained with the potentiometer. In voltammetry, the classification *working electrode* (WE) is to be reserved for the electrode at which the primary polarographic reaction occurs. The term will be used frequently to mean the platinum, gold, or other electrode involved.

Many times it is desirable to employ a three-electrode polarographic system. Here one speaks of an opposite working electrode, or *auxiliary electrode*. If, for example, an oxidation polarogram is being studied at a rotating platinum electrode (RPE), the auxiliary electrode (AUX) is the cathode of the polarographic cell. Depending on the requirements of the study, the AUX may be: (1) an inert electrode dipping directly in the test solution, (2) a shielded electrode (sintered glass partition, etc.), or (3) a conventional half-cell with salt bridge. The potential of the AUX is rarely of interest. It merely serves to complete the electrochemical cell. The potential of the WE is then measured with respect to a reference half-cell, the third electrode of this system.

B. Types of Working Electrodes

A great variety of solid electrodes have been employed over the years. While incomplete, the following general classifications can be made:

a. Stationary electrodes
   1. Quiet solution
   2. Stirred solution
   3. Flowing electrolyte
b. Rotated electrodes
   1. Conventional speed
   2. Very high rotation speed
c. Vibrating systems

This classification indicates nothing about the physical form or shape of the individual electrodes. Such descriptions are considered in detail later. The size and shape of the electrode becomes important when mass transfer processes are examined in Chapters 3 and 4.

One could also classify working electrodes according to the electrode material. Preparative electrochemists from the beginning have employed a wide variety of metals. These are selected on the basis of price, durability, and, probably most important, from the viewpoint of their effect on the overall product yield and purity. The preparative reactions in many cases bear little resemblance to the voltammetry discussed in this book. Consequently, such electrode materials are described only where they have been used in small-scale voltammetry. The great bulk of solid electrode polarographic work has been accomplished at platinum, gold, carbon, and amalgam electrodes.

C. Reference Electrodes

The most frequently used reference electrode (REF) is the SCE. For certain purposes the silver–silver chloride (Ag/AgCl) is useful. Mercury–mercurous sulfate and lead amalgam–lead sulfate half-cells have been

1-4. POLARIZATION AND POLARIZED ELECTRODES

Without fail, an electrode through which a finite current is passing has a potential different from its zero current or equilibrium value. This difference is ordinarily called overvoltage or *overpotential* and is given the symbol η. It is observed experimentally using a three-electrode system, measuring the potential of the WE with respect to a reference half-cell. It can be defined as the difference in potentials when the electrode is at equilibrium and when it is sustaining a net anodic or cathodic reaction:

$$\eta = (E)_{\text{reaction}} - (E)_{\text{equilibrium}} \tag{1-6}$$

Thus defined, η is the total overpotential η_T, which experimentally may be composed of three quantities (24): ohmic overpotential η_o, concentration overpotential η_c, and activation overpotential η_a. Thus

$$\eta_T = \eta_o + \eta_c + \eta_a \tag{1-7}$$

Ohmic overpotential or *iR* drop results from the attempt to measure the potential of any working electrode in a cell of finite resistance. Electrochemists would be delighted to eliminate *iR* loss. Refined measuring techniques reduce its magnitude to relatively small proportions, and corrections can be applied through Ohm's law calculations.

Concentration polarization or, better, overpotential, η_c, exists in principle at every working electrode. With sustained electrolysis, concentrations at the electrode surface differ from that in the bulk of the solution [Eq. (1-3)] and the electrode potential must differ from its value with no current flow. It is worthwhile to note that "concentration polarization becomes nonexistent provided that the calculated potential is based on concentrations or activities of the potential determining substances at the electrode surface rather than in the body of the solution (23)."

Concentration overpotential is the very basis of voltammetry. A current–voltage curve is nothing more than the measurement of concentration-overpotential effects at an electrode. The *polarizable* electrode used in voltammetry (i.e., the WE) is one at which it is possible to obtain extreme evidence of concentration overpotential. When this high degree of η_c develops, limiting currents are obtained.

In the strictest sense there are no *nonpolarizable* electrodes (25). However, η_c is quite small at a large electrode, provided the concentrations of electroactive material are very large (and, further, that only small currents are passed). Such an electrode is said to be relatively nonpolarizable. Clearly such electrodes are valuable as reference electrodes in voltammetry, since their potentials change but little with current flow. In the SCE the excess of solid mercurous chloride and potassium chloride together with the reservoir of mercury prevent any appreciable concentration overpotential in either the anodic or cathodic direction.

It should be observed that reference electrodes can be made by maintaining large concentrations of electroactive species at many working electrodes. A large platinum foil dipping into 1 M hydrochloric acid containing ca. 0.2 M ferric and ferrous ions would exhibit only slight changes in potential with small currents. If connected to the test solution through a suitable salt bridge, this system could function as a reference electrode. Such systems are not very practical, but it is worthwhile to visualize other than conventional reference electrodes. This type of reference electrode embodies the principles of redox buffers set forth by Furman and co-workers (26).

Unfortunately the terms polarization and polarized electrode are associated with the idea of "irreversibility" in the electrode process. In part this is due to definitions of η as

$$\eta_T = E_{\text{irreversible}} - E_{\text{reversible}} \qquad (1\text{-}8)$$

Actually only the quantity η_a has real significance with respect to the reversibility of the electrode process. The activation overpotential is associated with a slow or rate-determining electron-transfer process. As we have seen, η_o and η_c are a natural result of the current flow. A detailed discussion of η_a is beyond the scope of this book, but reversible and irreversible electrode reactions will be treated in more detail later.

1-5. RELATION TO DROPPING MERCURY POLAROGRAPHY

Fundamentally, a solid electrode polarogram is identical with its counterpart at the dropping mercury electrode. Both are measurements depending ultimately on concentration variations at the electrode surface during the electrolysis period. In practice, only the dropping electrode presents a constantly renewable surface. This factor alone introduces significant differences and makes comparisons of experimental polarograms difficult.

The surface conditions of the dropping mercury electrode (DME) are highly reproducible. Excluding for the moment those cases where reaction with the mercury occurs, it is usually possible to consider the soluble electroactive system without reference to the electrode proper. This is rarely ever the case with solid electrodes. Probably no truly inert solid electrode exists. Since the surface is not renewable during the period of electrolysis, changes in the electrode itself are reflected in the current–voltage curve of the electroactive system in solution. It is this situation which has caused distrust of solid electrode techniques in the past. Recent work shows that it is possible to separate electrode surface effects from the solution electrochemistry or, at least, to minimize these factors to the point where they no longer seriously interfere.

In several instances it is possible to make direct comparisons at the DME and solid electrodes. One might expect closest correlation with mercury-plated solid electrodes. Marple and Rogers (*27*) studied mercury-plated platinum wires, both rotated and stationary. Rotated electrodes were not dependable, but satisfactory results were obtained with stationary wires in both quiet and stirred solutions. For the reduction of cadmium and zinc results were directly comparable with the DME; i.e., half-wave potentials agreed within about 10 mV. In the case of cobalt, an insoluble film resulted from the reduction. This gave rise to erratic behavior at slow rates of voltage change. At higher scanning rates the waves were more reproducible but distinctly different from those obtained at the DME. This is a clear case of the complications which arise when the surface is not renewable with each drop.

These workers also compared organic reductions at the two electrodes. Conditions are more favorable for solid electrodes where depositions do not occur. Closest agreement between the stationary and dropping mercury electrodes occurred with slow voltage scan rates. Under such conditions the half-wave potentials for the reduction of the three isomeric dinitrobenzenes were quite close using either electrode. The stationary electrode, however, showed a third reduction wave which was absent at the DME. At stationary electrodes, products of the primary electrolysis reaction can accumulate in the interface area, giving rise to added complications. Since mass-transport conditions at the two electrodes are entirely different, there is no direct correlation of limiting currents.

Cooke (*28*), working with rotated silver amalgam electrodes, showed that for ion–ion reductions such as Co(III)–Co(II), the fundamental electrode process at the DME and rotating amalgam electrode were essentially the same. In the case of deposition of metal on the mercury, half-wave potentials were complicated.

Tsukamoto et al. (*29*) have shown that hydrodynamic derivations lead to a familiar form for current–voltage curves at a rotated platinum electrode:

$$E_{\rm RPE} = E^\circ - \frac{RT}{nF} \ln \frac{K_2}{K_1} + \frac{RT}{nF} \ln \frac{i}{i_L - i} \qquad (1\text{-}9)$$

with

$$E_{1/2} = E^\circ - \frac{RT}{nF} \ln \frac{K_2}{K_1} \qquad (1\text{-}10)$$

where K_1 and K_2 are constants depending on the mass-transfer characteristics of the particular system. The current–voltage curve is a typical S-shaped polarographic wave.

Delahay (*30*) has derived the current–voltage characteristics for stirred solutions. The results differ only slightly from those in unstirred media. The investigations of the above workers will be dealt with in detail in later sections.

Julian and Ruby (*31*) compared the oxidation waves of *p*-aminophenols and various N-alkyl phenylenediamines at a quiet platinum electrode and the DME. Although minor differences were obtained, the two electrodes gave $E_{1/2}$'s within a few millivolts. Such close agreement should not be expected in all cases. Compounds which give rise to "anodic depolarization waves" at the DME or result in film formation (mercaptans, etc.) can hardly be expected to have identical voltammetry at solid electrodes.

Studies of reductions at the RPE by Ferrett and Phillips (*32*) show close correlation of $E_{1/2}$'s with the DME for a variety of reductions.

In summary it can be said that there are no a priori reasons for suspecting that solid and dropping mercury electrodes will give identical current–voltage characteristics for the same electroactive system. Close correlation is infrequent, owing primarily to the physical differences in the two-electrode systems. In general, one expects entirely different polarograms for one or more of the following reasons: (1) Solid electrode surfaces are not comparable to the renewable DME, (2) specific interactions of the Hg metal with electroactive species, and (3) accumulation of electrode products at stationary electrodes may produce "extra" electrochemical information.

REFERENCES

1. M. LeBlanc, *Z. Physik. Chem.*, **12**, 333 (1893).
2. E. Salomen, *Z. Physik. Chem.*, **24**, 55 (1897); **25**, 365 (1898).
3. W. A. Caspari, *Z. Physik. Chem.*, **30**, 89 (1899).

REFERENCES

4. F. Haber, *Z. Physik. Chem.*, **32**, 193, 271 (1900).
5. R. Russ, *Z. Physik. Chem.*, **44**, 641 (1903).
6. W. Nernst and E. S. Merriam, *Z. Physik. Chem.*, **52**, 235 (1905).
7. E. Brunner, *Z. Physik. Chem.*, **56**, 321 (1906).
8. H. F. Weber, *Wied. Ann.*, **7**, 536 (1879).
9. H. J. S. Sand, *Phil. Mag.*, **1**, 45 (1901).
10. Z. Karaoglanoff, *Z. Elektrochem.*, **12**, 5 (1906).
11. R. E. Wilson and M. A. Youtz, *Ind. Eng. Chem.*, **15**, 603 (1923).
12. S. Glasstone, *Trans. Am. Electrochem. Soc.*, **59**, 277 (1931).
13. S. Glasstone and G. D. Reynolds, *Trans. Faraday Soc.*, **28**, 582 (1932); **29**, 399 (1933).
14. H. A. Laitinen and I. M. Kolthoff, *J. Am. Chem. Soc.*, **61**, 3344 (1939).
15. H. A. Laitinen and I. M. Kolthoff, *J. Phys. Chem.*, **45**, 1061, 1079 (1941).
16. I. Zlotowski, *Roczniki Chem.*, **14**, 640, 651, 666 (1934).
17. L. B. Rogers, H. H. Miller, R. B. Goodrich, and A. F. Stehney, *Anal. Chem.*, **21**, 777 (1949).
18. E. M. Skobets and S. A. Kacherova, *Zavodsk. Lab.*, **13**, 133 (1947).
19. N. N. Atamanenko and E. M. Skobets, *Ukr. Khim. Zh.*, **22**, 771 (1957).
20. Y. K. Delimarskii and E. M. Skobets, *Zh. Fiz. Khim.*, **20**, 1005 (1946).
21. Y. K. Delimarskii and A. V. Gorodyskii, *Zh. Fiz. Khim.*, **33**, 137 (1959).
22. A. N. Frumkin, *J. Electrochem. Soc.*, **107**, 461 (1960).
23. I. M. Kolthoff and J. J. Lingane, *Polarography*, Wiley (Interscience) New York, 2nd ed., 1952.
24. J. O'M. Bockris, in *Electrical Phenomena at Interfaces* (J. A. V. Butler, ed.), Methuen, London, 1951.
25. P. Delahay, *Instrumental Analysis*, Macmillan, New York, 1957, p. 93.
26. N. H. Furman, C. E. Bricker, and B. J. McDuffie, *J. Wash. Acad. Sci.*, **38**, 5 (1948).
27. T. L. Marple and L. B. Rogers, *Anal. Chem.*, **25**, 1351 (1953).
28. W. D. Cooke, *Anal. Chem.*, **25**, 215 (1953).
29. T. Tsukamoto, T. Kambara, and I. Tachi, *Proc. 1st Intern. Polarog. Congr., Prague, 1951*, pp. 524–541.
30. P. Delahay, *New Instrumental Methods in Electrochemistry*, Wiley (Interscience), New York, 1954, p. 221.
31. D. B. Julian and W. R. Ruby, *J. Am. Chem. Soc.*, **72**, 4719 (1950).
32. D. J. Ferrett and C. S. G. Phillips, *Trans. Faraday Soc.*, **51**, 390 (1955).

2 SCOPE AND LIMITATIONS OF SOLID ELECTRODES

2-1. Background Processes And Potential Limits 19
2-2. Potential Ranges In Nonaqueous Media 29
2-3. Residual Currents 36
2-4. Cathodic Reactions 37
2-5. Anodic Oxidations 38
2-6. Sensitivity 38
2-7. Half-Wave Potentials 40
References 41

2-1. BACKGROUND PROCESSES AND POTENTIAL LIMITS

The potential limits to which an electrode can be operated define the utility in terms of the polarographic systems which can be investigated. One normally speaks of the appearance of a *background wave* as being the limit of useful range in either the anodic or cathodic direction. Potentials for background waves cannot be precisely defined. Limiting currents are not obtained and, therefore, no $E_{1/2}$ can be given. Instead, "decomposition potentials" are frequently used. A practical method used by several investigators (*1,2*) is to define the potential limit or "cutoff," where the background electrolyte yields a current in excess of an arbitrary small value, i.e., 1 μA.

Reversible potential limits for aqueous solutions can be defined in terms of two possible background processes—reduction of hydrogen ion and oxidation of water—as

$$2H^+ + 2e \rightarrow H_2 \tag{2-1}$$

$$2H_2O \rightarrow O_2 + 4H^+ + 4e \tag{2-2}$$

The potentials of these reactions can be expressed by

$$E = E^\circ_{H_2/H^+} - 0.059 \, pH \tag{2-3}$$

and

$$E = 1.23 - 0.059 \, pH \tag{2-4}$$

where the pressure of hydrogen and oxygen, respectively, are assumed to be 1 atm. Since $E^\circ_{H_2/H^+} = 0$ is the standard of convention, Eq. (2-3) reduces to the simpler form

$$E = -0.059 \, pH \tag{2-5}$$

Plots of Eqs. (2-4) and (2-5) give potential–pH curves which efficiently summarize the reversible oxidation–reduction limits for aqueous background electrolytes. The theoretical curves are *independent of electrode material or type*.

Several points are to be considered for the hydrogen ion–water system.

1. The background limits are valid provided only that H^+ and H_2O are the most readily reduced and oxidized species present. The oxidation limit given by Eq. (2-4) would never be reached in ordinary voltammetry with a background electrolyte of, say, 1 M potassium iodide and 1 M sulfuric acid. The anodic reaction

$$3I^- \rightleftharpoons I_3^- + 2e \tag{2-6}$$

proceeds readily at much less positive potentials (ca. +0.5 V vs. SCE) than that of reaction (2-4). The liberation of iodine would be the limiting process regardless of the theoretical curve. Confusion seldom exists on this point. For instance, it is widely recognized that anodic limits in hydrochloric acid are somewhat less positive than in sulfuric acid, owing to the relative ease of chloride ion oxidation.

2. The theoretical E–pH curve indicates nothing about the overvoltage (better, *activation overpotential*) for the reactions. Since overpotential varies with current density and, even more important, may be strikingly different according to the electrode material, *experimental determinations of potential ranges are required*.

3. Specific interactions between electrode metal and background electrolyte are not covered by Eqs. (2-4) and (2-5). Such reactions sometimes severely alter the anodic limits of solid electrodes.

The solid lines of Fig. 2-1 show the theoretical (reversible) oxidation–reduction limits calculated from Eqs. (2-4) and (2-5). To be of most practical value, the potential scale has been converted to SCE. The region between the two solid lines may be called the stability domain of water (*3*).

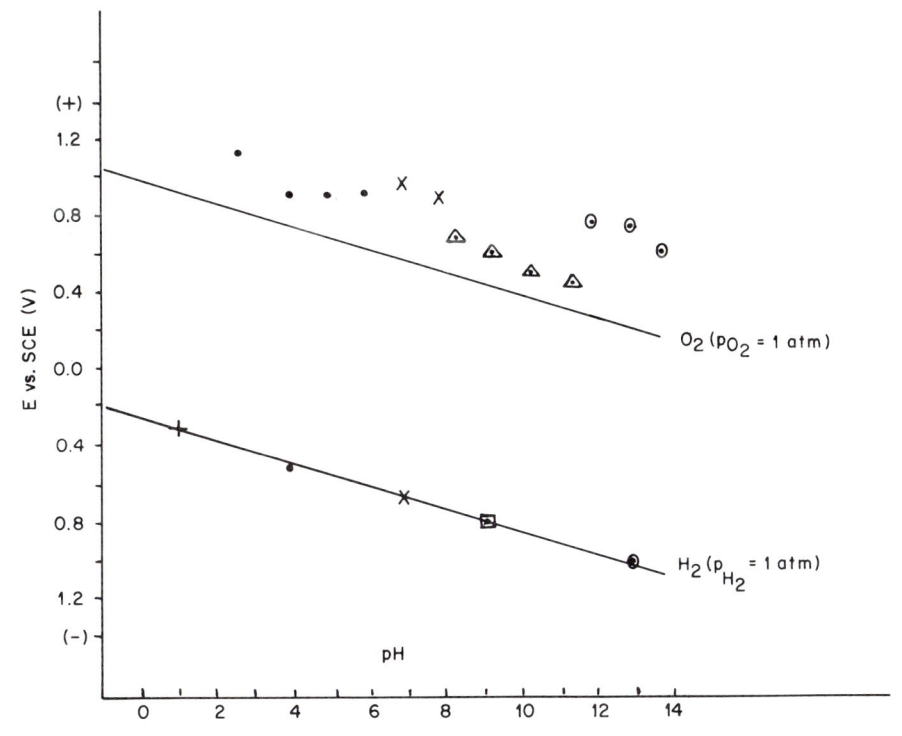

Fig. 2-1. Experimental potential limits of platinum electrodes. ●, Acetate buffer; ×, phosphate buffer; △, ammonia; ☐, borate buffer; ○, sodium hydroxide; +, perchloric acid. (Although exact current densities in Fig. 2-1 and Table 2-1 are unknown, all are probably in the general range $\mu A/cm^2$.)

Above the top line, evolution of oxygen is possible. Similarly, hydrogen may be evolved by electroreduction at potentials less positive than the lower line.

A. Platinum

The individual points in Fig. 2-1 show experimental limits for platinum electrodes taken from available literature data. (The data are taken from the solid electrode polarographic literature, not from overpotential studies.) It can be seen that the practical anodic limits for platinum generally exceed the theoretical values. This represents the well-known activation overpotential of oxygen on platinum. Specific electrolyte effects exist. Background electrolytes of the same pH but differing in buffer system do not necessarily have the same potential limits. This is due in part to anion effects on overpotential. In other cases reaction of the electrode with the background electrolyte may intervene.

The hydrogen overpotential on platinum is practically negligible. Hydrogen is evolved at essentially the theoretical limit (solid lower line,

TABLE 2-1

Potential Limits for Platinum Electrodes

Medium	Anodic	Cathodic
0.1 M HCl	+1.1	−0.3
0.1 M HClO$_4$	—	−0.31
0.1 M HAc, pH 2.7	1.12	—
2 M H$_2$SO$_4$	1.25	—
6 M HCl	0.97	−0.30
Acetate buffer, pH 4.0	0.90	−0.50
Acetate buffer, pH 5.0	0.90	—
Acetate buffer, pH 6.0	0.90	—
Phosphate buffer, pH 7.0	0.94	−0.70
Phosphate buffer, pH 8.0	0.88	—
0.05 M borox, pH 9.2	—	−0.80
NH$_3$, NH$_4$Cl buffer, pH 8.3	0.68	—
NH$_3$, NH$_4$Cl buffer, pH 9.3	0.60	—
1 M NH$_3$, pH 11.4	0.44	—
0.01 M NaOH, pH 12.0	0.75	—
0.1 M NaOH, pH 12.9	0.72	−0.91
1 M NaOH, pH 13.8	0.60	—
0.1 M NaClO$_4$	—	(−0.85)[a]
0.1 M KNO$_3$	—	−1.0
0.1 N KCl	1.0	—

[a] Unbuffered pH 7.

Fig. 2-1). There are few specific buffer effects. The lack of hydrogen overpotential is a serious disadvantage in the cathodic usage of platinum electrodes.

Table 2-1 contains more detailed information on platinum electrodes. These potentials should be considered only as approximate values for estimation of useful range. In acid solutions one probably should not plan to study any cathodic reactions beyond 0.0 V, although Table 2-1 shows, for instance, −0.3 V for the cutoff in 0.1 M hydrochloric or perchloric acids. In general, definition of a wave close to the background process is very poor. Frequently with platinum, reactions operating at

BACKGROUND PROCESSES AND POTENTIAL LIMITS 23

potentials close to anodic limits are also unfavorable, since surface oxidation may inhibit, or in some cases eliminate, the polarographic behavior. Details of surface oxide effects are considered in Chapter 7.

A point worth mentioning briefly here is that platinum electrodes which are subjected to potentials sufficiently cathodic that hydrogen ion reduction occurs, will adsorb the evolved hydrogen. Now, upon application of anodic potentials, the sorbed hydrogen is oxidized and gives rise to a considerable "anodic dissolution" current. The magnitude of this anodic current depends on the quantity of hydrogen adsorbed (for small quantities) and hence to the time of exposure to the cathodic potential. These currents are frequently of large magnitude and may give a false impression of an anodic cutoff. The situation is frequently met when the electrode is pretreated or cleaned at cathodic potentials. The hydrogen dissolution current decays with time and at most is only troublesome. If manual polarograms are run, the dissolution current is seen only as a transient. With recorded polarograms, well-defined peaks are obtained as discussed in Chapter 5.

B. Gold

Gold, shown in Fig. 2-2, presents an opposite picture with respect to hydrogen evolution. Between pH 4 and 10, the overpotential is of the order of 0.4–0.5 V. Gold electrodes are much more valuable than platinum for cathodic studies (1). The data on oxidation limits with gold are meager. In the cases studied, appreciable overpotential extends the anodic limit beyond the theoretical oxygen evolution line. In acid chloride media the anodic range is severely limited by oxidation of the metal to complex chlorides. The positive range in hydrochloric acid is about +0.6 V (4). Gold electrodes should not be used for anodic work in chloride-containing media.

More specific data are contained in Table 2-2. All values are referred to SCE. Cutoff potentials are given to the nearest 10 mV in the tables. Even this may be more than is justified by the nature of the measurements.

Surface oxidation of gold introduces complications. While the anodic limit of gold in 1 M perchloric acid is given in Table 2-2 as +1.5 V, Bauman and Shain point out the practical range extends to only +0.8 V (4). This is due to interference by gold oxide formation.

With all facts considered, gold is probably a better electrode material than platinum for general utility. Gold does not sorb hydrogen to any appreciable extent. After hydrogen evolution gold shows practically none of the hydrogen dissolution current associated with platinum. Oxide effects are about the same as with platinum and there is little to choose from in this respect. The one practical disadvantage of gold is that it is difficult to seal in glass, and this is probably responsible for the lesser popularity.

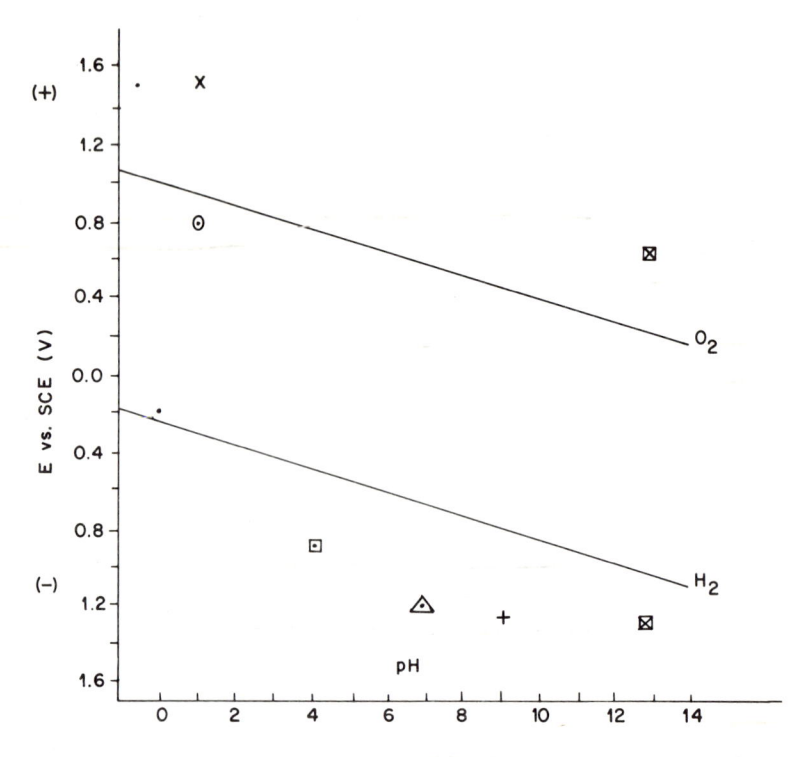

Fig. 2-2. Experimental potential limits of gold electrodes. •, Perchloric acid; ×, sulfuric acid; ⊙, hydrochloric acid; ⊡, acetate buffer; △, phosphate buffer; +, borate buffer; ⊠, sodium hydroxide. (Although exact current densities in Fig. 2-2 and Table 2-2 are unknown, all are probably in the general range $\mu A/cm^2$.)

C. Carbon (Graphite) Rod Electrodes

After a few initial studies with plain carbon rods, most work has been done with wax-impregnated graphite rods. While such electrodes sometimes suffer from high residual currents, their potential range is excellent (6). Table 2-3 summarizes the results of several investigators on potential ranges.

TABLE 2-2
Potential Limits for Gold Electrodes

Medium	Anodic	Cathodic
1 M HClO$_4$ (air-free)	+1.5	−0.2
0.1 M HClO$_4$, pH 1.0		−0.37
Acetate buffer, pH 4.0		−0.88
Phosphate buffer, pH 7.0		−1.19
0.05 M borax, pH 9.2		−1.25
0.1 M NaOH, pH 12.9		−1.28
0.1 M NaClO$_4$, pH 7[a]		−1.13

[a] Unbuffered.

Even WIGE's often have anomolous residual currents which apparently are due to slight variations in impregnation procedures. Elving and Smith have recently reported an improved technique for use with the WIGE (7). This consists of a momentary pretreatment of the electrode surface on a 0.003% solution of Triton X-100. This application of the wetting agent assures a completely wetted surface when immersed in the test solution and hence increases reproducibility of surface area. Elving and Smith give the anodic and cathodic range of this type electrode as 1.2 and −1.2 V vs. SCE, respectively, in both neutral and acidic media.

D. Pyrolytic Graphite Electrodes

The pyrolytic graphite electrode was introduced for aqueous solution usage by Beilby and co-workers (25) and Miller and Zittel (26). The

TABLE 2-3
Potential Ranges for Carbon Rod Electrodes

Type of electrode	Medium	Anodic	Cathodic
Graphite rod (untreated)	0.1 M HCl	+1.12	−0.32
	0.1 N H$_2$SO$_4$	1.19	−0.24
	0.1 M NaOH	0.68	−0.34
WIGE[a]	0.1 M KCl	1.0	−1.3
	HCl–KCl, pH 1.20	1.32	−1.28
	Phthalate buffer, pH 3.95	1.33	−1.30
	Phosphate buffer, pH 7.02	1.35	−1.38
	Borate buffer, pH 9.80	1.04	−1.61

[a] Wax impregnated graphite electrode.

useful range according to the latter authors is from +1.0 to −0.8 V vs. SCE in acidic chloride or nitrate media. Using a 1-μA deviation from the residual current as the cutoff limits, Elving et al. indicate a more restricted range in KCl–HCl of about +0.9 to −0.4 V vs. SCE. Further data on potential limits in several background media are given by these workers (37).

E. Carbon Paste Electrodes

Carbon paste electrodes, which are discussed in detail later, are made by mixing powdered graphite with an organic liquid such as Nujol to a fairly thick paste consistency (8,9). The paste is then packed in a depression in a holder (e.g., a Teflon rod). A platinum or copper wire in the bottom of the depression makes electrical contact, and the outer surface of the paste is the electrode proper.

Like their carbon rod counterparts, carbon paste electrodes have a wide range of anodic and cathodic utility. Table 2-4 summarizes some of the background media examined for anodic work. Nujol pastes (CE-NjP) have a very slightly greater anodic limit than bromonaphthalene pastes, but, in general, all pasting liquids give about the same range. Owing to an extremely low residual current the entire anodic range is available for study.

Determining cathodic potential limits for carbon pastes is somewhat complicated by the fact that a small residual wave is almost always obtained

TABLE 2-4

Anodic Limits for Carbon Paste Electrodes

Medium	E vs. SCE, V[a]
1 M KCl	+1.10
0.1 M KCl–HCl, pH 2.5	1.08
1 M HCl	1.02
0.1 M HAc–NaAc, pH 4.7	1.27
1 M Na$_2$SO$_4$	1.28
0.1 M H$_2$SO$_4$	1.30
B and R buffer,[b] pH 2.4	1.30
B and R buffer, pH 4.8	1.30
0.2 M NaOH	0.87

[a] When residual current exceeds 2 μA, CE–BnP (carbon electrode–bromonaphthalene paste).
[b] Britton and Robinson buffer.

TABLE 2-5

Cathodic Limits of Carbon Paste Electrodes

Medium	E vs. SCE, V[a]
1 M HClO$_4$	−0.90
1 M HCl	−0.90
1 M HAc–NaAc	−1.0
1 M KNO$_3$	−1.1
1 M KCl	−1.1
1 M NaClO$_4$	−1.1
1 M NH$_3$–NH$_4$Cl	−1.2
1 M NaOH	−1.4

[a] CE–NjP (Carbon electrode–nujol paste).

on cathodic background scans. This appears to be oxygen reduction, which is very difficult to remove even after thorough deaeration. (It can be *almost* eliminated by using alkaline sulfite background.) Apparently the small amount of dissolved oxygen is adsorbed on the carbon paste surface giving rise to the appreciable residual current. The saving factor is that this residual wave has a long flat plateau, and reduction waves of electroactive species can be superimposed without great difficulty. The magnitude of this residual current is about 2 μA. Thus cutoff limits are specified at the point where the sudden increase of current indicates the onset of the real background reduction. Such cathodic limits are given in Table 2-5.

F. Boron Carbide Electrodes

In a search for new inert electrode systems the boron carbide (B$_4$C) electrode was introduced (*10,11*). While pure B$_4$C is a nonconductor, Norbide (Norton Company, Worchester, Mass.) has negligibly low resistance for voltammetric work. The electrode is characterized by extreme inertness, low residual currents, and a wide potential range. Details on this electrode are discussed later, but Table 2-6 shows the potential ranges vs. SCE in a variety of background media.

There is no evidence to suspect that the background processes at B$_4$C differ from those observed at platinum or gold electrodes. No stripping currents are observed when the electrode has been held at high anodic potentials for prolonged periods of time and then scanned in a cathodic direction. Pretreatment of the electrode at either high positive or negative

TABLE 2-6
Potential Range of B$_4$C Electrodes

Medium	Anodic	Cathodic
1 M H$_2$SO$_4$	+1.13	−0.9
1 M HCl	1.0	−0.9
0.4 M KCl	1.0	−1.6
0.2 M NaOH	0.4	−1.7
0.1 M Na$_2$CO$_3$	0.6	−1.6

potentials has no effect on later polarograms. In short, the B$_4$C electrode approaches the concept of an inert electrode to a much higher degree than do the noble metal electrodes. It is, however, difficult to find a completely satisfactory method to mount the electrode. There is further uncertainty in the reliability of various batches of the commerically available B$_4$C rods. Some batches tested have shown abnormally high residual currents.

G. Miscellaneous Electrodes

The tungsten hemipentoxide electrode (THPE) developed by Jordan and Jiminez (*12*) is useful primarily in the anodic region. The range in 0.1 M perchloric acid was reported to be from −0.25 to +1.85 V vs. SCE. A residual current of less than 0.5 μA was observed over this span. In 0.1 M sodium perchlorate the cathodic region is extended to about −0.55 V with about the same anodic limit. Finally, in 0.1 M sodium hydroxide the range is shifted markedly. The cathodic cutoff occurs at about −1.7 V, but dissolution of the electrode in alkali limits the anodic range to −0.9 V. The pronounced shift is caused by this change of background processes—in acid and neutral solution, ordinary oxygen evolution is the anodic background reaction. The cathodic reactions in all three media were interpreted as being reduction of the oxide-coated electrode. The extreme anodic limit of the THPE makes it an interesting electrode, but further details on its usage are lacking.

Jordan and Jiminez also worked with a mercury-plated rotated tungsten wire whose cathodic range extended to about −2.0 V vs. SCE. The anodic limit was, of course, imposed by the anodic dissolution of mercury. Other mercury-plated electrodes have about the same cathodic range.

While the mercury chloride film anode (MCFA) developed by Kuwana and Adams (*13,14*) is not properly a solid electrode, its utility is in the anodic oxidation of organic compounds, and thus its potential range is of interest. In the usual background electrolyte of 0.2 M potassium chloride and 10^{-3} M hydrochloric acid the MCFA cuts off at about 1.4 V vs. SCE using conventional voltammetric scanning. It is interesting to note that *p*-nitroaniline oxidizes with a well-defined peak polarogram at the MCFA, whereas the wave is ordinarily merged with background at platinum. Several other moderately difficult organic oxidations occur surprisingly well as the MCFA (*14*).

2-2. POTENTIAL RANGES IN NONAQUEOUS MEDIA

Nonaqueous voltammetry employing solid electrodes is truly in its infancy. Excluding fused-salt systems and various water–organic solvent systems which have been used from time to time, the most important solvent system is acetonitrile. The potential limits of this and a few other systems are discussed individually below.

A. Acetonitrile

The potential range of acetonitrile is very large and has initiated a great deal of interest in this solvent. Kolthoff and Coetzee reported the potential span to be from $+1.8$ V to -1.5 V vs. the ordinary aqueous SCE using the rotating platinum electrode (RPE) with 0.1 M sodium perchlorate as supporting electrolyte (*15*). Popov and Geske (*16*) employed 0.1 M lithium perchlorate as the supporting electrolyte and a acetonitrile silver–silver nitrate (0.01 M) reference electrode. (The potential of this reference electrode is ca. $+0.3$ V with respect to the aqueous SCE. A more detailed discussion of reference electrodes in acetonitrile is given later.) Calculated on the basis of the aqueous SCE, Popov and Geske found a range of $+2.5$ to -3.0 V at the RPE in the lithium perchlorate medium. The cathodic reaction is quite clearly the reduction of lithium ion, as evidenced by lithium metal deposition on the electrode. The anodic reaction is not so readily understood. Schmidt and Noack studied the electrolysis of acetonitrile-containing silver perchlorate using a platinum anode (*20*). The proposed anodic reaction was the formation of perchlorate free radicals followed by reaction with the solvent as

$$ClO_4^- \rightarrow \cdot ClO_4 + e$$
$$\cdot ClO_4 + CH_3CN \rightarrow H^+ + ClO_4^- + \cdot CH_2CN$$
$$2 \cdot CH_2CN \rightarrow NCCH_2\!-\!CH_2CN$$

In an already classical paper Maki and Geske examined the anodic oxidation of 0.1 M lithium perchlorate in acetonitrile, carrying out the electrode reaction in the resonance cavity of an electron paramagnetic resonance (EPR) spectrometer (21). Using an anodic current of 3 mA, resonance signals were obtained after about 1 min of generation. The spectrum consisted of four equally spaced lines, which was interpreted as a radical species containing a single chlorine atom, perhaps the perchlorate radical. Russell has suggested that chlorine dioxide is the free-radical species involved in this case (46).

It is clear that the presence of appreciable quantities of water in the acetonitrile will give rise to a different limiting anodic reaction: the oxidation of water to hydrogen ion and oxygen. Indeed, this reaction has been used for coulometric generations in acetonitrile by Streuli and Hanselman (22,23). The presence of water was shown by Popov and Geske to give rise to surface oxidation of the platinum with resultant cathodic stripping patterns (16). The water content of well-purified acetonitrile is considerable. Acetonitrile twice distilled from phosphorous pentoxide ordinarily is about 1 mM in water, as determined by Karl Fisher titration. In ordinary work with acetonitrile, one should not be surprised to find small background currents from water contamination.

In a later paper concerned with the oxidation of tetraphenylborate ion, Geske indicates an anodic cutoff in acetonitrile -0.1 M lithium perchlorate of about $+2.1$ V vs. aqueous SCE (17). Small differences are to be expected from changes in the solvent purity and also the supporting electrolyte. Loveland and Dimeler, working with 0.5 M sodium perchlorate, found considerable residual currents at potentials more positive than $+2.0$ V vs. aqueous SCE. This residual current was a small wave with $E_{1/2}$ of ca. $+2.05$ V vs. SCE. They employed large-scale controlled potential oxidation of the background solution to eliminate the impurities (18). Lund indicates the anodic limit of acetonitrile with 0.5 M sodium perchlorate to be about $+2.5$ V when calculated vs. aqueous SCE (19).

A nonaqueous carbon paste electrode has been developed which is useful in acetonitrile. Background currents of less than 3 μA were obtained between $+1.1$ and -0.7 V vs. SCE with this electrode. The same carbon paste is usable in nitromethane and propylene carbonate but not dimethyl sulfoxide, benzonitrile, or dimethylformamide. Details of the paste composition and its behavior are given in the original literature (43).

A number of purification procedures have been reported for acetonitrile. One which has given consistently good results in the writer's laboratory involves drying commercial grade acetonitrile over calcium hydride, followed by two distillations from phosphorous pentoxide. A third distillation is made from calcium hydride. The middle 60% is collected from each distillation. From reagent-grade acetonitrile, a treatment with calcium hydride followed by a single distillation from phosphorous pentoxide suffices. Using tetraethyl ammonium perchlorate (0.1 M) as supporting electrolyte, this grade of acetonitrile ordinarily has a potential range of about $+2.0$ to -1.8 V vs. SCE at a stationary platinum electrode.

B. Benzonitrile

Because benzonitrile differs from acetonitrile in having delocalized electrons in the phenyl ring capable of interaction with the C≡N system, and the phenyl group affords possible steric interactions, Larson and Iwamoto made a fundamental comparison of polarographic reductions in benzonitrile and acetonitrile together with other nitriles (24). Benzonitrile has been found to be an interesting solvent for anodic voltammetry (38).

Using 0.1 M tetraethylammonium perchlorate (TEAP) as supporting electrolyte, the anodic range of benzonitrile extends to approximately $+2.0$ V vs. aqueous SCE before the background cutoff occurs at a stationary platinum electrode. With only a single distillation from phosphorous pentoxide, the residual currents were ordinarily less than 1–2 μA up to $+1.0$ V.

The effect of water on the background polarograms was investigated briefly. No more than about 1% water can be dissolved in benzonitrile. Table 2-7 is a numerical representation of the background current as a

TABLE 2-7

Effect of Water in Benzonitrile on Background Current

	H$_2$O–benzonitrile mixture, %		
E vs. SCE, V	0[a]	0.5	1.0
+0.9	2.0[b]	2.0	2.0
1.1	2.5	2.5	3.0
1.3	3.0	3.5	4.0
1.5	4.0	6.0	7.0
1.7	8.0	11.5	16.5
1.9	15.5	21.5	28.5
2.1	25.5	36.0	44.0

[a] No water added.
[b] Current in microamperes at indicated E_{app}.

function of applied voltage for three different amounts of water in benzonitrile–0.1 M TEAP. It is clear that the water has a decided effect on the nature of the anodic background process but, in the region less than +1.0 V, the change in overall current is negligible. Cyclic voltammetry of benzonitrile–0.1 M TEAP containing considerable water using a rapid sweep rate shows a cathodic peak at about +0.5 V. The nature of this process is not clear at this time.

Substitution of lithium perchlorate for TEAP has relatively little effect on the anodic range of benzonitrile. Other supporting electrolytes have not been examined. The cathodic range of benzonitrile at platinum is about −1.8 V vs. SCE.

C. Dimethyl Sulfoxide

Dimethylsulfoxide (DMSO) has certain useful properties as a solvent for voltammetry. It may be purified by vacuum distillation of the bulk material, collecting the middle 60%. It can be stored over molecular sieves. The anodic range is limited to only about +0.9 V at a stationary platinum electrode. The cathodic range extends to −1.8 V vs. SCE, with TEAP as supporting electrolyte. Kolthoff and Reddy report a range of +0.7 to −1.85 V at a RPE using either TEAP or sodium perchlorate background (44).

D. Propylene Carbonate

Propylene carbonate (4-methyldioxolone-2) is an excellent solvent with particularly desirable characteristics for quantitative electroanalytical measurements. Propylene carbonate (PC) has a low vapor pressure, high dielectric constant (69.0 at 23°C), relatively low toxicity, and high chemical stability. It is not very hygroscopic and noncorrosive. A technical bulletin by Jefferson Chemical Company covers much of the literature on physical and chemical properties, applications and potential uses of PC (39). PC may be purified by reduced pressure distillation at 120–130°C. A double distillation, collecting the middle 60% fraction, gives sufficient purification, and the product can be stored over molecular sieves with little apparent decomposition. The potential limits at a stationary platinum electrode are about +1.7 to −1.9 V vs. SCE. The background current at platinum is exceptionally low within this range. A relatively high solution resistance is overcome by using 0.25 M TEAP as supporting electrolyte.

One of the distinct advantages of PC is the high viscosity (1.67 centistokes at 38°C). This means that natural convection is greatly diminished relative to a solvent such as acetonitrile. Thus linear diffusion conditions can be maintained for periods as long as 40–50 sec. This is particularly advantageous in quantitative chronopotentiometry. It has also been found that PC is an excellent solvent for the electrogeneration of cation radicals. A summary of its utility in voltammetry has appeared recently (40).

E. Methylene Chloride

Bard and co-workers found methylene chloride very suitable especially for oxidation of hydrocarbons. With 0.2 M tetrabutylammonium perchlorate as supporting electrolyte, using a current density of 0.5 mA/cm^2, the potential limits at platinum were $+1.80$ to -1.70 V vs. SCE (47,48).

F. Nitro Compounds

Drushel and Miller used a 30:70 by volume mixture of nitrobenzene and methanol which was 0.2 M in hydrochloric acid. This mixture was used to solubilize petroleum samples for aliphatic sulfide voltammetry. The anodic range of this medium extends to about $+1.0$ V vs. an internal Ag/AgCl electrode (29,30).

Nitrobenzene itself has recently been found to be a very satisfactory solvent for anodic generation of hydrocarbon cation radicals. The nitrobenzene was vacuum-distilled at 60°C and then passed through an alumina column. TEAP is not soluble enough and tetrapropylammonium perchlorate is a satisfactory supporting electrolyte. The range at a stationary platinum electrode is from about $+1.6$ to -0.7 V vs. SCE (41).

Voorhies and Schurdak reported the use of nitromethane in solid electrode chronopotentiometric studies (42). The solubility of supporting electrolytes is somewhat limited, but tetramethylammonium chloride (TMAC) was satisfactory. The anodic limit at platinum was about $+0.9$ V. With 0.1 M magnesium perchlorate, the range was extended to about 2.2 V. The cathodic range is controlled by the water content and with dry nitromethane (ca. 0.03% water) and TMAC was about -0.9 V. These potentials are for a chronopotentiometric current density of 0.3 mA/cm^2 and involved the use of an aqueous -20% methanol normal calomel electrode. The potential of this electrode is ca. $+20$ mV vs. aqueous SCE, so the general ranges listed above are *approximately* those of the usual SCE.

G. Miscellaneous Solvent Systems

Alcohols were examined briefly by Lord and Rogers (2), who found that the anodic limit of platinum in 100% ethanol containing 0.1 M lithium chloride was about +0.75 V vs. SCE. Surprisingly, a lower cutoff was obtained with 20% ethanol–water. Gold had a slightly more positive range in the pure alcohol, ca. 1.1 V. Adams and Voorhies found about +0.7 V as the anodic limit at a RPE in 95% ethanol containing a very small amount of sulfuric acid (27). Nicholson reports the usable range of a stationary platinum electrode in methanolic HCl (0.1 M in HCl, 2.1 M in H_2O) is +0.9 to −0.2 V vs. SCE (28).

Delimarskii and Abarbarchuk studied some metal ion reductions at platinum in pyridine containing 1% potassium thiocyanate, but no potential ranges are given (31). Fujinaga and co-workers have investigated the decomposition products and background processes of N,N'-dimethylformamide (DMF) (45). DMF does not appear to be particularly promising for anodic studies.

H. Purity Criteria for Nonaqueous Solvents

It can be seen in the preceding paragraphs that the ordinary criteria of various nonaqueous solvents are the absence of extraneous background currents and the potential range. An often-used but sometimes misleading yardstick is that the more highly purified the solvent, the greater its potential range. The range obviously depends on the supporting electrolyte and, not so obviously, on the electrode material and its surface condition. Furthermore, the tolerance for background currents may depend on the particular measurement carried out. If analytical applications are the object, then obviously low backgrounds and sharp cutoff potentials are desirable. If, on the other hand, the general nature of a particular electrode process is to be examined, perhaps one can operate at moderately high concentrations of electroactive species so that the background current is a negligible fraction of the net observable current. It is probably always unsafe, however, to draw too many conclusions (from voltammetry alone) on a electrode process which occurs too close to the background cutoff. This is especially true of anodic oxidations.

Figure 2-3 provides an excellent example of a solvent-supporting electrolyte system which would be totally unsatisfactory for analytical voltammetry—or for studying the nature of electrode processes. The

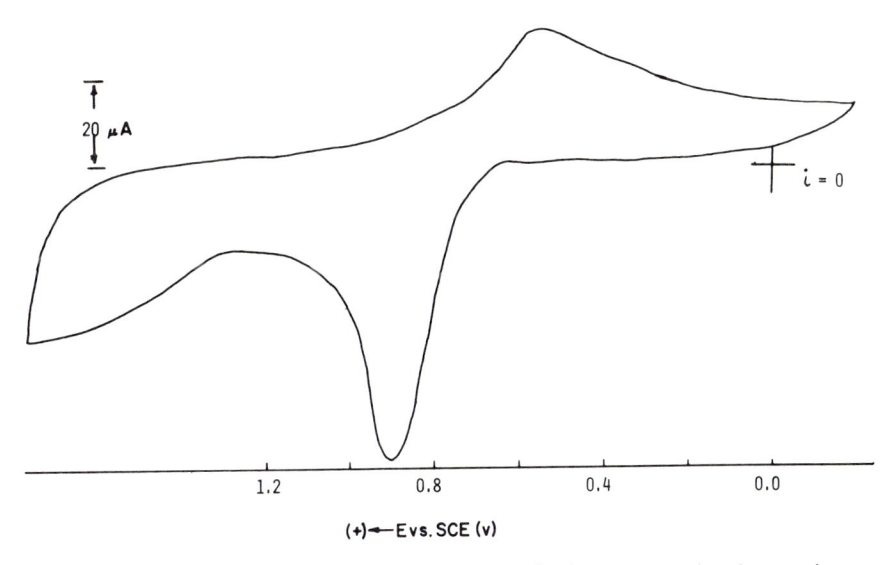

Fig. 2-3. Background cyclic polarogram of the nitrobenzene–tetrabutylammonium perchlorate solvent system.

system is highly purified nitrobenzene with tetrabutylammonium perchlorate as supporting electrolyte. The cyclic polarograms (cyclic voltammetry is discussed in Chapter 5) show large anodic and cathodic currents in the middle of the potential range of interest. (This electroactive system was found to be an unidentified constituent of the tetrabutylammonium perchlorate—background polarograms with tetrapropylammonium perchlorate show no extraneous current in this region.) However, the purpose for which this solvent system was intended was to anodically generate hydrocarbon cation radicals for ERP studies. The cation radicals of 9,10-dimethyl- and 9,10-diphenylanthracene as well as perylene were generated at ca. +1.2 V (past the anomolous peak at 0.85 V in Fig. 2-3). The EPR spectra of these cation radicals were, in every way, identical with those obtained in other media. This discussion is not meant to imply that known impurities in solvent systems should be tolerated. Rather, it does show that the criteria by which one judges solvent–background electrolyte purity is often set by the usage one intends for the system. Wide potential ranges and minimum background currents are not invarient standards for purity of the solvent system. At the same time, it should be noted that all the potential ranges mentioned above should serve only as general guides. The values vary considerably with individual purifications and other experimental conditions.

2-3. RESIDUAL CURRENTS

The anodic polarogram of Fig. 2-4 represents the background polarogram of 0.1 M KCl at a RPE. It is observed that a finite current flow exists throughout the interval 0.2 to 1.0 V before the sharp rise marking the limiting background potential. This small, slowly increasing current is usually termed residual current. It is present in all current–voltage curves to varying degrees and, indeed, even in electrolyses of "pure" background solutions.

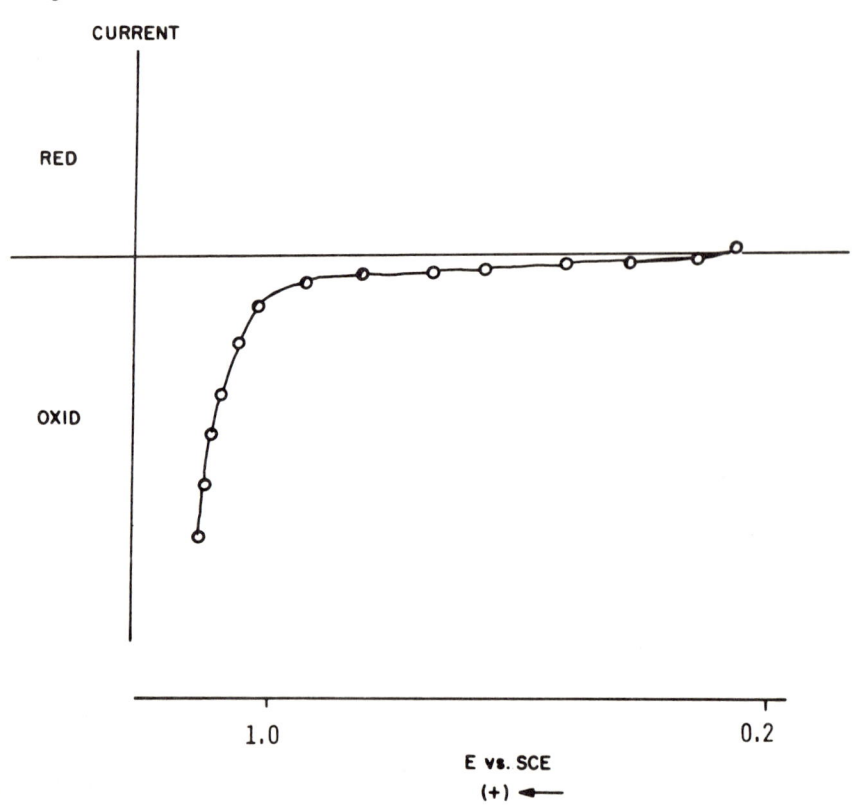

Fig. 2-4. Background polarogram of 0.1 M KCl at a rotated electrode.

Electrochemical rate theory provides that finite current flows exist at potentials considerably removed from the point where the sharp rise in current occurs (*32*). This *Faradaic* current forms only a small portion of what is commonly called residual current at a solid electrode.

Another contribution to the residual current is the *charging* or *capacity* current resulting from charging the double layer associated with the electrode–solution interface. The order of magnitude of the double-layer capacity at Pt is about 20–40 $\mu F/cm^2$, and the charging current ordinarily is relatively unimportant except at high voltage sweep rates. With some solid electrodes, indications are that the capacity may be of the order of 100–200 $\mu F/cm^2$, and the charging current is of significance.

In the main, residual currents at solid electrodes result from extraneous processes primarily associated with the electrode surface. Thus, according to the past history of the electrode, the surface may have become oxidized or covered with a layer of sorbed gas. Potentials applied to these electrodes may then give rise to currents from dissolution of the oxides or gas films. Furthermore, a "clean" electrode may show residual currents upon formation of these films. Such currents make up the bulk of the residual current at a solid electrode.

In addition, no background electrolyte is normally free of traces of oxidizable and reducible impurities. These substances contribute to the overall residual current. To summarize, residual current is composed primarily of four components: (1) expected Faradaic current from the electrochemical reaction, (2) charging current, (3) variable currents from electrochemical reactions of the electrode surface, and (4) trace currents from electroactive impurities in solution. A more detailed discussion of residual currents will be given in later sections. It suffices to note at this point that residual currents may limit the useful range of a solid electrode by masking the primary process. Untreated graphite electrodes may have residual currents of such magnitude as to be unusable over a considerable potential range (*2,5*).

2-4. CATHODIC REACTIONS

The major portion of the reductions studied at solid electrodes have been simple metal ion–metal processes. Silver ion is an example which has been used by many investigators for a variety of theoretical studies. The reductions of lead, thallium, cadmium, mercury, and copper ions have received similar attention. Ion–ion reductions are quite common, including ferric–ferrous, ferri–ferrocyanide, permanganate, and chromate. The reduction of iodine and bromine at platinum electrodes has been widely investigated. The importance of dissolved oxygen in physiological systems has prompted a large amount of work on electroreduction of oxygen at platinum and gold. A considerable number of aromatic organic molecules can be reduced in nonaqueous media at platinum electrodes.

Mercury-plated platinum electrodes have been used in only a few organic reduction processes. Copper- and silver-plated wires have received attention. Quinone is reducible at platinum, and the hydroquinone–quinone system has been studied widely. The lack of organic reduction work is unfortunate, for solid electrodes should have decided value in the analysis of flowing samples. They are suited for in-line monitoring operations.

2-5. ANODIC OXIDATIONS

Some of the ion–ion processes mentioned above have been examined anodically but are of minor importance. By far the most valuable area of solid electrode voltammetry is the application to organic oxidations. Here the solid electrode has no peer—most of the anodic reactions cannot be carried out with the dropping electrode.

To date, several broad classes of organic compounds have been investigated, including phenols, aromatic amines and diamines, sulfur compounds, and aromatic hydrocarbons. It is to be expected that much more information on the anodic reactions of organic compounds will be forthcoming.

2-6. SENSITIVITY

The total sensitivity of an electrode is a function of the limiting current which can be obtained under a given set of operating conditions. The useful or real sensitivity may be considerably different. It is best defined in terms of a "signal to noise" ratio, where the residual current corresponds to a noise level or unwanted portion of the total current response. It is not uncommon to find sensitivities expressed as the ratio residual current/limiting current (i_r/i_L) (5).

Since i_L is usually directly proportional to the electrode area, it is theoretically possible to increase i_L to any desired level by using large electrodes. A little reflection indicates that i_r is usually correspondingly increased by this technique. Practical area limits are soon reached beyond which no real increase in sensitivity is achieved. Rotating the electrode or stirring the solution is an effective method of increasing real sensitivity.

Kolthoff and co-workers have presented a very practical sensitivity evaluation. Various-sized electrodes are identified by their limiting currents obtained in the oxidation of thallous ion in 0.1 M sodium hydroxide. Using this well-characterized electrode reaction, the limiting current for a given electrode is expressed in microamperes per millimole per liter (33–35). For a rotated electrode the speed of rotation also needs to be specified.

Other electrode reactions are useful for comparison standards. The oxidation of ferrocyanide in KCl supporting electrolyte can be used with confidence provided the KCl concentration is 1 M or more and the ferrocyanide concentration greater than ca. 1.5 mM. The oxidation of o-dianisidine in 1 M H_2SO_4 has been found to be an extremely reproducible system. The oxidation of iodide ion is often used.

The sensitivity comparison is often used for the determination of n_T, the total number of electrons transferred in an unknown electrode reaction. This is done by comparing known and unknown systems at the same electrode and reducing the limiting currents to a common concentration basis as i_L/C. These comparisons have proved very useful at rotated electrodes, as shown by Geske (36).

The concept can be used at stationary electrodes. Here the electrode area is constant and one can compare the specific limiting current (SLC), which is the current obtained in microamperes for a 1 mM solution per cm² of electrode surface. The units of SLC then are $\mu a/mM/cm^2$. The electrode area may be that measured from the electrode dimensions (geometric area) or it may be obtained from a voltammetric measurement, where all other factors in the pertinent equation are known (electrochemical area).

Some care must be applied in using such SLC comparisons at stationary electrodes, as the following data show. Table 2-8 is a comparison of the SLC's for ferrocyanide oxidation (a known n_T = one electron) and the oxidation of N,N-dimethylaniline at pH 2.4. The purpose of the series of experiments was to obtain an idea of n_T for the dimethylaniline process. In the case of ferrocyanide the SLC is independent of the electrode type and configuration within about 4–5%. However, the SLC of the dimethylaniline depends very markedly on the electrode configuration, i.e., on the mass-transport conditions which supply and remove electroactive

TABLE 2-8
SLC's of Ferrocyanide and Dimethylaniline at Various Stationary Electrodes[a]

	SLC, μA/mM/cm²	
Electrode[b]	Ferrocyanide[c]	Dimethylaniline[d]
A	43.5	109
B	44.3	129
C	39.7	61.75
D	42.6	86.0

[a] Voltage scan rate, 200 mV/min.
[b] Electrode A, a circle of Pt foil inset in a Teflon plug to give upward linear diffusion configuration; electrode B, 12-gage Pt wire, extending vertically downward; electrode C, rectangular Pt foil, in vertical plane; electrode D, carbon paste electrode, linear diffusion configuration.
[c] In KCl background.
[d] In Britton and Robinson buffer, pH 2.4

species from the interface area. The SLC varies by a factor of 2 or more and no valid interpretation of n_T is possible in this case. Such complications are widespread in organic reactions and care should be used in interpretation of results. More consistent information in the dimethylaniline case can be obtained at rotated electrodes. It should be noted additionally that comparisons at stationary electrodes should be run with the same voltage scan rate (preferably at controlled potential with a three-electrode system) and approximately equal concentrations of the two electroactive species.

2-7. HALF-WAVE POTENTIALS

The half-wave potential ($E_{1/2}$) of a solid electrode polarogram is defined as usual as the potential where $i = i_{L/2}$. For a thermodynamically reversible electrode reaction, the $E_{1/2}$ can be correlated directly with the standard potential. For stationary electrodes where peak-type polarograms are obtained, $E_{p/2}$ is a close counterpart of $E_{1/2}$ at a rotated electrode.

The vast majority of solid electrode processes are irreversible, and no strict thermodynamic interpretation of the $E_{1/2}$ or $E_{p/2}$ is possible. This is only a minor limitation on the value of $E_{1/2}$ measurements. As in dropping mercury polarography, the $E_{1/2}$ or $E_{p/2}$ serves to characterize the electrode reaction and is frequently independent of concentration within experimental error.

Half-wave potentials at solid electrodes may not be as accurate as at the dropping electrode for several reasons. First, lack of reproducibility of surface conditions gives rise to distortions in the polarogram. Second, the shape of recorded current–voltage curves may vary with the rate of change of applied voltage. On the other hand, if these factors are held invarient, it is possible to obtain reproducibility in $E_{1/2}$ and $E_{p/2}$ measurements of ± 1–2 mV. It is the writer's opinion, however, that these values should be reported at most only to ± 5 mV. The reproducibility of successive measurements, as indicated above, may well exceed these limitations.

REFERENCES

1. I. M. Kolthoff and J. Jordan, *J. Am. Chem. Soc.*, **74**, 4801 (1952).
2. S. Lord and L. B. Rogers, *Anal. Chem.*, **26**, 284 (1954).
3. M. Pourbaix, *Thermodynamics of Dilute Aqueous Solutions* (J. N. Agar, (transl.)), Arnold, London, 1949.
4. F. Bauman and I. Shain, *Anal. Chem.*, **29**, 303 (1957).
5. V. F. Gaylor, A. L. Conrad, and J. H. Landerl, *Anal. Chem.*, **29**, 224 (1957).
6. J. B. Morris and J. M. Schempf, *Anal. Chem.*, **31**, 286 (1959).
7. P. J. Elving and D. L. Smith, *Anal. Chem.*, **32**, 1849 (1960).
8. R. N. Adams, *Anal. Chem.*, **30**, 1576 (1958).
9. C. Olson and R. N. Adams, *Anal. Chim. Acta*, **22**, 582 (1960).
10. T. R. Mueller, C. Olson, and R. N. Adams, *Proc. 2nd Intern. Polarog. Congr.*, Cambridge, *1959*, p. 198.
11. T. R. Mueller and R. N. Adams, *Anal. Chim. Acta*, **23**, 467 (1960).
12. J. Jordan and L. R. Jiminez, unpublished data, 1960.
13. T. Kuwana and R. N. Adams, *J. Am. Chem. Soc.*, **79**, 3609 (1957).
14. T. Kuwana and R. N. Adams, *Anal. Chim. Acta*, **20**, 51, 60 (1959).
15. I. M. Kolthoff and J. F. Coetzee, *J. Am. Chem. Soc.*, **79**, 870 (1957).
16. A. I. Popov and D. H. Geske, *J. Am. Chem. Soc.*, **80**, 1340 (1958).
17. D. H. Geske, *J. Phys. Chem.*, **63**, 1062 (1959).
18. J. W. Loveland and G. R. Dimeler, *Anal. Chem.*, **33**, 1196 (1961).
19. H. Lund, *Acta. Chem. Scand.*, **11**, 491 (1957).
20. H. Schmidt and J. Noack, *Z. Anorg. Allgem. Chem.*, **296**, 262 (1958).
21. A. H. Maki and D. H. Geske, *J. Chem. Phys.*, **30**, 1356 (1959).
22. C. A. Streuli, *Anal. Chem.*, **28**, 130 (1956).
23. R. B. Hanselman and C. A. Streuli, *Anal. Chem.*, **28**, 916 (1956).
24. R. C. Larson and R. T. Iwamoto, *J. Am. Chem. Soc.*, **82**, 3239, 3526 (1960).
25. A. L. Beilby, W. Brooks, and G. L. Lawrence, *Anal. Chem.*, **36**, 22 (1964).
26. F. J. Miller and H. E. Zittel, *Anal. Chem.*, **35**, 1866 (1963).
27. R. N. Adams and J. D. Voorhies, *Anal. Chem.*, **29**, 1690 (1957).
28. M. M. Nicholson, *J. Am. Chem. Soc.*, **76**, 2539 (1954).

29. H. V. Drushel and J. F. Miller, *Anal. Chim. Acta*, **15,** 389 (1956).
30. H. V. Drushel and J. F. Miller, *Anal. Chem.*, **29,** 1456 (1957).
31. Y. K. Delimarskii and I. L. Abarbarchuk, *Zavodsk. Lab.*, **16,** 929 (1950).
32. See, for example, P. Delahay, *New Instrumental Methods in Electrochemistry*, Wiley (Interscience), New York 1954, pp. 32–41, 218–227.
33. I. M. Kolthoff and J. Jordan, *J. Am. Chem. Soc.*, **74,** 382 (1952).
34. I. M. Kolthoff, J. Jordan, and A. Heyndrickx, *Anal. Chem.*, **25,** 884 (1953).
35. I. M. Kolthoff and J. Jordan, *Anal. Chem.*, **25,** 1833 (1953).
36. D. H. Geske, *J. Am. Chem. Soc.*, **63,** 1062 (1959).
37. L. Chuang, I. Fried, and P. J. Elving, *Anal. Chem.*, **36,** 2426 (1964).
38. S. Goodwin, K. Darlington, B. Mayrath, and R. N. Adams, unpublished data, 1963.
39. *Propylene Carbonate*, Tech. Bull., Jefferson Chemical Co., Houston, Texas.
40. R. F. Nelson and R. N. Adams, *J. Electroanal. Chem.*, **13,** 184 (1967).
41. L. Marcoux, P. Malachesky, and R. N. Adams, *J. Am. Chem. Soc.*, **89,** 5766 (1967).
42. J. D. Voorhies and E. J. Schurdak, *Anal. Chem.*, **34,** 939 (1962).
43. L. S. Marcoux, K. B. Prater, B. Prater, and R. N. Adams, *Anal. Chem.*, **37,** 1446 (1965).
44. I. M. Kolthoff and T. B. Reddy, *J. Electrochem. Soc.*, **108,** 980 (1961).
45. T. Fujinaga, K. Izutsu, and F. Kamiyama, *Rev. Polarog. (Kyoto)*, **13,** 90 (1966).
46. C. D. Russell, *Anal. Chem.*, **35,** 1291 (1963).
47. J. Phelps, K. S. V. Santhanam, and A. J. Bard, *J. Am. Chem. Soc.*, **89,** 1752 (1967).
48. A. J. Bard, private communication, 1967.

3 MASS TRANSFER TO STATIONARY ELECTRODES IN QUIET SOLUTIONS

3-1. Introduction 43
3-2. Linear Diffusion To Plane Electrodes 45
3-3. Cylindrical Diffusion 61
3-4. Diffusion To Spherical Electrodes 62
3-5. Convection In Unstirred Solution 63
References 64

3-1. INTRODUCTION

In this chapter the concern will be with processes whereby electroactive material is supplied to a working electrode and how these *mass-transfer processes* determine the current–voltage response.

Consider again why limiting currents are obtained in voltammetry. The general reaction

$$Ox + ne \rightleftharpoons Red$$

actually is composed of the individual processes

$$(Ox)_{bulk} \rightarrow (Ox)_{electrode} \qquad (3\text{-}1)$$

$$(Ox)_{electrode} + ne \rightarrow (Red)_{electrode} \qquad (3\text{-}2)$$

and

$$(Red)_{electrode} \rightarrow (Red)_{bulk} \qquad (3\text{-}3)$$

Examining, for the moment, only the first two equations, these are processes with finite rates and, in an electrical sense, they are reactions in series. The overall flow of electrons, which is the voltammetric current, can be limited by either of these series reactions. (Other series reactions which can alter the current are often present, but these are neglected in the present case.)

Now, the rate of the electron-transfer reaction (3-2) is an exponential function of applied potential; i.e., for a cathodic reaction the rate increases with increasing negative potentials—for an anodic reaction with increasing positive potentials. At applied potentials sufficiently different from the $E^{\circ\prime}$ of a given system (see Chapter 1), the rate of the electron-transfer reaction is at its maximum. In the usual case this rate is very rapid compared with that of process (3-1). Thus the over-all current is limited by the slower of the two series processes, in this case the mass transfer of Ox to the electrode. Such electrode reactions are said to be mass-transfer-controlled. They are also frequently called reversible processes. Reversible processes are sometimes referred to as Nernstian.

If the rate of mass transfer in the above situation could be increased to a very high value (by appropriate stirring techniques), so that it becomes comparable to or greater than the electron transfer rate, then the latter becomes limiting. The kinetics of the electron transfer now can be studied. Frequently, the electron-transfer rate is already relatively slow. Such reactions are called charge-transfer-controlled or irreversible. The terms reversible and irreversible are extremely poor choices, since, as seen above, the limiting rate and hence the "reversibility" can be altered as desired by such a variable as stirring.

Sometimes neither of the above processes are as slow as a chemical transformation involving the electroactive species. Such chemical reactions as ionic dehydration, acid–base dissociation, etc., may be the series limiting reaction. In this case the electrode reactions give rise to what are called kinetic currents—or it is said that the reaction is kinetically controlled.

In this and the next chapter electrode reactions limited only by mass transfer will be discussed. Three mass-transfer modes are normally encountered: *migration*, *convection*, and *diffusion*.

Mass transfer by migration is a result of the force exerted on charged particles by an electric field. Migration effects in laboratory-scale voltammetry serve no useful purpose. In the presence of a large excess of background or supporting electrolyte, migration of electroactive material is minimized to an extent where it can be neglected. Background electrolytes are used in voltammetry almost without exception, and such conditions are assumed throughout subsequent discussions.

Convection essentially means stirring of electroactive material to the electrode. It arises from thermal, mechanical, or other disturbances of the solution. It also develops when density variations appear near the electrode surface. Mass transport by convection is important in voltammetry. Processes in stirred solution depend largely on forced convection.

Diffusion is perhaps the most widely studied means of mass transport. In fact, the object of many voltammetric studies is to have diffusion the only process operative. Diffusion exists whenever concentration differences are established. Since a concentration gradient develops as soon as electrolysis is initiated, diffusion plays a part in every practical electrode reaction.

The next few chapters will describe the mass-transport conditions at most of the working electrode systems outlined in Chapter 1, Section 1-3, B. These descriptions in turn will define the limiting current behavior and current–voltage characteristics at particular electrodes. Detailed and authoritative treatments of the mass-transfer problems have been given (1). The intent of this chapter is to illustrate the problems in a somewhat practical fashion with experimental results.

3-2. LINEAR DIFFUSION TO PLANE ELECTRODES

Consider a plane electrode immersed in a solution containing Ox which is reduced typically as

$$\text{Ox} + ne \rightleftharpoons \text{Red}$$

The solution contains an excess of background electrolyte and the cell is free of mechanical and thermal disturbances; i.e., migration and convection are absent. The placement of the electrode is not defined, with one exception—the cell dimensions are very large with respect to the electrode size. Under these circumstances it is possible to obtain mass transfer by so-called *semiinfinite linear diffusion* (1). (Several other restrictions on the process will become apparent during the discussion. In particular, the geometric characteristics of practical linear diffusion electrodes will be defined later.)

If we carry out some electrolyses at constant potential with this somewhat ambiguously described electrode, three experimental facts are noted*:

* To conform to general practice, reduction processes are considered in this and most examples to follow. A slightly modified symbolism will be used for concentrations. The superscripts e and b are used to indicate concentrations at the *electrode* surface and in the *bulk* of solution, respectively. The subscript O refers to oxidant forms and R to the reductant species. This modification requires very little extra writing but seems to the writer to be more explicit than the commonly employed systems, which are somewhat confusing when set in different forms of print.

(1) The current is proportional to C_O^b, (2) the current is proportional to the area of the electrode, and (3) the current decreases with time of electrolysis. A derivation of the current–time curve from linear diffusion theory can be made, and the final result must fit these experimental facts.

The number of moles of a substance which diffuse past a given cross-sectional area of A cm² in a time dt is proportional to the concentration gradient of the diffusing species:

$$\frac{dN}{dt} = KA \frac{\partial C_O}{\partial x} \tag{3-4}$$

The proportionality constant K can be identified as the diffusion coefficient D_O:

$$\frac{dN}{dt} = D_O A \frac{\partial C_O}{\partial x} \tag{3-5}$$

Equation (3-5) is called Fick's first law, and it relates diffusion rates with concentration. A very useful modification of this equation is obtained if it is put on a unit area basis as

$$\frac{dN}{A\,dt} = D_O \frac{\partial C_O}{\partial x} \tag{3-6}$$

Equation (3-6) defines the *flux* of material to be the number of moles diffusing per unit time through unit area. The flux is most frequently given the symbol q; thus

$$q = \frac{dN}{A\,dt} = D_O \frac{\partial C_O}{\partial x} \tag{3-7}$$

Now, if one considers the electrolysis over a period of time, as an observer standing close by the electrode surface, it is evident that C_O and hence $\partial C_O/\partial x$ must vary and, in fact, decrease with time, since C_O is being consumed at the electrode. Thus it becomes necessary to know C_O as both a function of distance from the electrode and time. In this connection C_O should be written most properly as $C_O(x, t)$.

Now, the change in C_O with time, between two planes at distances x and $x + dx$ from the electrode surface ($x = 0$), is clearly the difference between the number of moles of Ox entering at plane $x + dx$ and leaving at plane x, if the process is considered on a unit area of flux basis. Thus

$$\frac{\partial C_O}{\partial t} = \frac{q(x + dx) - q(x)}{dx} \tag{3-8}$$

LINEAR DIFFUSION TO PLANE ELECTRODES

The right side of Eq. (3-8) can be identified with $\partial q/\partial x$ as $dx \to 0$, thus:

$$\frac{\partial C_O}{\partial t} = \frac{\partial q}{\partial x} \tag{3-9}$$

But the flux already has been given by Eq. (3-7) as

$$q = D_O \frac{\partial C_O}{\partial x}$$

Thus

$$\frac{\partial q}{\partial x} = D_O \frac{\partial^2 C_O}{\partial x^2} \tag{3-10}$$

Substituting for $\partial q/\partial x$ from Eq. (3-9) gives

$$\frac{\partial C_O}{\partial t} = D_O \frac{\partial^2 C_O}{\partial x^2} \tag{3-11}$$

which is Fick's second law, the fundamental equation for linear diffusion. For its solution, the following conditions apply in the present case:

at $t = 0$: $C_O^e = C_O^b$

at $t > 0$: $C_O^e = 0$ since $E_{app} \ll E^{o\prime}$

Also,

$$C_O \to C_O^b \quad \text{as} \quad x \to \infty$$

A standard treatment of this problem, as it exists for heat transfer, can be transposed to the electrolysis situation. The solution, in terms of $C_O(x, t)$ is

$$C_O(x, t) = C_O^b \frac{2}{\pi^{1/2}} \int_0^{x/2D^{1/2}t^{1/2}} e^{-y^2} \, dy \tag{3-12}$$

The integral $(2/\pi^{1/2}) \int_0^Z e^{-y^2} \, dy$ is called the error function of Z (erf Z). The value of erf Z depends only on Z. The integral is evaluated numerically and can be found in standard tables. A very abbreviated summary which suffices for this discussion is given in Table 3-1. It is observed that as $Z \to \infty$, the value of erf $Z \to 1$, and, in fact, does so rapidly. For $Z > 2$, erf Z, for all practical purposes, is equal to unity. From the point of view of acquiring a practical picture of the process of linear diffusion, it is worthwhile removing some of the abstract cloaks of erf Z.

TABLE 3-1

Z	erf Z
0	0.0000
0.2	0.2227
0.4	0.4284
0.8	0.7421
1.0	0.8427
1.2	0.9103
1.6	0.9764
2.0	0.9953
2.5	0.9996
3.0	0.999
∞	1.000

We can write Eq. (3-12) as

$$C_O(x, t) = C_O^b \cdot \text{erf } Z \tag{3-13}$$

where

$$Z = \frac{x}{2D^{1/2}t^{1/2}}$$

According to Table 3-1, Z is simply a dimensionless number. This is verified quickly by inserting the proper dimensions of cm, cm²/sec, and sec for x, D, and t, respectively. Now, Z must be a relatively small quantity, or else erf Z is practically unity. In that case, since erf Z is simply a multiplier of C_O^b, $C_O(x, t)$ would be essentially unchanged. This is contrary to the entire electrolysis picture. Suppose a small-sized observer were stationed a reasonable distance from the electrode surface, say at $x = 4 \times 10^{-3}$ cm. Observing after $t = 10$ sec and using $D = 1 \times 10^{-5}$ cm²/sec, these values can be substituted in the expression for Z to give

$$Z = \frac{4 \times 10^{-3}}{(2)(1 \times 10^{-5})^{1/2}(10)^{1/2}} = 0.20$$

From the table for $Z = 0.20$, erf Z is approximately 0.22 and C_O (4 × 10⁻³ cm, 10 sec) = $0.22 C_O^b$, which is an entirely sensible result.

Rearranging Eq. (3-13) to the form

$$\frac{C_O(x, t)}{C_O^b} = \text{erf } Z$$

illustrates how it is used to calculate the concentration–distance profile for selected values of the time of electrolysis (3). For the consumption

LINEAR DIFFUSION TO PLANE ELECTRODES 49

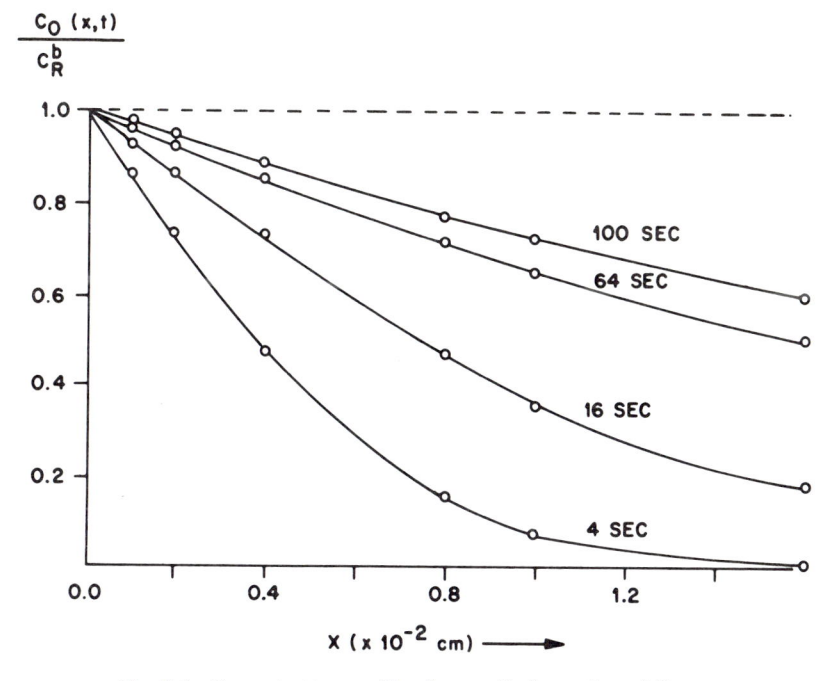

Fig. 3-1. Concentration profiles for anodic formation of Ox.

of Ox in a reduction, this is usually plotted as the ratio $C_O(x, t)/C_O^b$ vs. distance. For the production of Red, the expression corresponding to Eq. (3-13) is

$$C_O(x, t) = C_O^b \, \text{erfc} \, Z$$

where erfc Z, the error-function complement, is

$$\text{erfc} \, Z = 1 - \text{erf} \, Z$$

Thus it is easy also to draw the concentration–distance profile for the species formed during the electrolysis. This is shown in Fig. 3-1, which illustrates an anodic oxidation at a solid electrode. With the process

$$\text{Red} \rightleftharpoons \text{Ox} + e$$

the ratio $C_O(x, t)/C_R^b = 1 - \text{erf} \, Z$, since Ox is the product in an oxidation. The concentration–distance profile is plotted for several times of electrolysis. The particular times were obviously chosen for their convenient square roots.

Alternatively, one can assign a time and calculate the distance, x, at which $C_O(x)$ is essentially equal to C_O^b (returning to a reduction process). This distance is commonly called the thickness of the diffusion layer and is seen to expand into the solution as electrolysis proceeds. It is somewhat arbitrary, depending on the limit one sets for the equality of $C_O(x)$ and C_O^b, but an approximate value for 10 sec using the diffusion coefficient above would be about 0.3 mm (with C_O at 0.3 mm still less than C_O^b by about 3%).

Returning now to the actual current flow, the instantaneous current will be directly proportional to the flux at $x = 0$:

$$i_t = nFAq(0, t) = \left(nFAD\frac{\partial C_O}{\partial x}\right)_{0,t} \tag{3-14}$$

The value of the concentration gradient at the electrode surface, $(\partial C_O/\partial x)_{0,t}$, can be obtained by differentiating Eq. (3-12) or its equivalent (3-13) and evaluating at $x = 0$. This gives the remarkably simple result

$$\left(\frac{\partial C_O}{\partial x}\right)_{x=0} = \frac{C_O^b}{\pi^{1/2} D_O^{1/2} t^{1/2}} \tag{3-15}$$

which, when substituted in Eq. (3-14) gives the final expression for the instantaneous current at a plane electrode under semiinfinite linear diffusion control (4):

$$i_t = \frac{nFAD_O^{1/2} C_O^b}{\pi^{1/2} t^{1/2}} \tag{3-16}$$

Comparing this result with the original experimental observations it can be seen that the current is directly proportional to bulk concentration and the area of the electrode and that it decreases inversely with $t^{1/2}$. In terms of experimental utility, Eq. (3-16) can be rewritten as

$$i_t = KC_O^b \quad \text{or} \quad (i_t t^{1/2}) = K' C_O^b \tag{3-16a}$$

which indicates a possible application to analysis. Also,

$$A = \frac{(i_t t^{1/2})\pi^{1/2}}{nFD_O^{1/2} C_O^b} \tag{3-16b}$$

Hence the "electrochemical" area of an electrode can be determined provided n and D are known. Finally, since

$$D_O = \left(\frac{i_t t^{1/2} \pi^{1/2}}{nFAC_O^b}\right)^2 \tag{3-16c}$$

LINEAR DIFFUSION TO PLANE ELECTRODES

linear diffusion currents are well suited to the evaluation of diffusion coefficients. The determination of D values is discussed in detail in Chapter 8. Measurements in which $it^{1/2}$ curves are observed are called chronoamperometric methods.

Since the instantaneous current in Eq. (3-16) is purely diffusion-controlled, it is often given the symbol i_d. The term i_t is somewhat preferable, since it emphasizes the transient nature of the instantaneous current. The units of Eq. (3-16) are:

i_t = instantaneous current, amperes
F = coulombs
n = number of electrons involved in overall electrode reaction, i.e., n_T
A = area, cm²
D_O = diffusion coefficient, cm²/sec
C_O^b = bulk concentration, moles/ml
t = time, sec

Figure 3-2 illustrates the type of electrode at which semiinfinite linear diffusion conditions can be realized approximately. It consists of a circular metal-foil electrode sealed horizontally across a glass tube. The mantle restricts diffusion to an upward mode and eliminates peripheral contributions. The entire lower end may be bent in an inverted V shape if downward diffusion is to be examined—or it can be bent horizontally. The mantle need be really only a few millimeters extension. The usual duration of electrolysis is short and the mantle need be only several times longer than the maximum thickness of the diffusion layer. From the practical viewpoint, when the mantle is too long, air pockets and bubbles

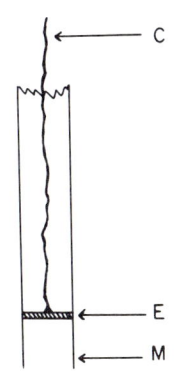

Fig. 3-2. Simple linear diffusion electrode. C, copper wire connection; E, circular foil electrode (Pt, Au, etc.); M, mantle or shield (glass, Teflon, etc.).

are difficult to dislodge. Sealing a smooth, horizontal foil in the glass tube is not so easy a task as the diagram implies. Another approach is to seal a foil on the end of the glass tube and build up a cylindrical mantle by painting ceresin wax till a ridge 4–5 mm high is obtained. Such a mantle is naturally not perfectly perpendicular to the foil surface but seems to work reasonably well.

Precise studies of linear diffusion to solid electrodes were first made by Laitinen and Kolthoff (5,6). Equation (3-16) was examined with a variety of electrodes like the one pictured in Fig. 3-2. In the oxidation of ferrocyanide, excellent agreement with theory was obtained provided upward diffusion was maintained. Discrepancies with the other electrodes were traced to density gradients produced during the electrolysis. The oxidation product, ferricyanide, is less dense than the ferrocyanide.

Very exacting studies were undertaken by von Stackelberg et al. in an evaluation of polarographic diffusion coefficients (7). Their excellent data on diffusion coefficients as a function of supporting electrolyte can be used with confidence for precise calculations of electrode area, peak currents, etc., in linear diffusion techniques. The original paper should be consulted for details of the electrode construction.

An experimental evaluation of linear diffusion is usually made by checking the constancy of the product $it^{1/2}$. This should be carried out at controlled potential, rather than merely applying a voltage to the cell from a polarographic bridge. Obviously the current decreases with time and the iR loss shifts the real potential of the working electrode accordingly. The applied potential should be sufficiently anodic or cathodic, as the case may be, to be past the rising portion of the corresponding peak polarogram (see Chapter 5). If the electrode reaction is well separated from other possible electrode processes, an applied voltage of several hundred millivolts beyond the peak potential is satisfactory, and the iR effect is a minor problem. Otherwise, potentiostatic control of the electrode should be maintained. Using shielded boron carbide electrodes Mueller et al. were able to get quite satisfactory results with ferrocyanide oxidation in potassium chloride media (8–10). Table 3-2 shows the constancy of $it^{1/2}$ for a typical case. The applied voltage was derived from a Leeds and Northrup Model E electrochemograph. With this particular instrument, convenient values can be read every 3 sec from the chart paper, which accounts for the selection in the first column of the table. Similar results can be obtained with any recording polarographic instrument, but care should be taken that at least 1-sec pen response is available on the recorder. An $it^{1/2}$ curve should always be run on the background and subtracted

TABLE 3-2
Constancy of $it^{1/2}$ Ferrocyanide Oxidation in 0.5 M KCl

t, sec	$it^{1/2}$ [a]
12	50.7
15	50.3
18	50.2
21	50.1
24	49.9
27	50.0
30	50.2
33	50.2
36	50.2
39	50.3
42	50.4
45	50.5
60	51.1

[a] Units are $\times 10^{-6}$ A-sec$^{1/2}$.

point by point from the curve of the electroactive species. While these corrections are negligible in some cases, this fact should be checked experimentally.

Mueller found that repetitive determination of $it^{1/2}$ on the same solution, or on solutions of the same composition using the same electrode, showed variations of about 0.5%. Using different electrodes, the over-all reproducibility was about $\pm 5\%$. With an operational-amplifier type of controlled potential polarograph it was verified that the simple applied voltage method was satisfactory for systems like ferrocyanide, where a wide range of applied voltages can be used.

In practice the attainment of $it^{1/2}$ constancy is sensitive to horizontal positioning of the shielded electrode. If, with a given shielded electrode, linear diffusion conditions apparently are not obtained, the electrode positioning should be varied and several $it^{1/2}$ curves run until constancy is obtained. With recording polarographs the current for the first 5 sec may not be in accord with Eq. (3-16), owing to recorder lag. Beyond 45 sec, $it^{1/2}$ generally increases as a result of natural convection, although many workers have obtained constancy for longer periods. Morris and Schempf showed that their WIGE with restricting glass mantle obeyed $it^{1/2}$ constancy up to about 200 sec (11). Other workers have maintained linear diffusion conditions for long periods (5–7).

Zimmerman carried out very exhaustive studies of linear diffusion at practical solid electrodes (36). Carbon paste and platinum electrodes with mantle-type shields consistently showed positive deviations in the $it^{1/2}$ curves which were appreciable after about 60 sec. For times less than 30 sec, deviations were, in general, minor. However, to get true semiinfinite linear diffusion conditions it was necessary to use a shielded electrode almost identical to that originally built by von Stackelberg et al. (7). Useful directions for building such an electrode were given by Macero (37). The electrode used by Zimmerman employed gold foil as the electrode surface. With this electrode (which is particularly unsuitable for practical work) $it^{1/2}$ constancy for very long periods can be achieved. This is illustrated by the data of Table 3-3 for the reduction of silver ion in 0.2 M KNO_3 as supporting electrolyte.

TABLE 3-3

Constancy of $it^{1/2}$ for Silver Ion Reduction at the von Stackelberg Electrode[a]

Time, sec	i, μA × 10	$it^{1/2}$, μA-sec$^{1/2}$ × 10
15	346.1	1340
30	240.8	1319
45	195.1	1309
60	167.9	1301
120	119.0	1303
180	97.1	1302
240	84.0$_5$	1302
300	75.1	1301
360	68.6	1301
420	63.5	1301
540	56.0	1301
600	53.1$_5$	1302
660	50.7	1302
720	48.5	1301
780	46.6	1303
840	44.9	1301
900	43.3	1299
960	42.0	1301
1080	39.5$_6$	1300
1140	38.5$_1$	1300
1200	37.5$_3$	1300
1260	36.6$_1$	1300
1380	35.0$_2$	1301
1500	33.5$_9$	1301
Mean		1303
Std. dev. of mean		±1

[a] Electrode area (geometric) = 0.124 cm², applied potential + 0.250 V vs. SCE.

Very approximate linear diffusion conditions can be obtained with unshielded electrodes; e.g., ca. 1-cm² vertical Pt foils and unshielded platinum disks and carbon paste electrodes frequently show $it^{1/2}$ "constancy" over short periods of electrolysis. Bard has shown, however, that unshielded electrodes deviate considerably from linear diffusion when examined by chronopotentiometric techniques (12). Lingane investigated planar, unshielded, circular disk electrodes for both chronoamperometric ($it^{1/2}$ curves) and chronopotentiometric conditions (40). With these electrodes the $it^{1/2}$ equation is of the form

$$i_{(t)} = \frac{nFAD^{1/2}C^b}{\pi^{1/2}t^{1/2}} \left[1 + a\left(\frac{DT}{r^2}\right)^{1/2} \right]$$

A theoretical analysis by Soos and Lingane (41) showed that the constant, a, was 2.26 and experimental evaluations on a variety of systems gave excellent agreement with $a = 2.12 \pm 0.11$. This modification of Eq. (3-16) can be applied to circular foil electrodes via extrapolation of data to $t^{1/2} = 0$.

Actually, unshielded carbon paste electrodes correspond to the geometry described above. In practice they show agreement with Eq. (3-16) for short periods of electrolysis. This is illustrated by the data of Table 3-4 for the oxidation of 4-methylcatechol in ca. 1 M sulfuric acid. The carbon paste electrode had an apparent area of 0.32 cm² and the reproducibility

TABLE 3-4

$it^{1/2}$ Constancy at Unshielded Circular Carbon Paste Electrodes[a]

	Surface 1		Surface 2		Surface 3	
t, sec	i, μA	$it^{1/2}$	i, μA	$it^{1/2}$	i, μA	$it^{1/2}$
2.25	69.3	104	70.6	106	71.0	106
3.00	60.0	104	60.7	105	61.0	106
3.75	53.3	103	54.2	105	54.3	105
4.50	48.8	103	49.6	105	49.5	105
5.25	45.1	103	45.8	104	45.8	104
6.00	42.2	103	42.8	105	42.8	105
6.75	40.0	104	40.3	105	40.2	104
7.50	37.9	104	38.3	105	38.4	105
8.25	36.3	104	36.7	105	36.7	105
9.00	34.7	104	35.1	105	35.1	105

[a] Oxidation of 4-methylcatechol in 0.945 M sulfuric acid, $E_{app} = +0.70$ V (potentiostatic control), new carbon paste surfaces prepared after each run.

TABLE 3-5

$it^{1/2}$ Constancy with Unshielded Platinum Button Electrode[a]

t, sec	i, μA	$it^{1/2}$
1.55	40.2	50.2
2.33	32.7	49.9
3.11	28.0	49.5
3.89	25.0	49.3
5.44	21.2	49.5
6.98	18.7	49.4
7.77	17.8	49.6
9.31	16.2	49.5
11.66	14.5	49.7
13.21	13.7	49.8
15.54	12.7	50.0
17.88	11.7	49.7
19.42	11.3	49.8
21.76	10.7	49.9
23.3	10.3	50.0
27.2	9.6$_5$	50.3
31.1	9.0	50.3
35.0	8.5	50.3

[a] Oxidation of 2×10^{-3} M N(CH$_3$)$_2$—C$_6$H$_4$—OCH$_3$ in propylene carbonate, supporting electrolyte 0.25 M tetraethylammonium perchlorate, Beckman platinum button electrode, $E_{app} = 0.85$ V (potentiostatic control).

of the $it^{1/2}$ curves is shown on three fresh electrode surfaces. Table 3-5 shows that a platinum button electrode (commercial Beckman-type electrode) has $it^{1/2}$ constancy to within ±1–2% for reasonable periods of time. This electrode is that used by Lingane for the experimental studies mentioned above. The data of Table 3-5 were obtained in propylene carbonate, which has a high viscosity and works well for $it^{1/2}$ studies.

The *approximate* application of Eq. (3-16) or the Lingane modified form may be made to other planar electrodes, but it should not be extended to wire electrodes. With wire electrodes the pseudo-$it^{1/2}$ curve never attains any degree of constancy in the time interval required with a recording polarograph.

One of the practical aspects of $it^{1/2}$ curves is the determination of electrode areas through Eq. (3-16b). The area so determined is designated the electrochemical area, which, under certain circumstances, may differ

LINEAR DIFFUSION TO PLANE ELECTRODES 57

considerably from the measured (geometric) area. Using the D values for ferri- and ferrocyanide in 0.5 M potassium chloride given by von Stackelberg et al. (7), the area of a B_4C electrode was determined by alternate oxidation of ferrocyanide and reduction of ferricyanide. The values of 0.373 and 0.367 cm² obtained differ by only 2%. A comparison of geometric and electrochemical areas for several B_4C electrodes shows the surface roughness differences as illustrated in Table 3-6.

TABLE 3-6
Surface Area of B_4C Electrodes

Electrode	Geometric area, cm²	Electrochemical area,[a] cm²
Polished	0.36	0.35$_3$
Rough A	0.18	0.19$_3$
Rough B	0.44	0.53$_1$
Rough C	0.7	0.94$_2$

[a] Via $it^{1/2}$ curves for ferrocyanide oxidation in 0.5 M KCl.

Again, using the von Stackelberg-type electrode, Zimmerman demonstrated how precisely electrochemical electrode areas can be determined from $it^{1/2}$ curves (36). Some of the data are shown in Table 3-7. Since

TABLE 3-7
$it^{1/2}$ Measurement of Electrode Areas

Series	System	$it^{1/2}/C^b$ [a]	Area, cm² [b]
1	Ferrocyanide, 2 M KCl	172.4	0.1263
1[c]	Ferrocyanide, 2 M KCl	172.4	0.1263
2	Ferrocyanide, 2 M KCl	172.3	0.1262
2[c]	Ferrocyanide, 2 M KCl	173.4	0.1270
3	Ferrocyanide, 2 M KCl	171.6	0.1257
2[c]	Ag^+ in 0.1 M KNO_3	274.6	0.1276
4	Ag^+ in 0.1 M KNO_3	271.6	0.1265
Mean			0.1265
Std. dev. of mean			±0.0004

[a] Average of three or more determinations.
[b] Based on D values of von Stackelberg et al. (7).
[c] Indicates the same electrode with a solution of different concentration.

the areas were calculated using the D values for ferrocyanide and silver ions given by von Stackelberg et al., they attest to the reliability of these data.

Electrochemical areas can be determined chronopotentiometry. This technique embodies semiinfinite linear diffusion and will be discussed later.

Equation (3-16) is derived on the basis that $C_O^e \to 0$. The potential of the electrode must be controlled at a value sufficiently negative (for a reduction) that this condition is met—or the applied voltage must be great enough to allow a potential "swing" during electrolysis. However, it is possible to get diffusion-controlled current with E_{app} such that C_O^e is other than zero but constant. Under these conditions it can be shown that the final expression for the current is (13)

$$i'_t = \frac{nFAD_O^{1/2}(C_O^b - C_O^e)}{\pi^{1/2}t^{1/2}} \qquad (3\text{-}17)$$

The current i'_t is obviously less than the maximum i_t of Eq. (3-16). Since the term $C_O^b - C_O^e$ is not an experimental quantity, Eq. (3-17) cannot be applied directly to area measurements or D-value determinations.

Further applications of the transient currents of linear diffusion are limited. Although the maximum i_t is directly proportional to bulk concentration, which is an obvious prerequisite for analytical usage, few applications have appeared.

Linear diffusion methods are used, of course, in a more conventional polarographic sense. A linear voltage sweep is applied to the cell and the resulting current–voltage curve is recorded as a so-called peak polarogram. This technique is described in Chapter 5 in detail and is sometimes called linear potential sweep chronoamperometry. Many workers prefer to bypass linear diffusion and use rotated electrodes where steady-state currents are obtained.

An interesting technique which eliminates any attempt to measure the instantaneous current was devised by Booman et al. (14). It was applied to cylindrical electrodes and is discussed in the next section. Briefly, however, one integrates the total charge transferred during a very short interval of time after initiation of electrolysis. The charge integral is a linear function of concentration of electroactive species.

Crittenden and co-workers used wire electrodes and employed short periods of electrolysis to avoid convention. This same technique should be applicable on a larger time scale using a shielded linear diffusion electrode. Figure 3-3 shows some experimental data on the oxidation

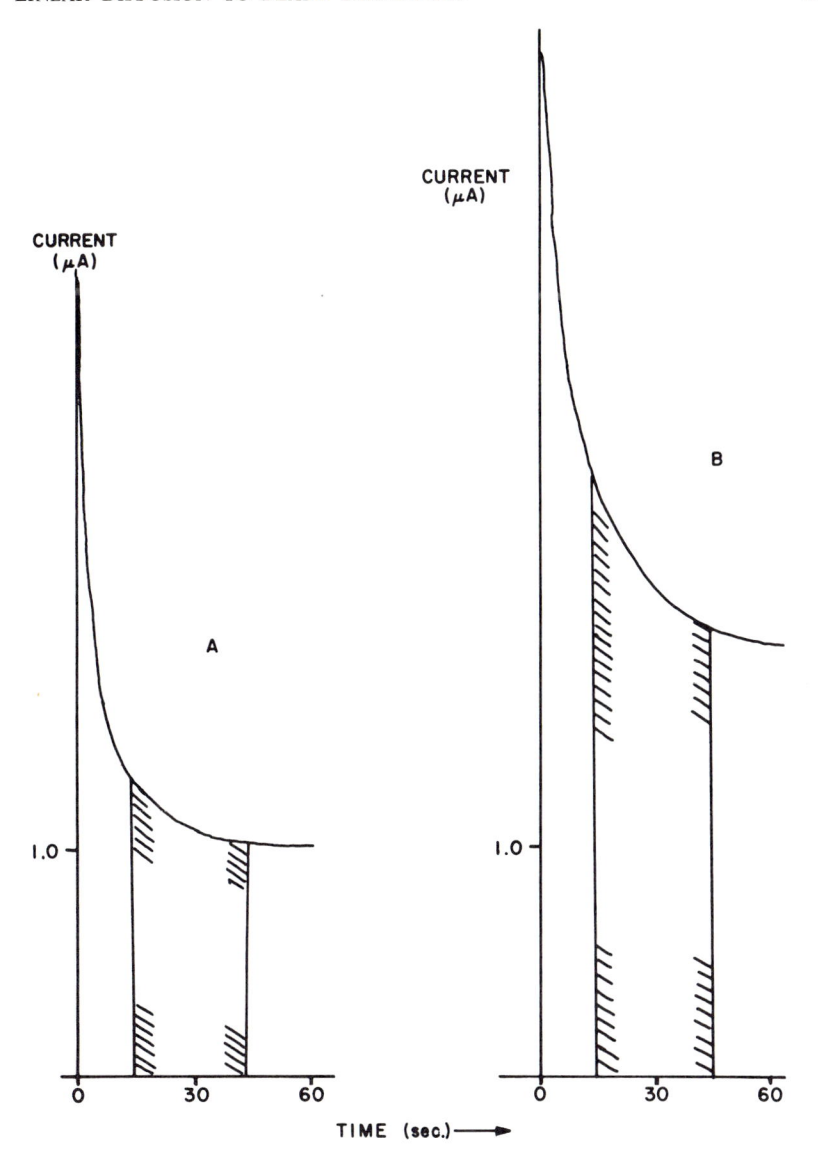

Fig. 3-3. Current–time curves for oxidation of 3,4-diaminotoluene. A, 8.0×10^{-5} M 3,4-diaminotoluene; B, 1.6×10^{-4} M 3,4-diaminotoluene. In 0.1 M hydrochloric acid, shielded Pt-foil electrode similar to that of Fig. 3-2.

of 3,4-diaminotoluene in 0.1 M hydrochloric acid at a shielded Pt foil electrode (similar to that of Fig. 3-2). This compound was chosen because it is typical of a variety of aromatic compounds whose oxidations are partially controlled by charge transfer and which tend to produce films on the electrode surface in addition to other complications. The applied potential of +0.52 V vs. SCE was chosen from previous polarographic runs to be such that $C_R^e = 0$; i.e., the maximum instantaneous current was obtainable at this E_{app}. The potential was applied from an Electrochemograph and the current decay with time recorded, as indicated, for a total period of 60 sec. Curves A and B of Fig. 3-3 are for diaminotoluene concentrations of 8.0×10^{-5} and 1.6×10^{-4} M, respectively. Assuming recorder lag to make the first 15 sec unreliable, the charge was integrated over the period 15–45 sec by simply counting squares on the chart paper under the curve. The areas for curves A and B are 221.9 and 434.2 units, respectively. Since area B should be exactly twice that of A, this is an error of about 2%. Actually, this is an extremely gratifying result in view of the fact that the system is operating far from linear diffusion control. Table 3-8 shows that the product $it^{1/2}$ for curve A is seriously increasing even in the short interval between 15 and 45 sec. Convection should be small with a shielded electrode operating properly. Most likely the cause can be ascribed to the nature of the diaminotoluene oxidation. This and similar molecules show abnormal specific limiting currents and are known to have follow-up chemical reactions producing products which are also electroactive at the applied potentials. Thus at a stationary electrode the currents are higher than expected. The approximate agreement between the two charge integrals is a result of compensating effects. When the relative concentrations differ by a factor of 4, the charge integrals show a discrepancy of some 15–20%.

TABLE 3-8

$it^{1/2}$ Variation of Curve A, Fig. 3-3

t, sec	i, μA	$it^{1/2}$
15	1.28	4.95
18	1.23	5.20
21	1.17	4.36
24	1.13	5.53
27	1.11	5.76
36	1.07	6.42
42	1.05	6.80
45	1.04	6.97

In any event, the technique above is not a practical one for such systems. Less ambiguous results can be obtained with much more ease using voltage scanning techniques at either stationary or rotated electrodes. The material was presented because it illustrates several points of interest in linear diffusion studies. Especially with aromatic organic systems one must exercise considerable care in selecting applicable techniques for routine analysis. Precise data on such compounds are only beginning to appear, and it is particularly unwise to generalize or predict any specific behavior.

Integration of $it^{1/2}$ curves is considerably more involved than implied in the crude experiments illustrated above. Osteryoung and co-workers utilized current integration upon application of a potential step. Their approach was completely different in nature from that of Crittenden et al. in that they essentially were concerned with the variation of charge passed as a function of time (*42,43*). From a plot of charge Q vs. $t^{1/2}$ one can determine kinetic parameters of the electrode reaction and information about adsorption. The original literature should be consulted for details of the procedure (*42–44*). Osteryoung, Lauer, and Anson have employed integration of peak polarograms (see Chapter 5) for similar purposes (*44–46*). Very recently Pool et al. have again employed $it^{1/2}$ integration to examine the surface oxidation of platinum electrodes (*49*).

3-3. CYLINDRICAL DIFFUSION

The equations for the instantaneous current at a cylindrical electrode are considerably more complex than that for linear diffusion. Equation (3-18) shows this:

$$i_t = nFAD_0C_O^b \frac{1}{r}\left[\frac{1}{\pi^{1/2}\phi^{1/2}} + \frac{1}{2} - \frac{1}{4}\left(\frac{\phi}{\pi}\right)^{1/2} + \tfrac{1}{8}\phi \cdots \right] \quad (3\text{-}18)$$

where $\phi = D_0 t/r^2$ and r = radius of the cylinder. This form of the expansion is applicable for small values of the dimensionless quantity $D_0 t/r^2$. In practice, Eq. (3-18) is not as formidable as it seems at first glance. When $D_0 t/r^2$ is quite small, all but the first term inside the brackets can be dropped. Noting the substitution for ϕ, it can be seen that the result is then identical with the linear diffusion equation. Typically, if $r = 0.1$ cm, the linear diffusion approximation is satisfactory for $t > 10$ sec.

For long periods of electrolysis the equation can be written in other forms which show that the current approaches zero as t increases. It does so very slowly, however, because it turns out the current decreases inversely with the logarithm of time (*15*).

A wire electrode extending vertically downward into solution approximates cylindrical diffusion. A small glass bead sealed on the bottom can be used to eliminate end effects. Berzins and Delahay studied the deposition of cadmium on Pt with such an electrode (*16*). Using an oscillographic technique with its accompanying short period of electrolysis they applied the linear diffusion approximation. Laitinen and Kolthoff in their early studies of Pt wires showed that a fairly steady current was obtained after about 2–3 min. Convection effects were believed to be prevalent (*5*).

Nicholson used a rigorous approach in a study of the oxidation of organic sulfides at Pt wires, taking into account the curvature. She concluded that convection effects were minor using a relatively slow voltage sweep to obtain peak polarograms (*17*).

Booman et al. in their charge integration work mentioned earlier, used ordinary open-end Pt wires of 0.4 mm radius and 5.5 mm length (*14*). For short electrolysis times the experimental results agreed with Eq. (3-18) using only the first term, i.e., linear diffusion approximation. (The diffusion-controlled charge is directly related to i_t.) Usually, however, they included the first three terms of the equation in their calculations. They obtained excellent linearity of total charge vs. concentration for the reduction of hydrogen ion and ferricyanide, as well as the oxidation of ferrocyanide and iodide ions. The integration was performed with an electronic integrating network. Details are to be found in the original paper and the thesis of Booman (*18*). Beilby and Crittenden also applied the integration technique to Pt-foil electrodes (*19*).

3-4. DIFFUSION TO SPHERICAL ELECTRODES

Solid spherical electrodes per se are seldom used. Nevertheless, several practical working electrodes operate under conditions where spherical diffusion plays a part.

It is possible to derive the i–t characteristics for spherical diffusion by methods analogous to those of the previous sections (*15*). For the maximum instantaneous current one has

$$i_t = 4\pi r n F D_O C_O^b + \frac{nFAD_O^{1/2}C_O^b}{\pi^{1/2}t^{1/2}} \qquad (3\text{-}19)$$

where r is the radius of the sphere and the other symbols have their usual significance. Conditions of electrolysis are as before—migration and convection are absent and E_{app} is adjusted for $C_O^e = 0$ for values of $t > 0$.

The current is composed of a constant term plus a transitory portion. The latter is identical with that obtained in linear diffusion. After long periods of electrolysis the current becomes approximately equal to the constant term. It is difficult to prevent convection with prolonged electrolysis, and Laitinen and Kolthoff (5) were unable to verify Eq. (3-19) in aqueous solution. With gels, where convection was minimized, closer approach to theory was observed.

For very short periods of electrolysis (ca. <10 sec) the current is controlled mainly by the transient term. Linear diffusion equations can be applied to the sphere within this time interval without too much error. This approximation is also a function of the size of the sphere.

Unshielded rod type electrodes, where only the end is used (carbon rods, WIGE, etc.), as well as pool configuration carbon paste electrodes, probably have some (hemi) spherical diffusion contributions. The deliberate use of *solid* spherical electrodes can hardly be recommended.

3-5. CONVECTION IN UNSTIRRED SOLUTION

Natural or free convection develops spontaneously in any solution undergoing electrolysis since density differences develop near the electrode. This mode of mass transfer is of particular importance at vertical electrodes in unstirred solution.

Theoretical treatments of natural convection have been given by Levich (20) and Agar (21). Wagner has presented a thorough treatment of convection processes at vertical electrodes (22). Natural convection control differs from diffusion in at least one important aspect: Limiting currents are theoretically proportional to the *1.25th power* of the concentration. Experimental verification has been found in several instances of metal depositions on vertical plate electrodes (22–24).

Ibl et al. (25) examined the deposition of Cd, Tl, and Cu on vertical plate cathodes in unstirred solution. The height of the electrodes varied from 0.1 to 4 cm. Provided concentrations exceeded ca. $5 \times 10^{-3} M$, the limiting currents were in good agreement with natural convection theory. At lower concentrations considerable deviations from the theoretical were obtained. For instance, limiting currents were found to be proportional to $C^{0.5}$ to $C^{0.7}$ as opposed to the theoretical $C^{1.25}$.

In view of the above observations on natural convection, it is interesting to examine the available data on vertical wire electrodes in unstirred solution. Laitinen and Kolthoff (5) showed that the current decreased rapidly at first but soon reached a relatively steady state in 1–3 min. The steady-state limiting currents were directly proportional to concentration. Such behavior has generally been reported for wire electrodes. In contrast to the data with plate electrodes, Ibl et al. found limiting currents directly proportional to concentration for vertical wires (25).

Paulopoulos and Strickland (26) indicated that limiting currents at vertical wires, especially in the case of metal depositions, were proportional to $C^{1.2}$. An exponent of 1.2 was also found for the reduction of ferric ion in 1.5 M sodium perchlorate as the background electrolyte. The limiting currents were measured after a 3-min wait for the so-called steady-state value. O'Brien and co-workers have studied natural convection in metal depositions using interferometer methods (47,48).

Ibl has derived an equation for the limiting current at a wire electrode which shows $C^{1.25}$ dependence. In addition, the equation accounts for the density effects responsible for natural convection in the practical voltammetric sense, i.e., for the reduction or oxidation of an electroactive species in the presence of supporting electrolyte. While this density effect is a difficult one to calculate in the absolute sense, Ibl has shown remarkable agreement between his calculated limiting currents and those determined experimentally by various workers at wire electrodes (27). It should be emphasized that those currents are steady-state convection currents and not those obtained by relatively rapid scan voltammetry at wire electrodes.

For a thorough coverage of details on natural convection effects, the reader is referred to the contribution of Ibl and Muller (28–32) and the other references listed (33–35,38,39,47,48).

It is evident that some care should be taken when using stationary electrodes in unstirred solution for quantitative measurements. In particular, vertical plate or foil electrodes should probably be avoided, since it appears difficult to reproduce natural convection conditions. Stationary wire electrodes seem to reach their steady-state convection condition more easily. It is clear that the best usage of stationary electrodes lies in rapid scanning techniques, where the short time interval minimizes natural convection effects.

REFERENCES

1. For complete derivations of many of these transport processes, see P. Delahay, *New Instrumental Methods of Analysis*, Wiley (Interscience), New York, 1954, and the literature cited later.
2. I. M. Kolthoff and J. J. Lingane, *Polarography*, Vol. 1, Wiley (Interscience), New York, 2nd ed., 1952, p. 21.

REFERENCES

3. G. W. C. Milner, *The Principles and Applications of Polarography and Other Electroanalytical Processes*, Longmans, London, 1957, pp. 32, 33.
4. For detailed solution see Refs. *1–3*.
5. H. A. Laitinen and I. M. Kolthoff, *J. Phys. Chem.*, **45**, 1062 (1941); *J. Am. Chem. Soc.*, **61**, 3344 (1949).
6. H. A. Laitinen, *Trans. Electrochem. Soc.*, **82**, 289 (1942).
7. M. von Stackelberg, M. Pilgram, and V. Toome, *Z. Elektrochem.*, **57**, 342 (1953).
8. T. R. Mueller, C. L. Olson, and R. N. Adams, *Proc. 2nd Intern. Polarog. Congr.*, Cambridge, *1959*, p. 198.
9. T. R. Mueller and R. N. Adams, *Anal. Chim. Acta*, **23**, 467 (1960).
10. T. R. Mueller and R. N. Adams, *Anal. Chim. Acta*, 482 (1961).
11. J. B. Morris and J. M. Schempf, *Anal. Chem.*, **31**, 286 (1959).
12. A. J. Bard, *Anal. Chem.*, **33**, 11 (1961).
13. See Ref. *1*, p. 55, and Ref. *3*, p. 34.
14. G. L. Booman, E. Morgan, and A. L. Crittenden, *J. Am. Chem. Soc.*, **78**, 5533 (1956).
15. Ref. *1*, p. 69.
16. T. Berzins and P. Delahay, *J. Am. Chem. Soc.*, **75**, 555 (1953).
17. M. M. Nicholson, *J. Am. Chem. Soc.*, **76**, 2539 (1954).
18. G. L. Booman, thesis, Univ. Washington, Seattle, 1956.
19. A. L. Beilby and A. L. Crittenden, *J. Phys. Chem.*, **64**, 177 (1960).
20. B. Levich, *Acta Physicochem. USSR*, **17**, 257 (1942); **19**, 117, 133 (1944).
21. J. N. Agar, *Discussions Faraday Soc.*, **1**, 26 (1947).
22. C. Wagner, *Trans. Electrochem. Soc.*, **95**, 161 (1949).
23. C. R. Wilke, M. Eisenberg, and C. W. Tobias, *J. Electrochem. Soc.*, **100**, 513 (1953).
24. G. H. Kevlegan, *J. Res. Natl. Bur. Std.*, **47**, 156 (1951).
25. N. Ibl, K. Buob, and G. Trumpler, *Helv. Chim. Acta*, **37**, 2251 (1954).
26. T. Paulopoulos and J. D. H. Strickland, *J. Electrochem. Soc.*, **104**, 116 (1957).
27. N. Ibl, *Electrochim. Acta*, **1**, 3 (1959).
28. N. Ibl, *Proc. 8th Meeting C.I.T.C.E.*, Madrid, *1956*, Butterworth, London, 1958, p. 174.
29. N. Ibl and R. H. Muller, *J. Electrochem. Soc.*, **105**, 346 (1958).
30. N. Ibl, *Chimia (Aaraw)*, **9**, 135 (1955).
31. N. Ibl, *Helv. Chim. Acta*, **37**, 1149 (1954).
32. N. Ibl and R. Muller, *Z. Elektrochem.*, **59**, 671 (1955).
33. G. Wranglin, *Acta Chem. Scand.*, **12**, 1143 (1958).
34. K. Asada, F. Hine, S. Yoshizawa, and S. Okada, *J. Electrochem. Soc.*, **107**, 242 (1960).
35. E. J. French and C. W. Tobias, *Electrochim. Acta*, **2**, 311 (1960).
36. J. Zimmerman, Ph.D. thesis, Univ. Kansas, Lawrence, 1964.
37. D. J. Macero, Ph.D. thesis, Univ. Michigan, Ann Arbor, 1958.
38. M. G. Fouad and N. Ibl, *Electrochim. Acta*, **3**, 233 (1960).
39. M. G. Fouad and T. Gouda, *Electrochim. Acta*, **9**, 1071 (1964).
40. P. J. Lingane, *Anal. Chem.*, **36**, 1723 (1964).
41. Z. G. Soos and P. J. Lingane, *J. Phys. Chem.*, **68**, 3821 (1964).
42. J. H. Christie, G. Lauer, R. A. Osteryoung, and F. C. Anson, *Anal. Chem.*, **35**, 1979 (1963).
43. J. H. Christie, G. Lauer, and R. A. Osteryoung, *J. Electroanal. Chem.*, **7**, 60 (1964).

44. F. C. Anson, *Anal. Chem.*, **38,** 54 (1966).
45. R. A. Osteryoung, G. Lauer, and F. C. Anson, *Anal. Chem.*, **34,** 1833 (1962).
46. R. A. Osteryoung, G. Lauer, and F. C. Anson, *J. Electrochem. Soc.*, **110,** 926 (1963).
47. R. N. O'Brien and C. Rosenfield, *Nature*, **187,** 935 (1960).
48. R. N. O'Brien, *Nature*, **201,** 74 (1964).
49. K. H. Pool, J. G. Smith, and A. L. Crittenden, *Anal. Chem.*, **38,** 1242 (1966).

4 MASS TRANSFER BY FORCED CONVECTION

4-1. Nernst Diffusion Layer 67
4-2. Hydrodynamics And Forced-Convection Electrodes 71
4-3. Stationary Electrodes In Flowing Solution 76
4-4. Rotated Disk Electrodes 80
4-5. Applications Of RDE To Electrode Kinetics And Mechanisms . . 92
4-6. Mass Transport With Turbulent Flow 102
4-7. Rotated Wire Electrodes . . . . . : 104
4-8. Vibrating Wire Electrodes 107
References 107
Special Bibliography On Rotated Disk Electrodes 110

Mass-transfer processes in stirred solution are difficult to evaluate. The transport conditions depend to a high degree on the cell geometry and mode of stirring. Even when the conditions are known precisely, solutions of the corresponding equations may be difficult. Many practical solid electrode techniques utilize stirred solutions or rotated electrodes, so it is necessary to examine the existing knowledge. In this chapter, after first considering the "diffusion-layer" approach, attention is focused on the few instances where a quantitative hydrodynamic treatment can be given. Migration effects again will be ignored by considering only solutions which contain an excess of indifferent electrolyte.

4-1. NERNST DIFFUSION LAYER

One of the first approaches to mass transfer in electrode processes was given by Nernst in 1904 (*1*). Nernst assumed a stationary thin layer of solution in contact with the electrode. Within this layer it was postulated that diffusion alone controlled the transfer of substances to the electrode.

Outside the layer diffusion was negligible and the concentration of electroactive material was maintained at the value of C^b by convective transfer. This hypothetical layer has become known as the Nernst diffusion layer, δ_N.

The Nernst diffusion-layer hypothesis has been criticized in terms which imply that it is a wrong approach to the problem. Actually the basic treatment is acceptable and one can hardly hold Nernst responsible for shortcomings in modern fluid hydrodynamics. It is now clear that one of the real errors in Nernst's treatment was the assumption that the liquid within the layer of thickness δ_N is motionless. Experimental evidence on the flow of liquids past solid surfaces shows this to be untrue. In addition, Nernst assumed the concentration varied linearly with distance within δ_N or, in terms of the concentration gradient,

$$\frac{\delta C_O(x)}{\delta x} = \frac{C_O^b - C_O^e}{\delta_N} \tag{4-1}$$

The concentration gradient is not linear in forced convection situations. (Note that concentrations are a function of distance x but not time, since steady-state conditions prevail in stirred solutions.) Figure 4-1 is a pictorial representation of the Nernst diffusion layer. The most serious objection to the Nernst treatment lies in its purely qualitative nature. While the concept of the thickness δ_N is very useful experimentally, it allows no predictions as to the variation of δ_N with solution viscosity, diffusion coefficient, etc.

Setting aside the electrochemistry for a moment, the fundamental problem in stirred solutions is that one is dealing with a fluid moving past a solid surface. Exactly at the surface of the solid the fluid flow velocity U is zero. Some distance away, measured in the x direction normal to the solid surface, the flow velocity has a value U_o, characteristic of the bulk of the solution, unaffected by the solid body. Between these extremes there exists, as a result of the viscous properties of the liquid, a thin layer extending out from the solid body, in which there is a large variation of fluid velocity. The region in which this high velocity gradient is observed is called the *boundary layer* or *hydrodynamic boundary layer*. To define the exact distance from the solid to the point where this viscous resistance to flow is no longer felt has little significance. Practically speaking, the thickness of the boundary layer (usually given the symbol δ_o) is that distance in which the main change in velocity gradient appears. The boundary layer exists with all real liquids at ordinary flow velocities.

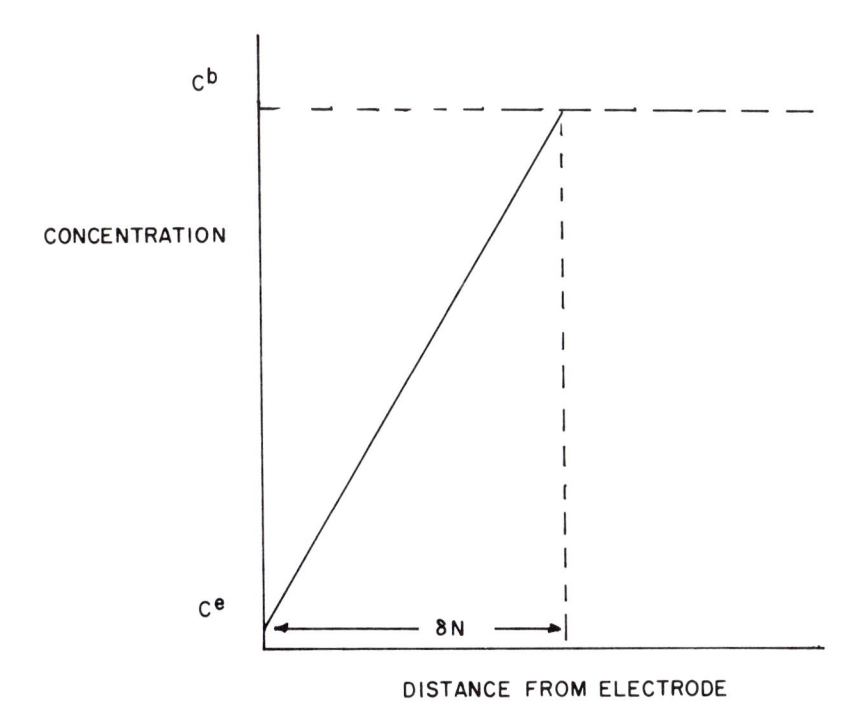

[Figure: plot of CONCENTRATION vs DISTANCE FROM ELECTRODE, showing a linear rise from c^e at the electrode to c^b at distance δ_N, then flat at c^b.]

Fig. 4-1. Representation of Nernst diffusion layer.

If now the moving fluid contains solute, and if conditions are such that the concentration of this solute varies throughout the solution, the effect of molecular or ionic diffusion is added to the situation. If, in the electrochemical case, a potential is applied so that the concentration of solute at the electrode surface is maintained at zero, then solute is transported to the electrode by diffusion as a result of the concentration gradient. Further, the solute molecules are carried along by the physically moving liquid. These two mass-transport processes, molecular or ionic diffusion and convective transfer, constitute *convective diffusion*. Both of these mass-transport processes exist simultaneously in forced convection. One or the other may predominate in magnitude, and in forced convection it can be visualized that diffusion has the minor contribution if one considers the entire solution. However, in a restricted region very close to the electrode both diffusion and convection play major roles and any rigorous treatment must consider both.

Hydrodynamic treatment of the problem leads to the concept of a thin *diffusion boundary layer* close to the electrode surface. The largest change of concentration occurs in this thin layer. The liquid flow velocity within this diffusion boundary layer is not zero except at $x = 0$, i.e., at the electrode surface. Viscosity changes must be taken into account in this layer. Any transition from pure diffusion to convection must take place continuously. The diffusion boundary layer also has no *exactly* defined thickness—it is simply a depth which it is convenient to define as the region within which the *maximum change in concentration occurs*. It can be shown that this diffusion-boundary-layer thickness, δ, is proportional to the physical properties of the solution (flow velocity, viscosity) as well as the value of D. In principle each electroactive species has its own value of δ. The value of δ is several times smaller than that of the hydrodynamic boundary layer, δ_o, in which the overall forces of viscous flow past a surface are encountered. Finally, there exists some distance from the electrode where the concentration is identical with C^b.

The hydrodynamic treatment then leads to a physical picture not totally unlike that of the Nernst hypothesis. There are several important differences which will become apparent in the following discussion.

Returning to the Nernst derivation, assuming the validity of a linear concentration gradient [Eq. (4-1), the rate of transfer of electroactive substance per unit area is given by

$$r = D_O \frac{C_O^b - C_O^e}{\delta_N} \tag{4-2}$$

The current flow then is

$$i = nFAD_O \frac{C_O^b - C_O^e}{\delta_N} \tag{4-3}$$

From the usual conditions of E_{app} such that $C_O^e = 0$, the limiting current is given by

$$i_L = \frac{nFAD_O C_O^b}{\delta_N} \tag{4-4}$$

where A is the area of the electrode and δ_N the thickness of the diffusion layer.

This equation has been widely applied to limiting currents in stirred solution. There are distinct advantages as well as inconsistencies in its use.

First, the Nernst derivation indicates that i_L is proportional to concentration. *Provided experimental conditions are maintained constant*, this relation is valid and serves as the basis for all analytical applications.

Furthermore, Eq. (4-4) is applicable to flowing or stirred solutions, whether the stirring be mechanical or by electrode rotation. It is also essentially independent of the cell geometry. This is so because the hypothetical quantity δ_N actually embodies parameters described by the hydrodynamic conditions existing in a given cell. When examined from the hydrodynamic standpoint, δ_N inherently contains such quantities as D, the diffusion coefficient, ν, the kinematic viscosity, ω, the angular rotation velocity in the case of a rotated electrode, and L, the characteristic dimension of the electrode. It is evident then that the equation can only qualitatively describe the limiting current for many situations.

On the other hand, Eq. (4-4) predicts that any means of reducing the value of δ_N should increase i_L. Experimentally, values of δ_N are calculated from Eq. (4-4) by measuring i_L where the other quantities are known (2,3). The thickness varies from ca. 0.05 cm for quiet solution to 10^{-3} cm or less according to the type of stirring (2,4,5). Empirical relationships between δ_N and the rate of stirring have been widely studied and can be summarized by the expression

$$\delta_N = \frac{B}{U^n} \qquad (4\text{-}5)$$

where B is a constant for a given set of conditions and U represents the velocity of the liquid flow. Experimental values of n usually fall between about 0.4 and 1.0 (6,7). Such relationships have little practical value. The important point is that i_L can be enhanced by increased stirring. Furthermore, it is clear that precise control of stirring rates is essential to quantitative analytical work.

4-2. HYDRODYNAMICS AND FORCED-CONVECTION ELECTRODES

The application of hydrodynamics has provided very valuable information about several practical solid electrode systems. This section is intended only as a mere introduction to the methods and techniques as applied to specific situations. Authoritative treatments are given in the references. In particular, the English translation of Levich's monograph *Physicochemical Hydrodynamics* has appeared recently (8).

A hydrodynamic approach is followed whenever the *details* of fluid mass transfer are examined. In a forced-convection electrolysis, if one determines the dependence of the limiting current on the solution flow velocity, viscosity, cell geometry, etc., one deals with the hydrodynamics of the situation. Provided the flow is laminar at the surface of interest, the limiting current equation probably can be obtained with certainty, and even in the case of turbulent flow considerable information is available.

To reduce the number of variables which must be examined, so-called dimensionless groups may be used. These dimensionless groups or numbers are simply characteristic collections of parameters whose dimensions cancel. In *heat transfer* from a fluid medium to a probe or surface, the Nusselt number (N_{Nu}) is given by

$$N_{Nu} = \frac{q°L}{k(T - T°)}$$

where $q°$ = heat-transfer rate, cal/cm²/sec
L = characteristic dimension of the surface along which fluid travels, cm
k = thermal conductivity of fluid, cal/cm/sec/deg
T = temperature, deg

The ° indicates that the quantity is taken with respect to the surface of interest (here the surface of the heat probe immersed in the fluid). A quick glance indicates that N_{Nu} is dimensionless.

By judicious choice of parameters, several such dimensionless numbers can be defined which are particularly useful in heat-transport problems. A completely analogous set of dimensionless groups can be written for mass transport to an electrode surface. For mass transport the Nusselt number (most often called the Sherwood number, N_{Sh}, when used in mass transport) becomes

$$N_{Sh} = \frac{jL}{D(C^b - C^e)}$$

where j = flux or mass-transfer rate, moles/cm²/sec
L = length or characteristic dimension of electrode along which solution flow occurs, cm
D = diffusion coefficient of electroactive material, cm²/sec
C^b, C^e = concentrations, in bulk of solution and at electrode, respectively, moles/cm³

N_{Sh} is dimensionless and is seen to embody familiar parameters of voltammetry. The equation for N_{Sh} may be compared, for instance with Eq. (4-3) describing the Nernst-diffusion-layer treatment when it becomes evident that δ_N is not a *simple* dimension but a composite function of various hydrodynamic factors. Other dimensionless numbers may be formed with simplicity. The Schmidt number, N_{Sc}, is

$$N_{Sc} = \frac{\nu}{D}$$

i.e., the ratio of the kinematic viscosity (cm²/sec) to the diffusion coefficient. Table 4-1 summarizes the dimensionless groups. It can be seen

TABLE 4-1

Dimensionless Groups Used in Heat and Mass Transport

Heat transport[a]	
Nusselt No.	$N_{Nu} = \dfrac{q°L}{k(T - T°)}$
Prandtl No.	$N_{Pr} = \dfrac{\nu s \rho}{k}$
Reynolds No.	$N_{Re} = \dfrac{UL}{\nu}$
Mass transport	
Sherwood (or Nusselt) No.	$N_{Sh} = \dfrac{jL}{D(C^b - C^e)}$
Schmidt No. (or Prandtl No.)	$N_{Sc} = \dfrac{\nu}{D}$
Reynolds No.	$N_{Re} = \dfrac{UL}{\nu}$

[a] In addition to quantities defined in the text, the symbols have the meaning: s, specific heat of fluid, cal/g/deg; ρ, density of fluid or solution, g/cm³; U, solution velocity, cm/sec.

there are analogous numbers for both heat and mass transport (other dimensionless groups exist but are not included in the table). Specific, authoritative treatments of dimensionless groups in electrochemistry have been given (8–10).

Frequently the solutions of the equations for heat-transfer problems can be transferred with ease to their mass-transport counterparts.

As a typical example, the limiting current for a forced-convection electrode can be formulated in terms of some general function Z of the bulk and surface concentration, solution velocity, characteristic length of the electrode, etc., as

$$i_L = Z(C^b, C^e, U, L, D, \nu) \tag{4-6}$$

On the other hand, the limiting current is related to the flux j at the electrode surface by

$$i_L = nFAj \tag{4-7}$$

From Table 4-1 it is seen that the flux at the electrode surface is related to N_{Sh} through

$$j = \frac{N_{\text{Sh}} D(C^b - C^e)}{L} \tag{4-8}$$

The limiting current then can be expressed as

$$i_L = nFA \frac{N_{\text{Sh}}(C^b - C^e)}{L} \tag{4-9}$$

If the solution of the pertinent differential equation for heat transfer involving the Nusselt number is available, the transformation to the mass-transport case is direct. As often as not, the heat-transport solution is available, and this has facilitated the solution of electrochemical problems.

It should be mentioned that the nomenclature in this area is overlapping. This results from the fact that those electrochemists working with topics like dimensionless groups also tend to be well versed in hydrodynamics. Thus the Schmidt and Prandtl numbers are used almost interchangeably in mass-transport discussions, and the hydrodynamic boundary layer mentioned previously is frequently known as the Prandtl boundary layer. The symbols used here are frequently simplified (e.g., $N_{\text{Re}} \equiv \text{Re}$).

The dimensionless numbers are interwoven by relationships of the form:

For heat transport: $N_{\text{Nu}} = Z(N_{\text{Re}}, N_{\text{Pr}})$
For mass transport: $N_{\text{Sh}} = Z(N_{\text{Re}}, N_{\text{Sc}})$

Even where exact solutions of the mass-transfer problem are not available it is possible to experimentally vary the parameters embodied in the dimensionless numbers and simplify the overall transport picture. This approach is well illustrated by the study by Eisenberg et al. on rotating cylindrical electrodes (11). Other examples, both in natural and forced convection, have been given (12,13). Ibl has shown the rather amazing correlation of data taken from such diverse experiments as (1) limiting currents in metal deposition at vertical plate electrodes (natural convection), and (2) nonelectrolytic dissolution of organic acids and salts, when plotted in dimensionless form (10).

It is worthwhile to acquire some feeling for the physical interpretation and orders of magnitude of the various dimensionless groups. The Schmidt or Prandtl number is seen to be the simple ratio of kinematic

HYDRODYNAMICS AND FORCED-CONVECTION ELECTRODES 75

viscosity to the diffusion coefficient. (The kinematic viscosity, ν, is the ratio of viscosity to density. Viscosity is expressed in poise (P) $\equiv$ g/cm/sec, density in g/cm^3, and the dimensions of ν are cm^2/sec.) N_{Sc} (N_{Pr}) is seen to be independent of flow velocity or characteristic dimension. It can be considered as a ratio which reflects those properties of the liquid which characterize *velocity* transport (i.e., viscosity) with pure *diffusional* transport (the diffusion coefficient, D). In ordinary aqueous media $\nu = 0.01$ cm^2/sec, while D values are $\sim 10^{-5}$ cm^2/sec. Thus commonly $N_{Sc} \simeq 10^3$. This relation is valid for aqueous media with ordinary quantities of supporting electrolytes. Thus ν of 0.1 M KCl or 1 M H$_2$SO$_4$ is still ca. 0.01 cm^2/sec and $N_{Sc} \simeq 10^3$. Since D decreases inversely with ν, as ν increases, N_{Sc} tends to increase as ν^2. For very viscous liquids N_{Sc} can approach 10^5–10^6. As the simple ratio form of N_{Sc} shows, high values of N_{Sc} means convective mass-transport predominates. In gases where ν and D have the same order of magnitude and therefore $N_{Sc} = 1$, the situation is quite different.

The Nusselt or Sherwood number can be given a qualitative picture as follows. On a unit-area basis, the actual mass-transfer rate in a process controlled both by convective and diffusional transport is given by the flux j. If only diffusional control were present, the mass-transfer rate would be proportional to $D(C^b - C^e)$ divided by a characteristic dimension through which diffusion occurs [see Eq. (4-3)]. From Table 4-1, N_{Sh} is seen to reflect the contribution of diffusional transport to the actual mass transport which exists in the mixed convective-diffusion process.

The Reynolds number is the only dimensionless group listed in Table 4-1 which contains the solution flow velocity U. It most directly reflects the experimental variables. Thus, if a rotated electrode is used, it indicates the variation with rotation speed. N_{Re} varies over wide ranges according to the stirring regime. N_{Re} can vary from about unity or less to 10^4–10^5 under not too unusual experimental conditions.

The general expression for the rate of change of concentration at any point in a solution which is controlled by convective diffusion is given by

$$\frac{\partial C}{\partial t} = (\text{diffusion}) - (\text{convection}) \qquad (4\text{-}10)$$

In terms of a three-dimensional picture with flow velocities u, v, and w, in the x, y, and z directions, respectively, this can be written as

$$\frac{\partial C}{\partial t} = D\left(\frac{\partial^2 C}{\partial x^2} + \frac{\partial^2 C}{\partial y^2} + \frac{\partial^2 C}{\partial z^2}\right) - \left(u\frac{\partial C}{\partial x} + v\frac{\partial C}{\partial y} + w\frac{\partial C}{\partial z}\right) \qquad (4\text{-}11)$$

In forced convection $\partial C/\partial t = 0$; i.e., a steady-state process is considered. This is the starting point for rigorous solutions of the mass-transfer problems in convective diffusion. Solutions for a few specific electrode situations have been developed. No attempt will be made here to describe the details of these treatments. The results of interest in voltammetry will be considered briefly in terms of working electrodes. For complete derivations the reader is referred especially to the monograph by Levich.

4-3. STATIONARY ELECTRODES IN FLOWING SOLUTION

A. Plate Electrode

Exact solutions for i_L have been given for several electrodes which are fixed in position with solution flowing past the electrode. The first of these is a plate electrode set in a laminar flow of liquid (6,8,14). It is assumed that the effect of the edges can be neglected and that no turbulence exists near the electrode. The limiting current according to Levich is

$$i_L = 0.68 nFDC^b b \left(\frac{v}{D}\right)^{1/3} \left(\frac{Ul}{v}\right)^{1/2} \qquad (4\text{-}12)$$

where l is the length of the plate in the direction of liquid flow and b is the width or dimension normal to the flow direction. It is interesting to note that Eq. (4-12) is of the form

$$i_L = \text{const.} \ (N_{\text{Sh}})^{1/2} (N_{\text{Re}})^{1/2} \qquad (4\text{-}13)$$

Collecting terms in Eq. (4-12) one obtains the expression

$$i_L = 0.68 nFC^b D^{2/3} bl^{1/2} U^{1/2} v^{-1/6} \qquad (4\text{-}14)$$

With this electrode the diffusion-layer thickness increases with $l^{1/2}$. Hence the maximum flux is inversely proportional to $l^{1/2}$ and the mean value of i_L is not directly proportional to electrode area. Trümpler and Zeller have made an experimental evaluation of the plate electrode with laminar flow (15). Recently Wranglén and Nilsson have studied both laminar and turbulent flow at horizontal plate electrodes. In the laminar region an i_L dependence on $U^{1/2}$ was found with a sharp increase as turbulent flow developed (77).

The closest practical analog in solid electrode voltammetry is probably a foil electrode immersed in a stirred solution. Foil electrodes are often used in end-point-detection systems. No quantitative studies of such electrodes have been made. When used, for instance, in amperometry, they tend to give linear response of i_L with C^b. It is doubtful that even approximately laminar flow would exist in the practical applications.

B. Conical Microelectrode

In a comprehensive study of voltammetry at solid electrodes, Jordon et al. constructed a conical platinum electrode for which a rigorous limiting current equation was available and which was suitable, by virtue of its geometry and dimensions, for comparisons with ordinary voltammetric measurements (*16*). The electrode was fabricated by sealing 0.05-cm-diameter platinum wire (3 cm long) in Pyrex capillary tubing. With the aid of diamond and Carborundum abrasive wheels the glass and platinum was ground down to a smooth cone much like a pencil point. The finished cone had a slant height of ca. 0.04 cm with a surface area of ca. 0.003 cm².

The limiting current equation for this electrode was adapted from the corresponding heat-transport case and has the form

$$i_L = 0.77 nFAC^b D^{2/3} U^{1/2} \nu^{-1/6} l^{-1/2} \qquad (4\text{-}15)$$

where *l* is the slant height of the cone and the other symbols are as defined previously. Equation (4-15) is of the same form as Eq. (4-13). Further, the area of the cone can be written as $\pi r l$, where r is the radius of the right cone. Then Eq. (4-15) contains a term in $l^{1/2}$ and, as expected, is seen to be identical in form to Eq. (4-14) for the plate electrode.

Jordan and co-workers used this conical electrode with several cell configurations. Flow velocities up to about 200 cm/sec were obtained with a gravity flow tube where the conical electrode was positioned at the bottom of the flow pipe. In another form the entire electrolysis cell was rotated (*16,17*). Flow velocities up to 500 cm/sec were possible with a circulating flow assembly as pictured in Fig. 4-2. Solution containing the electroactive material was kept at constant temperature in the large reservoir with conventional temperature control. This solution was continually forced by an impeller pump through a flow tube about 10 cm in diameter and 45 cm in length. To aid in preserving laminar flow, a flow "straightener" was inserted near the bottom of the flow tube. This was a piece of Lucite 2 cm high drilled so as to provide a bundle of 170 parallel channels. The wall spacings between channels were about 0.2 cm thick. The conical electrode was positioned near the top of the tube facing the flow stream. The reference electrode was positioned out of the path. Flow velocities were measured via static and impact pressure tubes made

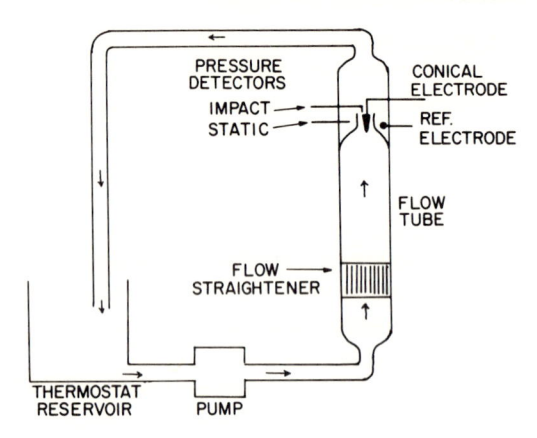

Fig. 4-2. Circulating-flow electrolysis apparatus.

of stainless-steel hypodermic needles. These were connected to a suitable manometer.

The conical electrode showed i_L directly proportional to C^b within 1–2%. In the reduction of oxygen and ferric ion log plots of i_L vs. flow velocity showed slopes of 0.45 ± 0.05, which is in good agreement with the theoretical $U^{1/2}$ dependence. In the case of oxygen, this relationship was verified over a range of flow velocities from 25 to 500 cm/sec.

Jordan and co-workers have given the name hydrodynamic voltammetry to this style of measurement, in which careful control of electrode design and convection conditions is employed. Using this approach they point out the possibilities of such systems as flow-velocity indicators, a reversal of the usual applications of voltammetry. Thus, provided C^b is constant, Eq. (4-15) takes the form

$$i_L = kU^{1/2} \tag{4-16}$$

By providing a suitable calibration flow, unknown velocities of solution could be measured. Another application of hydrodynamic voltammetry which has been very successful in the hands of Jordan's group is the measurement of heterogeneous rate constants and mechanisms of electron transfer processes (82–84).

C. Tubular Electrodes

Bazán and Arvia studied tubular electrodes past which solution was forced in a carefully controlled fashion (78). The electrodes were set in a vertical plane and the flow system was analogous to that shown in Fig. 4-2. The working electrode was an isolated annular section of the outer wall or a similar section of the inner axial tube. In general, the limiting currents were found to be proportional to $U^{1/2}$, $D^{2/3}$, and $\nu^{-1/6}$, as expected. The dependence of i_L on the characteristic length is complicated by factors involving the annular separation and total flow pattern length.

These studies were criticized by Ross and Wragg, who pointed out uncertainties as to full development of the hydrodynamic flow at the electrode surfaces. In a study in which the electrochemistry was secondary to elucidation of the subtleties of the hydrodynamics, these workers found i_L to be proportional to the average flow velocity, $U_m^{0.33}$, for laminar flow and to $U_m^{0.58}$ in the turbulent region (*79*). Gurinov and Gorbachev also have studied tubular electrode systems (*89*).

Blaedel et al. developed a tubular platinum electrode (TPE) for practical voltammetric measurements (*80*). A seamless platinum tube (10-mil wall thickness) was sealed in glass carefully to give a smooth glass–platinum inner wall. Flow rates up to ca. 40 ml/min were accomplished by simple gravity feed at constant head. The most recent design uses platinum tubing press fitted into Plexiglas channels and a peristaltic pump for flow-rate control (*81*). The limiting current equation for the TPE is

$$i_L = 5.306 \times 10^5 \, n D^{2/3} X^{2/3} V^{1/3} C^b$$

where X is the length of the platinum tube in the direction of flow and V is the volume flow rate in cm³/sec. If X is given in cm, V in cm³/sec, and the other terms in their usual units, i_L is in milliamperes. With the glass-sealed electrodes i_L was found to be proportional to $V^{0.335}$ for flow rates less than about 10 ml/min. Above this value (measured up to 40 ml/min) a $V^{0.47}$ dependence was observed showing some evidence of turbulent flow. The Plexiglas cells showed $i_L \propto V^{0.349}$ for moderate flow rates and all the data are in excellent accord with the theoretical $V^{1/3}$ in the nonturbulent region. An average exponent of 0.617 was found for the length factor, again in good agreement with theory. The i_L was also shown to be independent of the tubular radius in accord with theory.

Although earlier studies indicated currents could be detected for C^b as low as 10^{-8} M, practical current measurements are limited to concentrations about 10^{-5} M. The practical design and utility of the TPE show that it deserves further usage in hydrodynamic voltammetry.

D. Miscellaneous Electrode Shapes

A variety of electrode shapes set in a flow stream have been used for in-line polarographic measurements. An early example is the so-called "bypass" electrode of Müller (*85*). Although hydrodynamic treatments obviously were not attempted, the i_L was found proportional to C^b and log of the flow rate.

Stráfelda and Kimla studied spherical platinum electrodes in controlled flow systems (*86,87*). The dependence of i_L on the electrode geometry and flow rate were investigated. With ferrocyanide oxidation as a test system, i_L was found to be proportional to $V^{1/3}$. Recent studies of Stráfelda, Kimla, and co-workers have been concerned with dropping mercury electrodes in flowing solution (*88*). Arvia's group has studied conical and disk electrodes in laminar flow systems for hydrodynamic voltammetry applications (*117–119*).

4-4. ROTATED DISK ELECTRODES

The most practical form of electrode for which a completely rigorous hydrodynamic treatment can be given is the rotated disk electrode (RDE). The mass-transport equations for the RDE were given by Levich in 1942 (*18*). These papers in *Acta Physicochim. URSS* are in English. The RDE received almost no attention, except for continuing studies by the Moscow group, until the 1950s, when work from England and the United States began to appear. In 1953, when the literature survey for this monograph was initiated, only some six or seven publications dealing with RDE's existed. At present the number is well over 100. The RDE shows great promise in routine analytical applications, but its most important use to date is in the study of electrode mechanisms. The limiting current equations and some applications of the RDE are discussed in this section. In addition, to provide a consolidated source of information, all known RDE publications are listed chronologically in a special bibliography at the end of the chapter. (Although the bibliography is thorough, it is not exhaustive, and RDE papers having little connection to electrochemistry are not necessarily included. It should be noted that all papers in *Zh. Fiz. Khim.*, exist in translated form as *Russ. J. Phys. Chem. English Transl.* after 1959. Similarly, *Elektrokhimiya* exists translated as *Soviet Electrochem.* since its inception in 1965. The volume and year numbers are identical to the Russian citations; only the page numbers change.)

The theory of a RDE applies to a thin plane surface, so large in diameter that the edges may be neglected with respect to the total surface. This plane is rotated with constant angular velocity about an axis perpendicular to the plane. In practice this takes the form of a disk 1 mm to several centimeters in diameter rotated at constant speed in a laboratory-sized vessel (e.g., 50- to 1000-ml beaker or similar cell).

A physical picture of the fluid flow to a RDE is as follows. As the disk rotates, liquid in an adjacent thin layer acquires the rotational motion of the disk. Liquid thus entrained has a tangential velocity and, from the centrifugal force, also develops a radial velocity away from the center of the disk. This flow pattern, which moves liquid horizontally out and away from the center of the disk, requires an upward axial flow to replenish liquid at the disk surface.

These effects are well illustrated by the photographs of Fig. 4-3. In these experiments a green dye (liquid food color) was carefully pipeted into the bottom of a beaker so as to form a dye layer at the bottom of an otherwise clear solution. A conventional RDE was immersed slightly below the liquid surface. A sequence of photographs were taken before and after starting the RDE. In the first picture, before starting the rotation, only a small amount of dye is randomly distributed in the solution by accidental mixing. In the second picture, shortly after stirring was initiated, the solution under the disk has already begun to acquire a vertical movement. Finally, in the third frame, a well-defined flow pattern upward to the disk surface is observed. These naive experiments are not intended to represent theoretical flow conditions, but they can be used to illustrate the general behavior of a RDE.

The hydrodynamic boundary layer, δ_0, can be given approximately as (8,91)

$$\delta_0 \simeq 3\left(\frac{\nu}{\omega}\right)^{1/2} \qquad (4\text{-}17)$$

where ω is the angular velocity of the disk. Within the thickness δ_0, the radial and tangential fluid velocities decrease as a function of distance y, measured vertically downward from the disk surface. At δ_0 the tangential velocity, according to Levich, has decreased to one-twentieth of its value at the disk surface. At distances from the disk $y > \delta$, one considers that only axial (vertical) motion exists. With aqueous solutions and rotation rates of 16 rps or ca. 1000 rpm, δ_0 is of the order of a few tenths of a millimeter. Physically, δ_0 can be pictured as the approximate thickness of the liquid layer dragged by the rotating disk.

It should be noticed that the foregoing discussion is only concerned with liquid flow and applies to a rotated disk whether it be used as an electrode or not. If concentration gradients are now included, the complete

Fig. 4-3. Liquid flow patterns at a rotated disk electrode.

ROTATED DISK ELECTRODES

convective diffusion problem can be solved. In the usual electrochemical case with an excess of supporting electrolyte, the boundary conditions for convective diffusion of electroactive material are $C = C^b$ as $y \to \infty$ and $C = 0$ at $y = 0$. The full details of this problem have been given by Levich (*18–20*). An especially readable and authoritative account is given by Riddiford (*91*).

The final results of these calculations in terms of a limiting current for a reaction controlled only by mass transfer (reversible) is given by Levich as

$$i_L = 0.62nFAC^b D^{2/3} v^{-1/6} \omega^{1/2} \qquad (4\text{-}18)$$

where ω = angular velocity of the disk given by $\omega = 2\pi N$, with N = rps
v = kinematic viscosity, cm²/sec
C^b = concentration of electroactive species, moles/liter
i_L = limiting current, mA

Concentrations in limiting current equations are frequently expressed in moles/cc, and equivalent forms of Eq. (4-18) are seen. If i_L is in amperes and C^b in moles/cc, the equation is identical with Eq. (4-18) but requires the factor 10^{-3} if C^b is in the more conventional form of moles/liter. Other forms of the equation, involving rpm, or with disk area in terms of πr^2, are sometimes used. In terms of disk radius r, Eq. (4-18) can be written

$$i_L = 1.95nFD^{2/3} v^{-1/6} \omega^{1/2} r^2 C^b \qquad (4\text{-}19)$$

Equations (4-18) and (4-19) hold for $E_{app} \ll E^{\circ\prime}$ (where $C_O^e = 0$).

Before examining experimental verifications of Eq. (4-18), it is necessary to discuss the physical form of the RDE. The theory to which the transport equations apply requires a disk of very large diameter and infinitely thin, i.e., a wafer-like construction. The large diameter is necessary to reduce the effect of the edges to a negligible level [Eq. (4-18) does not cover transport to the edges]. The disk must be quite thin or otherwise it corresponds to a rotating cylinder for which a different fluid-flow problem is encountered.

The form of RDE used by Hogge and Kraichman (*21,22*) corresponds to the above description and is shown in Fig. 4-4. The disks were mounted directly on the rotating shaft. All but the bottom surface of the disk must obviously be rendered nonconducting with wax or a similar coating. According to Riddiford et al., disks of this type may show the rotating cylinder effect with thicknesses greater than about 0.5 mm (*23–25*). Electrodes of this form are not simple to fabricate.

Other styles of RDE's are shown in Fig. 4-5. Electrode A has been

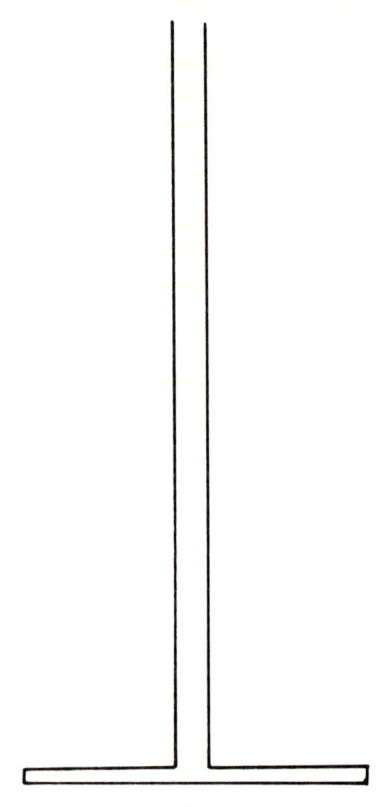

Fig. 4-4. Simple rotated disk electrode. All surfaces except bottom of disk covered with electrically insulated wax or coating.

widely used. In this design a stout piece of platinum, gold, or other metal wire is sealed into thick-walled glass tubing. The end is then ground flat to form a center disk with protecting rim. Such electrodes have been used for instance by Frumkin and Tedoradse (26) and Vielstich and Jahn (27). In a later study Jahn and Vielstich employed the glass rod type of

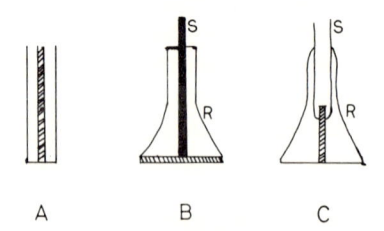

Fig. 4-5. Practical forms of rotated disk electrodes. S, shaft to motor connection; R, resin (epoxy or other forms); crosshatched areas indicate working electrode surfaces.

ROTATED DISK ELECTRODES 85

electrode with the glass edge tapered (28). Similar RDE's in which coating is effected by Teflon sheaths (29) or acrylic tubing have been described (30). Platinum wires can also be pressure-fitted into blocks or rods of Teflon or other inert material and then machined to a form like the glass rod RDE (31).

Cone-shaped electrodes, type B, where the disk occupies the entire end surface of the cone have been widely used (32–34). These electrodes have the advantage that edge effects are minimized and the main part of the axial shaft is integral with the disk proper. With proper machining the eccentricity should be small and, if present, is of minor importance, owing to the large electrode area. On the other hand, these electrodes pass rather high currents. High currents seldom make electrochemical measurements simpler, and, also, these currents ordinarily are too large to be handled by commercial polarographic equipment.

Azim and Riddiford (25) have combined the best features of the cone and wire RDE's in the design shown in Fig. 4-5C. The wire is soldered or otherwise fitted to the steel shaft. The shaft and wire are encapsulated with an epoxy resin and then machined to the cone configuration.

In the writer's laboratory RDE's of carbon paste have been used (35). Their design is shown in Fig. 4-6. Electrode A, with a small disk diameter, corresponds to the glass rod type A of Fig. 4-5. Electrode design B in Fig. 4-6 is the "double-disk" or ring-disk electrode, which will be discussed later. Other forms of carbon-type RDE's have been reported (31,57). Further details on the fabrication of electrodes are given in Chapter 9.

A serious question arose when studies by Blurton and Riddiford indicated that the cylindrical electrode (CEL) shapes (and other shapes) might deviate considerably from the theoretical requirements for a RDE (90,91). Expressed in the most succinct terms, the theoretical requirements are these. The region below the disk, $y = \infty$, should serve as an infinite fluid source. The region past the periphery of the disk, $y = 0$, $r = \infty$, should be an infinite sink for the fluid transport. These criteria are met by the hypothetical infinitely large, horizontal, rotating lamina set in an infinite volume. With the CEL, Blurton and Riddiford found that mixing of fluid flow occurred in the sense that flow patterns from above the $y = 0$ plane of the disk could interact with the transfer at and below $y = 0$. Using permanganate as a flow marker (analogous to the dye experiments shown in Fig. 4-3) they showed that very complex flow patterns resulted with several practical forms of RDE's (90). The CEL was particularly suspect in this connection. However, almost all practical RDE's have inert, insulating shrouds. The mixing effects are mediated somewhat in

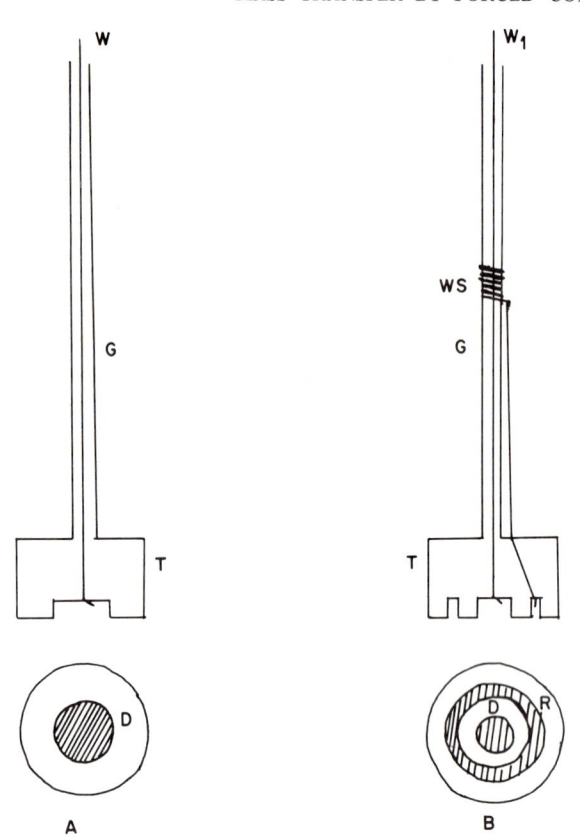

Fig. 4-6. Rotated disk electrodes of carbon paste. A, single disk: W, electrical connection to disk; G, glass tubing spindle, 6–8 mm diameter; T, Teflon holder, typical diameter 12 mm; D, carbon paste disk, typical diameter 2–8 mm. B, Ring disk or double disk: W_1, electrical connection to inner disk; WS, slip-ring electrical connection to outer ring (wire shielded where contact with solution is possible); T, Teflon holder, typical diameter 12 mm; D, carbon paste inner disk, typical diameter ca. 3 mm; R, carbon paste outer ring, typical width 1–1.5 mm.

favor of agreement with theory, since the shroud area decreases the effect of fluid transfer between upper and lower regions—at least in the region of the active disk surface.

Blurton and Riddiford found that the practical size requirements for agreement of experimental i_L with theory were somewhat less than those predicted from the hydrodynamic criteria.

Most of the RDE studies in the writer's laboratory had employed CEL's. Hence a thorough comparison of i_L's at CEL's and bell-shaped (BEL)

RDE's was made. This study was analogous to those of Riddiford and co-workers, i.e., after examining i_L at a CEL, the electrode was machined down on a lathe to a bell shape and the i_L's remeasured (92). The results showed there was practically no difference between the two electrodes within experimental error. For the CEL the average value of $i_L/R^{1/2}$ was 11.4 ± 0.3 compared to 11.3 ± 0.1 for the BEL over the range 1–40 rps. It should be emphasized that these electrodes had a fairly thick shroud of inert resin; the total radius of disk plus shroud was ca. 1 cm, while the actual disk radius was only 0.16 cm. Further, these studies were carried out at slightly higher Reynolds numbers (16–640 calculated for the outer edge of the electroactive surface). At these higher Reynolds numbers the shape factor is of slightly lesser importance. The results, in any event, were in accord with the findings of Blurton and Riddiford. It can be concluded that if cylindrical-shaped RDE's are used, one should ensure that a sufficiently thick insulating shield is provided to minimize fluid transfer between upper and lower layers.

The physical dimensions of the vessel in which a RDE is operated are important, since the theory requires an infinite volume of solution. A wide variety of actual conditions have been used. In the work by Hogge and Kraichman, the volume of solution was ca. 600 ml and the disk and vessel diameters were 2 and 9 cm, respectively (21,22). Using a disk of ca. 5 cm diameter rotated at 146 rpm, Gregory and Riddiford reported that the results were independent of vessel diameter, provided the latter was greater than 11 cm. They also showed that the height of the disk above the vessel bottom was immaterial if this distance exceeded 0.5 cm (23). Prater showed that a variation of the vessel size between a 100-ml beaker and a 9-liter lab bucket had practically no effect on i_L at the type of cylindrical RDE discussed above. The effect of immersion depth was also negligible at the cylindrical electrode (92). In most studies, where the information is available, it appears that the lower surface of the disk has been positioned just slightly below the liquid level. This is probably a practical approach dictated by the shaft eccentricity—any wobble is accentuated when the disk (and shaft) are immersed deep in the liquid. It is evident that attention must be paid to proper alignment of bearings, shaft hangers etc., in the mechanical design of RDE assemblies. These details are not as stringent as one might expect unless very high rotation velocities are used (20,000 rpm and greater), when even the natural resonance of the motorshaft may become a problem.

Several criteria have been cited as proof that a given RDE is functioning satisfactorily. At least some of these apparently can be verified with almost any form of RDE combined with vessel sizes and shapes that are minimal with regard to the theoretical limits mentioned above. Equation (4-18) first of all predicts that i_L is proportional to $\omega^{1/2}$. This was first verified by the study of Kabanov and co-workers for the reduction of oxygen and hydrogen ion at amalgamated RDE cathodes. The variation between experimental and theoretical values of i_L was less than 3% (36–38). Excellent verification of the linear dependence of i_L with $\omega^{1/2}$ for triiodide ion was obtained by Hogge and Kraichman (21–22). In fact, almost without exception, every publication on the RDE has verified this dependence of i_L vs. rotation speed for reversible reactions. Studies in the writer's laboratory have employed a carbon paste RDE of 4 mm diameter built in a Teflon shield of 1.1 cm diameter overall. This electrode shows excellent i_L vs. $\omega^{1/2}$ dependence when rotated in a 6-cm-diameter vessel of ca. 200 ml volume (39). These are hardly infinite size conditions. It must be concluded that the simple determination of linear dependence of i_L vs. $\omega^{1/2}$ is not a critical test of the performance of a RDE.

A slightly more informative test was carried out in the early studies of Siver and Kabanov (36). Using hydrogen and oxygen they showed that the values of i_L for these test species were in the ratio $(D_{H^+}/D_{O_2})^{2/3}$, as predicted by Eq. (4-18). Their limiting current ratios left no doubt as to the $\frac{2}{3}$ dependence of D, as opposed to the simple first-power ratio predicted by Nernst layer theory.

The real test of the RDE lies in checking experimental i_L's with theoretical values. To do this one must know the value of D under the experimental conditions of supporting electrolyte (as well as electrode area and n, the number of electrons transferred). A real difficulty now enters the picture. D values needed are often obtained from i_L data at a RDE. In any event they are evaluated by one or another of the mass-transfer-controlled electrochemical techniques. Further, electrode areas are frequently calculated from i_L of a "test" system such as ferricyanide ion, whose D value in turn was determined electrochemically. This circular method of acquiring data is a real problem in electrochemical studies. The pitfalls are not so apparent until results better than 5% are required.

The rather indirect condemnations of the performance of a RDE given above are really unwarranted. The RDE in almost any form is a highly successful electrode system which gives excellent results. The point of the discussion is that a simple evaluation like i_L vs. $\omega^{1/2}$ behavior is a necessary, but not sufficient proof that a given RDE is functioning according to theory. This check may be quite satisfactory if the RDE is to be used for analytical purposes, but if one wishes to determine D values or crosscheck other electrochemical parameters, a more rigorous examination of the RDE involving absolute values is necessary. It should be pointed out that the studies of Kabanov et al. and Hogge and Kraichman were not limited to the simple i_L vs. $\omega^{1/2}$ determinations but also involved comparing experimental and theoretical values of i_L.

Indeed, a careful check of the accuracy of limiting currents over a wide range of rotation rates and as a function of temperature led Gregory and Riddiford (23) to propose an important modification to Eq. (4-18). Studying the dissolution of zinc disks in aqueous iodine solutions (a process whose rate was previously shown to be determined only by mass transport of iodine to the disk surface) they were able to show that the unit rate constant k_T was directly proportional to the product $(D^{2/3}\nu^{-1/6}\omega^{1/2})$ within limits which allowed no questioning of the fundamentals of the disk behavior. However, the Levich equation (4-18) predicted i_L values which were consistently too large when compared with experiment. This was traced to an approximation in the numerical evaluation of the constant 0.62 in the Levich equation. Using additional terms in the expansion of the particular integral involved, Gregory and Riddiford concluded that the numerical constant 0.62 should be replaced by the quantity

$$\frac{0.554}{0.8934 + 0.316(D/\nu)^{0.36}} \quad (4\text{-}20)$$

This correction applies (within 1%) for values of D/ν in the range $0\text{-}4 \times 10^{-3}$. When $D/\nu = 10^{-3}$, the value of the denominator in Eq. (4-20) differs by about 3% from that originally used by Levich.

The corrective term can be used in terms of a successive approximation but it would appear that a single substitution in (4-20) with known D and ν values is sufficient to give reasonable correction. For instance, the constant 0.620 of Eq. (4-18) becomes 0.603 for ferricyanide ion in 1 M potassium chloride when $D = 0.763 \times 10^{-5}$ cm^2/sec and the kinematic viscosity $\nu = 0.853 \times 10^{-2}$ cm^2/sec are substituted in Eq. (4-20). The final result of the Gregory and Riddiford correction is, of course, to lower the value of the theoretical i_L. According to Levich, experimental values are more closely given by the original Eq. (4-18). This is attributed by Levich to additional corrections of comparable magnitude but opposite in sign. One such correction (not possible to evaluate quantitatively) is the edge effect, which becomes increasingly important at low rotation velocities (40). Recently Newson and Riddiford have pointed out that the correction to the Levich equation becomes less important as the Prandtl (Schmidt) number ($N_{Pr} = \nu/D$) increases. They studied triiodide ion reduction in solutions of varying sucrose concentration. With sucrose-free solution the correction amounted to 3% but decreased to 0.2% in 1.5 M sucrose (i.e., as N_{Pr} increased). Moreover, edge effects were apparently absent in these studies (*33*). Newman has recently proposed what appears to be a better correction for moderate Schmidt numbers and compares the new expression with the Levich and Riddiford formulations (*113*). While the corrective term is relatively small, it is by no means to be disregarded when precise values of D (or other quantities depending on i_L) are to be measured with the RDE.

It has been tacitly assumed in all the previous discussion that the RDE is applicable to any electrochemical system. Actually, the Levich equation and its counterparts apply strictly to a solution containing no more than three ionic species. Practical buffer systems, for instance, are usually more complex. If the electroactive species is uncharged there is no problem. If it is charged and in a complex supporting electrolyte the numerical constant in the Levich equation or the Riddiford corrected form can be expected to be very slightly altered (*8,91*). With the large excess of supporting electrolyte ordinarily used, the effective transference number of the electroactive species can be assumed zero and this correction neglected.

The precise measurement of D values requires a further examination of the fundamental RDE equation. Equation (4-18) at first glance predicts that $i_L = 0$ at $\omega = 0$. This, in fact, is not so, as has been pointed out by several workers (*23,29,35,44,45*). Actually the inapplicability of Eq. (4-18) in the absence of forced convection ($\omega = 0$) was indicated by Levich (*46*). Nevertheless, many plots in the literature of i_L vs. $N^{1/2}$ or $\omega^{1/2}$ show extrapolations to $i = 0$ at $N = 0$. The true value of i under stationary conditions is difficult to evaluate, since the various RDE's correspond to unshielded electrodes with upward diffusion in the absence of rotation. For instance, carbon paste RDE's in the writer's laboratory, when used in a stationary configuration, fail to meet exact linear diffusion criteria (i.e., $it^{1/2}$ is not constant at constant E_{app}). This is, of course, to be expected, since they, like most RDE's are simply flat, exposed electrodes with no restricting mantle for true linear diffusion. In any event, i at $\omega = 0$ is not zero but a finite value. Actually, as $\omega^{1/2}$ decreases, one expects natural convection to add to the small forced convection component resulting in i_L higher than that predicted by Eq. (4-18). A plot of i_L vs. $\omega^{1/2}$ therefore might be expected to show a slight upward bend at very low rotation rates (1–2 rps). Such effects have been noted in some instances (*32*).

It would appear then that precise measurements of quantities like D values can be accomplished best by measuring i_L at several values of ω (preferably all above perhaps 5–10 rps) and evaluating D from the slope of the i_L vs. $\omega^{1/2}$ plot. The averaging process of several points is a decided advantage and, in addition, the slope should be independent of the $\omega = 0$ contribution (*47*).

The behavior of the RDE discussed so far assumes that laminar flow exists near the disk surface. Such conditions can be maintained apparently up to Reynolds numbers between 10^4 and 10^5. The value of N_{Re} for a RDE is given by (*31,41*)

$$N_{Re} = \frac{r^2 \omega}{\nu} \qquad (4\text{-}21)$$

where r is the disk radius in cm and is the total radius of working plus nonworking area. Nonturbulent flow is favored, as can be seen by Eq. (4-21) by using a small disk size. This, however, must be countered against the edge effects discussed previously. The limits for laminar flow given above apply naturally to a smooth, well-centered RDE. The N_{Re} limit for laminar flow with many practical electrodes may be lowered by a factor of 10.

Under laminar flow conditions the thickness of the diffusion boundary layer, δ, at the RDE is given by

$$\delta = 1.61 D^{1/3} \nu^{1/6} \omega^{-1/2} \qquad (4\text{-}22)$$

In terms of the Gregory and Riddiford correction, the constant 1.61 becomes (*23,27*)

$$1.805 \left[0.8934 + 0.316 \left(\frac{D}{\nu} \right)^{0.36} \right] \qquad (4\text{-}23)$$

With the usual values of D and ν, the thickness of δ is only about 5% that of the hydrodynamic boundary layer, δ_o. The large concentration changes occur within a distance of ca. 10^{-3} cm from the electrode, whereas the viscous fluid flow effects extend out for a few tenths to 1 mm away, depending on rotation velocity. There is one very important point about the diffusion boundary layer—as seen in Eq. (4-22), δ contains no terms involving electrode geometry. The thickness is not a function of size of the electrode but is constant over the entire surface. This is completely different from a plate set in a flowing fluid (Section 4-3), where δ changes with distance along the plate. Indeed, this property of the RDE sets it apart from all other surfaces immersed in a flowing liquid. This type of reaction surface is termed *uniformly accessible* by Levich (*40*). Of considerable importance to electrochemistry is that on this uniformly accessible surface, the rate of an electrochemical reaction is everywhere the same (disregarding surface microinhomogeneities such as active sites). However, some important limitations on the uniformly accessible surface have been pointed out recently by Newman (*114*). The possibility of non-uniform current densities due to ohmic potential drop at currents $<i_L$ and other requirements for uniform accessibility merit close attention (*114–116*).

4-5. APPLICATION OF RDE TO ELECTRODE KINETICS AND MECHANISMS

Probably the most valuable applications of the RDE have been to investigations of the rates and mechanisms of electrode processes. Only a brief discussion will be given here. Some applications to specific situations will be discussed in detail in Chapter 8.

The distinction between electrode reactions controlled by mass transfer (m.t.) and charge transfer (c.t.) was made in Chapter 3. For a first-order electrode reaction which is controlled both by m.t. and c.t., Levich gives

$$i = \frac{nFADC^b}{1.61 D^{1/3} \nu^{1/6} \omega^{-1/2} + (D/k)} \qquad (4\text{-}24)$$

This relation can be written

$$i = \frac{nFADC^b}{\delta + (D/k)} \qquad (4\text{-}25)$$

Here C^b is the bulk concentration of electroactive species, A the electrode area, and k the heterogeneous rate constant at the given potential. The other symbols have their usual significance. Depending on the relative magnitudes of δ and the ratio D/k, Eq. (4-24) or (4-25) has limiting forms corresponding to control of the electrode process by c.t. or m.t.

If $\delta = 10^{-3}$ cm (a reasonable value under the experimental conditions used), then for values of $k > 10^{-1}$ cm/sec the term D/k can be dropped to give the familiar relation for the limiting current of a diffusion-controlled process at the RDE, i.e., Eq. (4-18). On the other hand, if, under the same experimental conditions, $k < 10^{-4}$ cm/sec, then δ is negligible with respect to D/k and one has

$$i = nFAkC^b \qquad (4\text{-}26)$$

APPLICATION OF RDE TO ELECTRODE KINETICS AND MECHANISMS 93

It is clear that the observed current in this case is independent of the m.t.—the current is entirely controlled by c.t. The rate constant can be evaluated from Eq. (4-25) and (4-26).

Next, $\log k_f$ and $\log k_b$ can be plotted as a function of electrode potential. According to the relations

$$k_f = k_f^\circ \exp\left(\frac{-\alpha n_a FE}{RT}\right)$$

and

$$k_b = k_b^\circ \exp\left[\frac{(1 - \alpha)n_a FE}{RT}\right]$$

where k_f = "forward" rate constant for cathodic direction. Two straight lines should be obtained. From the slopes of these the transfer coefficient, α, can be evaluated. The intersection, at the formal potential of the system, E_f°, gives the rate constant at the formal potential, k_s. Only one of the conjugate pair Ox–Red is in the bulk of solution, and measurements are made at high overpotentials, so the inverse reaction can be neglected.

Figure 4-7 shows current vs. $\omega^{1/2}$ behavior for the reduction of Fe(III) in 0.1 M hydrochloric acid at a carbon paste RDE (31). These data are typical of an electrode reaction controlled by both m.t. and c.t. The curves are of the form of Eqs. (4-24) and (4-26). At low applied potentials

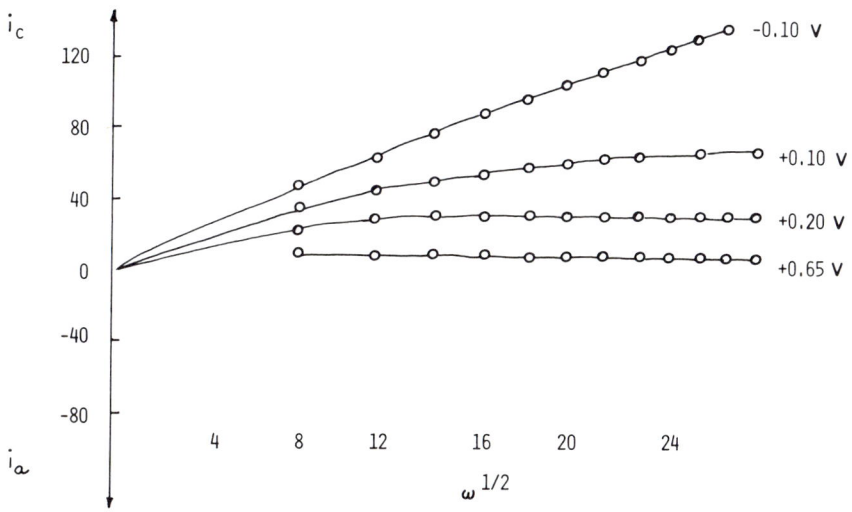

Fig. 4-7. Reduction of Fe(III) at a RDE. (Numbers indicate applied potential in volts vs. SCE.)

(+0.3 V), where the rate constant is small, the current is entirely c.t.-controlled, i.e., almost independent of $\omega^{1/2}$ [Eq. (4-26) applies]. As the reduction potential is made more favorable for the cathodic reaction, i is nonlinear with $\omega^{1/2}$, showing mixed control. Finally, at ca. -0.1 V the reaction is *almost* fully controlled by m.t., as evidenced by close to linear dependence of i on $\omega^{1/2}$.

It is possible to determine the order of the electrochemical reaction from RDE measurements. The order is actually reflected in the shape of i vs. $\omega^{1/2}$ curves such as those of Fig. 4-7 for c.t.-controlled reactions. Alternatively, the order can be found from the expression

$$m \ln \left[\frac{1 - (j_1 \delta_1 / C^b D)}{1 - (j_2 \delta_2 / C^b D)} \right] = \ln \frac{j_1}{j_2} \qquad (4\text{-}27)$$

where m = order of reaction
$j = 0.62 D^{2/3} \nu^{-1/6} \omega^{1/2} C^b$
$\delta = 1.61 D^{1/3} \nu^{1/6} \omega^{-1/2}$

The flux, j, is merely i/nFA. The thickness of the diffusion layer, δ, is evaluated from known D and ν values. The determination of m then involves measurement of i at two different rotation velocities and evaluation of the various parameters (42,43). Other techniques for evaluating reaction orders at RDE's have been given by Frumkin and Tedoradse (26). Vielstich and Jahn have measured charge transfer rates using the RDE (27). Holleck et al. have measured the kinetics of the second electron transfer in the reduction of aromatic nitro compounds at metal RDE's (94).

The study of kinetic and catalytic electrode processes was pioneered by the Czech school, principally by Brdicka, Koutecky, Wiesner, and coworkers. Reviews of the applications of various voltammetric techniques to such processes have been made (52,53). The RDE offers some particular advantages for studying such reactions. In particular, the reactions at the RDE take place under conditions of a stationary state which, in some instances, is helpful in deriving the rather complex equation describing such situations. The fundamental equations for kinetic and catalytic processes at the RDE were given by Koutecky and Levich (54,55).

A kinetic process may be one in which the rate of the electrode reaction and hence the current may be limited by the availability of electroactive species which derive from a homogeneous chemical reaction in the solution (precursor reaction). Such a reaction can be written

$$A \underset{k_2}{\overset{k_1}{\rightleftarrows}} B \qquad \text{solution reaction}$$

$$B \pm ne \rightarrow C \qquad \text{electron transfer}$$

The solution reaction is ordinarily considered to be pseudo-first-order. A typical example is the dissociation of a weak acid where the solution contains a large excess of the conjugate base. For such a kinetic process the limiting current at the RDE surface is given by (54,55)

$$i = \frac{nFADC^b}{1.61 D^{1/3} v^{1/6} \omega^{-1/2} [1 + (\sigma/1.61)(\omega^{1/2}/\alpha^{1/2})(D^{1/6}/v^{1/6})]} \quad (4\text{-}28)$$

Since

$$\frac{nFADC^b}{1.61 D^{1/3} v^{1/6} \omega^{-1/2}} = i_o$$

is the value of i_L which would be observed if no prior kinetic process were involved, Eq. (4-28) can be written

$$i = \frac{i_o}{1 + (\sigma \omega^{1/2} D^{1/6}/1.61 \alpha^{1/2} v^{1/6})} \quad (4\text{-}29)$$

The quantity σ is given by k_1/k_2, where k_2 may be written alternatively as $k_2[Z]$ for the pseudo-first-order reaction. The parameter α is related to the thickness of the so-called kinetic layer, δ_k, by $\delta_k = (D/\alpha)^{1/2}$. The physical significance of δ_k is that outside this layer the chemical reaction proceeds in an equilibrium fashion—only within the thickness δ_k from the electrode surface is the kinetic process evidenced. The derivations above assume all diffusion coefficients are equal. Dagonadze has examined the derivations for unequal D values and concluded that the results are not too significantly altered, although the mathematics becomes far more cumbersome (56). Vielstich and Jahn examined the dissociation of weak acids, a frequently investigated kinetic process. Rather than attempt to measure the rates from direct comparison of a kinetic current with a diffusion-controlled i_L, they pointed out the merits of plotting $i/\omega^{1/2}$ vs. i—a technique analogous to that used for kinetic processes in chronopotentiometry (27).

Equation (4-29) can be written

$$i = \frac{i_o}{1 + (\sigma \delta_k/\delta)} \quad (4\text{-}30)$$

and rearranged to give

$$i = i_o - \frac{\sigma \delta_k}{\delta} i$$

or

$$i = i_o - \left[\frac{(k_1/k_2)\delta_k}{\delta}\right] i \quad (4\text{-}31)$$

Dividing through by $\omega^{1/2}$ gives

$$\frac{i}{\omega^{1/2}} = \frac{i_o}{\omega^{1/2}} - \frac{\delta_k(k_1/k_2)}{\delta\omega^{1/2}} i \qquad (4\text{-}32)$$

$$\frac{i}{\omega^{1/2}} = \frac{i_o}{\omega^{1/2}} + \frac{\delta_k(k_1/k_2)}{1.61 D^{1/3}\nu^{1/6}} \qquad (4\text{-}32a)$$

Since δ, the usual diffusion-boundary-layer thickness, is given by $\delta = 1.61 D^{1/3}\nu^{1/6}\omega^{-1/2}$, the product $(\delta\omega^{1/2})$ equals $1.61 D^{1/3}\nu^{1/6}$ as shown in Eq. (4-32a). According to Dagonadze, δ_k may be expressed in terms of the diffusion coefficients and k_1, k_2. When this substitution is made in Eq. (4-32), an expression similar to that used by Vielstich and Jahn is obtained. A measure of the rate constants k_1 and k_2 is obtained from the slope of a plot of $i/\omega^{1/2}$ vs. i. Although it is assumed that k_{eq} (dissociation) is known for these reactions, it should be noted that Eq. (4-32) or its equivalents do not provide direct measurement from the slope, since δ_k includes these factors. Further details of the technique can be found in the original literature (27).

In the reduction of weak acids it is difficult to obtain flat limiting current plateaus on the RDE. Albery and Bell introduced a novel arrangement to circumvent this difficulty. Two RDE's, one in a standard strong acid solution and the other in a weak acid buffer, were arranged in a bridge network. This style of measurement allows more precise measurement of i_L and the technique offers promise in other situations where background reactions cause serious sloping of the i_L region (93).

Very recently a quantitative treatment of the so-called ECE reaction (an electron transfer followed by a chemical reaction with a subsequent electron transfer) has been given by Malachesky and co-workers. Variation of the apparent number of electrons transferred, n_{app}, with rotation rate allows a measurement of the rate of the intermediate chemical reaction. The theoretical treatment was tested at the RDE and rates for selected reactions were found to agree well with those measured by other techniques (58). This treatment is discussed further in Chapter 8 (Section 8-5). For further details on the application of the RDE to more complex kinetic and catalytic reactions the reader is referred to the monograph by Levich (55).

The so-called ring-disk electrode developed by the Frumkin school provides some of the most exciting prospects for studying electrode mechanisms. A typical design was shown in Fig. 4-6. The electrode consists of two separate electroactive surfaces, the inner disk and, separated

APPLICATION OF RDE TO ELECTRODE KINETICS AND MECHANISMS 97

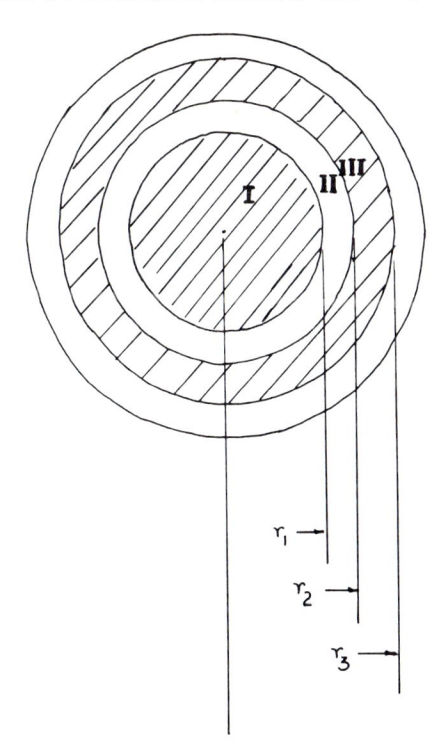

Fig. 4-8. Ring-disk electrode. (See text for explanation of dimensions.)

from it by an insulated band, an outer ring electrode. An enlarged diagram is seen in Fig. 4-8, where, using the nomenclature of Levich and Frumkin, the electrode can be divided into three zones of interest in the following fashion:

Zone I: $r < r_1$ disk, active surface

Zone II: $r_1 < r < r_2$ insulating spacer

Zone III: $r_2 < r < r_3$ ring, active surface

In practice the ring rotated disk electrode (RRDE) is connected so as to provide two separate voltammetric circuits, one to the disk and one to the ring. Current–voltage curves or limiting currents are measured at each surface. The normal operation is to measure an i_L at the ring as a function of E_{app} to the inner disk.

The general picture of the RRDE can be given qualitatively as follows. In zone I an electrochemical reaction can be caused to proceed by application of a potential E_1. If this reaction is designated as

$$A \rightarrow B$$

then product B is formed uniformly over the inner disk surface (uniformly accessible surface). The rate of formation of B is given by the flux derived from the Levich equation for a RDE. Now, recalling the flow pattern at the RDE, it is readily seen that product B will be swept outward from zone I, through the nonreactive area zone II, to zone III. The potential applied to zone III is such that C_B at the ring surface is kept at zero (i.e., a limiting current situation is set up for B). If B is a stable product, then presumably the only thing which can be accomplished is to apply a potential at the ring so as to carry out the reverse reaction:

$$B \rightarrow A$$

Depending on the geometric characteristics of the RRDE, a certain amount of B can be found in terms of the ring current i_R. This obviously is not equivalent to the disk current i_D. Only a certain fraction of i_D is observed at the ring. The fraction N which is equivalent to $|i_R/i_D|$ is a rather cumbersome function of r_1, r_2, and r_3. The original calculation of N by Ivanov and Levich (95) was modified by Bruckenstein and Feldman (96). Albery and Bruckenstein very recently have refined the calculation of N for RRDE's with thin rings and thin insulating gaps. They aptly designate N as the *collection efficiency* of the RRDE. Tables of N for common radius ratios of r_3/r_2 and r_2/r_1 are given by Albery and Bruckenstein (98). They observed excellent agreement between calculated N's and those observed experimentally.

In practice, experimental determination of N is relatively straightforward. One uses a simple electron transfer reaction with no chemical complications and measures $|i_R/i_D|$. (The ring current is inverse to i_D; hence i_R should properly be designated as $-i_R$.) Frumkin and Nekrasov reduced quinone at the disk and reoxidized the hydroquinone at the ring and found 38 ± 1% collection efficiency (48). N values were also determined by Frumkin et al. (51), via oxygen and cupric reductions. Albery and Bruckenstein found the oxidation of Br⁻ at the disk and reduction of Br_2 at the ring to be convenient. They also stripped silver anodically from the disk and reduced the Ag^+ at the ring to calculate N (98). Prater and Adams used several organic test systems for carbon paste RRDE's. Some typical experimental results for the collection efficiency of such an electrode with

APPLICATION OF RDE TO ELECTRODE KINETICS AND MECHANISMS 99

TABLE 4-2
Electrode Collection Efficiency

Electrode reaction at disk	N 10 rps	N 40 rps
Oxidation of o-dianisidine	0.455 ± 0.008	0.458 ± 0.006
Oxidation of hydroquinone	0.453 ± 0.003	0.451 ± 0.003
Reduction of quinone	0.463 ± 0.003	0.462 ± 0.003

$r_1 = 0.22$ cm, $r_2 = 0.29$ cm, and $r_3 = 0.44$ cm (a wide ring and wide gap) are given (102) in Table 4-2. The experimental determination may be, in many cases, as easy to accomplish as the corresponding measurement of the radii required for calculating N. The geometry of practical RRDE's is largely determined by the difficulty of fabrication, and most of the electrodes used to date have had radius ratios such that N was between 0.30 and 0.45.

The real utility of the RRDE is not realized in the above situations, where the product B is completely stable. If, however, B undergoes further reactions, then it might escape detection at ordinary disk electrode. The first situation one could consider is the case where B is capable of further heterogeneous electrochemical reaction; i.e., B is the intermediate oxidation state in a two-stage oxidation or reduction. Frumkin and co-workers suggested that by controlled stirring of the intermediate away from the area of reaction, it should be possible to detect it. This case was first considered by Ivanov and Levich (49) and is discussed in detail in Levich's monograph (50). A simplified summary of the expression for i_R for a two-stage reduction is (51)

$$A + n_A \rightarrow B$$

$$B + n_B \rightarrow P$$

$$i_R = \left(\frac{n_B}{n_A}\right) \frac{N i_D}{1 + (\delta_B/\delta_A)} \qquad (4\text{-}33)$$

Here i_R is the current measured at the ring. The ratio of electrons in the two stages of reduction, if unequal, appears as seen in Eq. (4-33). The first-order rate constant k for the disappearance of B is potential-dependent. The value δ_B is the diffusion-layer thickness for B and D_B has it usual meaning. When $k = 0$, rearrangement of Eq. (4-33) shows that i_R/i_D gives the ordinary collection efficiency (when B is stable and also assuming $n_A = n_B$).

The reduction of oxygen, which occurs in two stages via H_2O_2 as an intermediate, provides an excellent example of this application of a RRDE. The reactions are

$$O_2 + 2e + 2H_2O \rightarrow 2H_2O_2$$
$$H_2O_2 + 2e \rightarrow 2OH^-$$

Figure 4-9 shows the corresponding current i_D and i_R at the RRDE as obtained by Frumkin and co-workers (48,51). At potentials less cathodic than ca. -1.0 V (first oxygen wave), the H_2O_2 formed as intermediate B is stable—the applied potential is not sufficiently cathodic to carry out the second stage of reduction and $k = 0$. As it accumulates at the disk

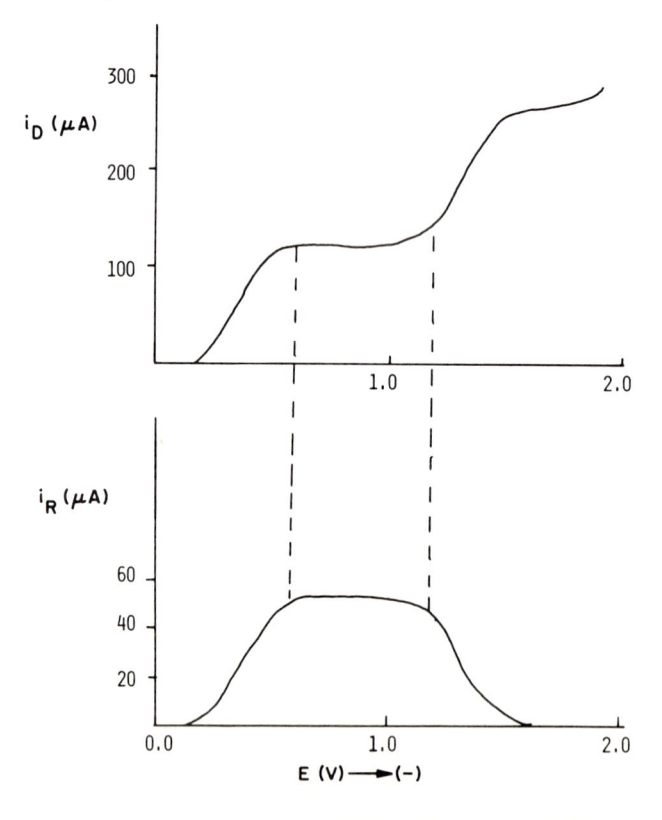

Fig. 4-9. Disk and ring currents for oxygen reduction at the RRDE. Top curve, disk current vs. E_{app}; bottom curve, ring current vs. E_{app}.

surface, it is swept out to the ring and detected as i_R. Thus i_R grows with increasing production of H_2O_2 up to ca. -0.6 V, where the limiting current i_D gives the maximum formation of H_2O_2. It levels off corresponding to the plateau of the first wave for i_D, and it can be seen that the fraction obtained as ring current is roughly 30–40% of i_D [actually $N = 0.37$–0.38 in the oxygen study (51)]. Now, at potentials more cathodic than ca. -1.0 V, the H_2O_2 becomes a reactive intermediate ($k > 0$) and i_R decreases as the rate of the second stage of reduction increases with increasing E. While the two-stage reduction of oxygen was a fairly well-established electrode process, the study confirms the utility of the RRDE to this type of electrode reaction. Several other examples of two-stage processes have been studied by Nekrasov and co-workers (103–105).

The collection efficiency is clearly the parameter associated with the RRDE, which can provide information on the kinetics of electrode processes. One can define N as modified by various competing processes which consume material in transit between the disk and ring. Actually this collection efficiency in the presence of kinetic complications, N_k, may only bear formal resemblance to the ordinary N. This is illustrated by the important case where the reactive electrogenerated intermediate undergoes homogeneous chemical reaction. The interpretation of N is far more subtle than it appears at first glance.

Bruckenstein and Feldman gave an approximate treatment of this problem using the approach of an average transit time between disk and ring (96). The rigorous solution by Albery and Bruckenstein showed that there are two limiting situations (101). When the chemical reaction is very fast, the amount of ring current is mainly limited by the kinetic decomposition. For slower reactions, the more ordinary situation involving competition between the ring and loss of material to the bulk of solution exists. The concentration profiles between these two extremes are very different. An important conclusion of the Albery and Bruckenstein work is that it is possible to calibrate RRDE's in terms of known first-order (or pseudo) rate constants. The electrode may then be used for unknown rate measurements. Some preliminary studies of this type have been carried out (106). First-order rate constants in the region 10^{-2} to 10^3 sec^{-1} should be accessible to the RRDE. Intermediates with half-lives less than ca. 10^{-3} sec would be undetectable with the RRDE. The original papers should be consulted for further details.

Albery et al. have also considered second-order kinetic cases and other uses of the RRDE (100,101). Much activity can be expected in the applications of the RRDE to homogeneous chemical reactions coupled to electron transfers. The only difficulty lies in the fabrication of electrodes with the desired geometries. This, in fact, is not a simple problem. Nevertheless, the application of the RRDE in a qualitative sense to detect unstable intermediates in complex organic electrode reactions has been very successful (*42*), and the possibility of empirical calibrations using known reaction rates may circumvent some of the problems.

Equations have been derived and tested for a ring electrode in which the "disk" consists of a coated inner portion and the electroactive surface is an annular ring only. Such an electrode is of considerable interest in that the limiting current to the ring surface (if the ring is small in width) is greater than that at an equivalent area of "pure disk" surface (*64,65,91*). Heusler and Schurig have developed a double-ring electrode for determining reactive intermediates (*107*).

Finally, it is possible to examine transient responses at the RDE (galvanostatic and potentiostatic methods). As pointed out by Hale and others, there are some advantages to using transient techniques with controlled convective mass transport (*91,108,109*). The reader is referred to the original literature for further details on these non-steady-state processes at the RDE (*108–111*).

4-6. MASS TRANSPORT WITH TURBULENT FLOW

Only laminar flow situations have been considered so far in Chapter 4. The RDE has served very well as a tool in liquid turbulence studies. Turbulent flow is a very complex situation and one about which expert opinions are still quite divergent (*59,91*). Only a brief description of some problems of interest to voltammetry is attempted here. At first thought, turbulence in electrochemical studies might be considered an experimental difficulty to be avoided at all costs. In cases where the highest rates of mass transport are desired, turbulence is often unavoidable. In general, in heterogeneous chemical reactions, turbulent conditions are desired for efficient mixing and shorter reaction times. Quantitative treatments of turbulent flow are therefore of great interest.

In turbulent flow there is an extremely random and chaotic motion of fluid near the solid body past which the liquid flows. Material is transported in an erratic fashion by turbulence eddies—not at all unlike random molecular diffusion in gases (*59*). Imposition of a concentration gradient as in the electrochemical case leads to a resultant transfer of electroactive species to the electrode surface. Levich and co-workers distinguish four rather distinct boundary layers of importance. While most of the mass transport is via the turbulence eddies, there is still a thin layer next to the electrode surface in which molecular diffusion predominates. One may characterize an effective turbulent flow diffusion coefficient D_{turb}, but it bears little resemblance to the usual D.

Turbulence develops at large values of the Reynolds number. In general mass-transport terms,

$$N_{Re} = \frac{UL}{\nu}$$

Since the kinematic viscosity of ordinary solutions is quite small, a large N_{Re} is easily encountered. For the RDE,

$$N_{Re} = \frac{r^2 \omega}{\nu}$$

and turbulence is encountered at N_{Re} of ca. 10^4–10^5. The onset of turbulence with a RDE is quite clearly a critical function of the centering and smoothness of the disk, wobble of the shaft, etc. Unless particular care is taken in the design of the entire RDE assembly it may well be true that turbulence will be encountered at $N_{Re} < 10^4$. Owing to the large variation in disk sizes, it is hardly practical to designate turbulence limits in terms of rotation velocities.

In the Levich theory of turbulent flow at a RDE there are two distinct differences when compared to laminar flow conditions. First, the value of i_L is proportional to ω or N to the power 0.8–0.95, as contrasted to the $\omega^{1/2}$ dependence in laminar flow. This difference in exponents from 0.5 to almost 1 is very noticeable. Further, $i_L \propto kD^{3/4}$, as opposed to $D^{2/3}$ in laminar flow.

The work of Bogotskaya has verified some of the predictions of the Levich theory of turbulence at the RDE (60). Using a large (radius = 3 cm) amalgamated copper RDE, the oxygen-reduction and hydrogen-discharge reactions were studied. Rotation velocities varied between 50 and 1500 rpm with corresponding Reynolds numbers between 10^4 and 10^5. The almost-unity dependence of i_L on rotation velocity was observed. As in the studies by Siver and Kabanov, the ratio (D_{H^+}/D_{O_2}) was measured by Bogotskaya. The observed ratio of 0.67–0.72 is close to the $\frac{3}{4}$ predicted by the Levich theory and considerably different from the value of unity given by some other theories of turbulent flow (59). Since a $D^{2/3}$ dependence is not expected in turbulent flow, it is not significant that the experimental values happen also to lie close to 0.67. Other studies supporting the Levich ideas on turbulent flow at the RDE were concerned with the apparent activation energies for the turbulent mass transport (61–63). Recent studies of turbulent flow at planar and cylindrical electrodes have been reported (77,112).

One point which emerges from these studies is that it does not appear to be too difficult to obtain reproducible values of i_L under turbulent conditions at the RDE. This is important in the measurement of heterogeneous rate constants via the RDE, where high mass-transport rates are needed. Since absolute values of i_L are not required, if they can be made reproducible, transitions between laminar and turbulent flow would seem to offer little difficulty in such studies. However, in all cases where the RDE is used to quantitatively measure i_L to calculate D, area, etc., it should be restricted to the laminar flow regime.

4-7. ROTATED WIRE ELECTRODES

The rotating platinum wire electrode was apparently first studied by Nernst and Merriam (2). It was introduced as an analytical tool by Laitinen and Kolthoff (66) in an effort to eliminate the waiting for steady-state currents with the stationary electrode. The electrode takes many physical forms. The construction and evaluation of various rotated electrodes will be given in Chapter 9.

The rotated wire is by far the most widely used solid electrode. It was observed from the beginning that limiting currents at this electrode were proportional to concentration. Any theoretical evaluation of i_L was generally in terms of the Nernst diffusion layer until Tsukamoto et al. (67) presented the results of their experimental hydrodynamic studies.

Their derivations assumed laminar flow at a RPE. The electrode consisted of a platinum wire fused in the end of a glass tube and bent 90° to the axis of rotation, i.e., the wire rotated in a horizontal plane. Details of their derivation of equation of i_L are to be found in the original paper. It is significant that a $D^{2/3}$ dependence was predicted under ordinary conditions.

A nonuniform dependence of i_L on rotation rate was observed. In the region 200–1600 rpm, $i_L = k(\text{rpm})^{1/3}$, while between 10 and 90 rpm, $i_L = k(\text{rpm})^{1/4}$. There appeared to be no discernible proportionality between i_L and rotation rate in the intermediate range 90–200 rpm. Tsukamoto et al. also derived and checked the dependence of i_L on the electrode geometry as a function of rotation rate.

It is difficult to access the reliability of limiting current data at rotating wire electrodes because of the extreme variation in actual stirring conditions with small changes in electrode configuration. Recent studies by Ferrett and Phillips with the horizontally revolving type of electrode showed values of i_L to vary as

$$i_L = k(\text{rpm})^{0.51-0.64}$$

for reductions of metal ions such as Cu^{2+}, Pb^{2+}, Tl^+, and Cd^{2+}. For oxygen reduction an exponent of 0.27 was obtained (68). While this electrode was of the horizontal-sweep type, only the leading edge moving in the direction of rotation was exposed as electroactive surface. They found that much more reproducible behavior was obtained by waxing the trailing portion of the electrode. This would suggest that laminar flow conditions are not achieved at such an electrode but rather that turbulence exists behind the electrode. It is difficult to imagine pure laminar flow at the RPE used by Tsukamoto et al.

On the other hand, Nightengale has demonstrated that RPE's in which the wire is rotated coaxially with the glass tube give i_L proportional to $D^{2/3}$, as predicted for laminar flow (69).

Eisenberg et al. (11), using a smooth nickel cylinder rotating about its axis in the center of a stationary, circular, cylindrical outer electrode found the following relation for the limiting current:

$$i_L = nFAC^b D^{0.64} V^{0.70} d_i^{-0.30} \nu^{-0.34} \tag{4-34}$$

where i_L = amperes
A = area of inner electrode, cm²
D = diffusion coefficient
V = peripheral velocity at the inner electrode, cm/sec
d_i = diameter of inner rotating electrode, cm
ν = kinematic viscosity, cm²/sec

These workers studied the ferro-ferricyanide oxidation–reduction reactions. The hydrodynamic conditions were far more rigidly defined than with the rotated wire systems used in analytical work. Arvia and Carrozza similarly found a $V^{0.70}$ dependence for electrolysis of $CuSO_4$ at rotated cylinder electrodes (76).

The preceding discussion indicates that correlations of i_L with rotation speed (and other hydrodynamic parameters) vary markedly with the individual wire electrode. It is perhaps most wise to summarize such effects in the form of a mass-transfer coefficient, as has been done by

Jordan and Javick (*70,71*). The limiting current for any electrode can be generalized in the form

$$i_L = nFAC_O^b m \qquad (4\text{-}35)$$

where *m* is the *mass-transfer coefficient* usually expressed in cm/sec. For the unique case of semiinfinite linear diffusion, *m* becomes

$$m_{\text{lin. diff}} = \frac{D^{1/2}}{\pi^{1/2} t^{1/2}} \qquad (4\text{-}36)$$

Substitution of Eq. (4-36) into Eq. (4-35) is seen to give Eq. (3-16), which expresses the instantaneous current at a plane electrode under linear diffusion conditions.

For rotated electrodes which involve diffusion plus convection, *m* is of a more complex nature, having the general form

$$m_{\text{rot}} = \text{const. } D^q V^p d^a v^s \qquad (4\text{-}37)$$

where D = diffusion coefficient
       V = solution velocity
       d = density of the solution
       v = kinematic viscosity

The exponents of the various parameters are determined by the exact hydrodynamic conditions existing with each electrode system and may indeed vary with seemingly identical electrodes, owing to experimentally unobservable variances in stirring [note Eq. (4-34) for a rotated cylinder].

From the analytical viewpoint it is fortunate that i_L is proportional to bulk concentration for *all rotated electrode systems.* Limiting currents can be used with confidence for quantitative work. To ensure reproducibility it is necessary to impose the following restrictions:

1. The geometry of the electrode and of the entire electrolysis vessel should be kept invariant. This is particularly important with respect to the wire electrode. Accidental bending of wire during cleaning operations, etc., may give rise to large variations in i_L.

2. The electrode should be rotated at reasonably high rates to ensure steady-state conditions. Since 600 rpm represents a convenient "gear-down" value for 1800-rpm synchronous motors, it would seem sensible to use 600 rpm as a standard rotation speed. In any event, the rotation rate should be held constant for all quantitative work.

Much work remains to be done on the best design of rotated electrode systems. The use of modifications such as waxed trailing edges, specially designed electrode shapes, etc., hold promise for greater precision in limiting current measurements. Without question, the RDE offers the most promise for precise measurements of limiting currents.

4-8. VIBRATING WIRE ELECTRODES

Harris and Lindsey (72–74) introduced the vibrating platinum-wire electrode for analytical applications. The electrode vibrates in a vertical (up and down) sense with constant frequency and an amplitude considerably larger than that electrode dimensions. A commercial massage-type vibrator can be used to power it (73).

Lindsey found that i_L varied linearly with frequency of vibration (at constant amplitude) in the range 1–40 Hz. Above 40 Hz, i_L became independent of frequency. Similar relations were found to hold for variation in the amplitude of vibration at constant frequency (100 Hz). Limiting currents were independent of amplitude above ca. 0.5 mm. These effects are qualitatively in accord with the rotation effects considered in the previous section.

No critical treatment of the mass-transfer conditions at vibrating wire electrodes has been attempted. It is evident, however, that i_L is proportional to bulk concentration within the frequency and amplitude limitations of about 60 Hz and 0.5 mm, respectively. To date, the vibrating electrode has been used mainly in amperometric titrations (74). Roberts and Meek employed a double vibrating platinum-wire system in an interesting study of the reduction of organic peroxides (75). Vibrating electrodes are used in a variety of physiological chemistry measurements.

REFERENCES

1. W. Nernst, *Z. Physik. Chem.*, **47**, 52 (1904).
2. W. Nernst and E. S. Merriam, *Z. Physik. Chem.*, **53**, 235 (1905).
3. I. M. Kolthoff and J. Jordan, *J. Am. Chem. Soc.*, **76**, 3843 (1954).
4. J. N. Agar, *Discussions Faraday Soc.*, **1**, 26 (1947).
5. S. Glasstone and A. Hickling, *Electrolytic Oxidation and Reduction*, Chapman and Hall, London, 1935, Chap. 3.
6. B. Levich, *Discussions Faraday Soc.*, **1**, 37 (1947). (Note the name B. Levich appeared in this and early English copy, the initials V. G. are used later.)
7. L. L. Bircumshaw and A. C. Riddiford, *Quart. Rev. (London)*, **6**, 157 (1952).
8. V. G. Levich, *Physicochemical Hydrodynamics*, Prentice-Hall, Englewood Cliffs, N.J., 1962.
9. J. N. Agar, *Discussions Faraday Soc.*, **1**, 26 (1947).
10. See especially, N. Ibl, *Chimia (Aarau)*, **9**, 135 (1955); *Electrochim. Acta*, **1**, 177 (1959).

11. M. Eisenberg, C. W. Tobias, and C. R. Wilke, *J. Electrochem. Soc.*, **101**, 306 (1954).
12. C. R. Wilke, M. Eisenberg, and C. W. Tobias, *J. Electrochem. Soc.*, **100**, 513 (1953).
13. C. S. Lin, E. B. Denton, H. S. Gaskill, and G. L. Putnam, *Ind. Eng. Chem.*, **43**, 2136 (1951).
14. P. Delahay, *New Instrumental Methods in Electrochemistry*, Wiley (Interscience), New York, 1954, p. 231.
15. G. Trümpler and H. Zeller, *Helv. Chim. Acta*, **34**, 952 (1951).
16. J. Jordan, R. A. Javick, and W. E. Ranz, *J. Am. Chem. Soc.*, **80**, 3846 (1958).
17. J. Jordan, *Anal. Chem.*, **27**, 1708 (1955).
18. B. Levich, *Acta Physicochim. URSS*, **17**, 257 (1942).
19. Ref. *8*, pp. 60–70.
20. V. G. Levich, *Soviet Electrochem., Proc. Conf. Electrochem.* (*English Transl.*) 4th Moscow, *1956*, **1**, 139 (1961).
21. E. A. Hogge and M. B. Kraichman, *J. Am. Chem. Soc.*, **76**, 1431 (1954).
22. M. B. Kraichman and E. A. Hogge, *J. Phys. Chem.*, **59**, 986 (1955).
23. D. P. Gregory and A. C. Riddiford, *J. Chem. Soc.*, **1956**, 3756.
24. D. P. Gregory and A. C. Riddiford, *J. Electrochem. Soc.*, **107**, 950 (1960).
25. S. Azim and A. C. Riddiford, *Anal. Chem.*, **34**, 1023 (1962).
26. A. N. Frumkin and G. Tedoradse, *Z. Elektrochem.*, **62**, 251 (1958).
27. W. Vielstich and D. Jahn, *Z. Elektrochem.*, **64**, 43 (1960).
28. D. Jahn and W. Vielstich, *J. Electrochem. Soc.*, **109**, 849 (1962).
29. Yu. K. Plesov, *Zh. Fiz. Khim.*, **34**, 623 (1960); *Russ. J. Phys. Chem.*, **34**, 296 (1960).
30. G. P. Lewis and P. Reutschi, *J. Electrochem. Soc.*, **67**, 65 (1963).
31. Z. Galus and R. N. Adams, *J Phys. Chem.*, **67**, 866 (1963).
32. E. A. Alkazyan and A. I. Fedorova, *Dokl. Akad. Nauk SSSR*, **86**, 1137 (1952); see also Ref. *8*, p. 311.
33. J. D. Newson and A. C. Riddiford, *J. Electrochem. Soc.*, **108**, 695 (1961).
34. G. R. Johnson and D. R. Turner, *J. Electrochem. Soc.*, **109**, 918 (1962).
35. Z. Galus, C. Olson, H. Y. Lee, and R. N. Adams, *Anal. Chem.*, **34**, 164 (1962).
36. Yu. G. Siver and B. N. Kabanov, *Zh. Fiz. Khim.*, **22**, 53 (1948).
37. B. N. Kabanov, *Zh. Fiz. Khim.*, **23**, 428 (1949).
38. Ref. *8*, pp. 308–310.
39. J. Zimmerman, D. Hawley, and R. N. Adams, unpublished studies, 1963.
40. Ref. *8*, p. 72.
41. Ref. *8*, p. 78.
42. Z. Galus and R. N. Adams, *J. Am. Chem. Soc.*, **84**, 2061 (1962); note error in Eq. (10) of this reference; the exponent of D should be 1/3.
43. Ref. *8*, pp. 72–76.
44. G. P. Dezider'ev and S. I. Berezina, *Dokl. Akad. Nauk SSSR*, **130**, 1270 (1960).
45. L. P. Kholpanov, *Zh. Fiz. Khim.*, **35**, 1538 (1961).
46. Ref. *8*, p. 65.
47. J. Zimmerman, Ph.D. thesis, Univ. Kansas, Lawrence, 1963.
48. A. N. Frumkin and L. N. Nekrasov, *Dokl. Akad. Nauk SSSR*, **126**, 1029 (1959).
49. Yu. B. Ivanov and V. G. Levich, *Dokl. Akad. Nauk SSSR*, **126**, 1029 (1959).
50. Ref. *8*, pp. 327–336.
51. A. N. Frumkin, L. Nekrasov, B. Levich, and Yu. Ivanov, *J. Electroanal. Chem.*, **1**, 84 (1959/60).
52. Ref. *14*, Chaps. 5, 7, 8.

REFERENCES

53. Z. *Elektrochem.*, **64**, 1-204 (1960), *Symposium on Fast Reactions in Solution, Hahnenklee/Harz, 1959.*
54. J. Koutecky and V. G. Levich, *Zh. Fiz. Khim.*, **32**, 1565 (1958).
55. Ref. *8*, pp. 345-357.
56. R. R. Dagonadze, *Zh. Fiz. Khim.*, **32**, 2437 (1958).
57. H. S. Swofford and R. L. Carman, *Anal. Chem.*, **38**, 966 (1966).
58. P. A. Malachesky, L. S. Marcoux, and R. N. Adams, *J. Phys. Chem.*, **70**, 4068 (1966).
59. Ref. *8*, Chap. 3.
60. I. A. Bogotskaya, *Dokl. Akad. Nauk SSSR*, **85**, 1057 (1952).
61. A. I. Fedorova and G. A. Vidovich, *Dokl. Akad. Nauk SSSR*, **109**, 135 (1956).
62. A. I. Fedorova, G. A. Vidovich, L. I. Boguslavskii, and V. D. Yukhtanova, *Soviet Electrochem. Proc. Conf. Electrochem.*, (*English Transl.*) *4th Moscow, 1956*, **1**, 153 (1961).
63. Ref. *8*, p. 319.
64. Ref. *8*, pp. 108, 313-315.
65. N. Ibl, *J. Electrochem. Soc.*, **108**, 610 (1961).
66. H. A. Laitinen and I. M. Kolthoff, *J. Phys. Chem.*, **45**, 1079 (1941).
67. T. Tsukamoto, T. Kambara, and I. Tachi, *Proc. 1st Intern. Polarog. Congr., Prague, 1951*, p. 525.
68. D. J. Ferrett and C. S. G. Phillips, *Trans. Faraday Soc.*, **51**, 390 (1955).
69. E. R. Nightengale, *Anal. Chim. Acta*, **16**, 493 (1957).
70. J. Jordan, *Anal. Chem.*, **27**, 1708 (1955).
71. J. Jordan and R. A. Javick, *Electrochim. Acta*, **6**, 23 (1962).
72. E. D. Harris and A. J. Lindsey, *Nature*, **162**, 413 (1948).
73. A. J. Lindsey, *J. Phys. Chem.*, **56**, 439 (1952).
74. E. D. Harris and A. J. Lindsey, *Analyst*, **76**, 647, 650 (1951).
75. E. R. Roberts and J. S. Meek, *Analyst*, **77**, 43 (1952).
76. A. J. Arvia and J. S. W. Carrozza, *Electrochim. Acta*, **7**, 65 (1962).
77. G. Wranglén and O. Nilsson, *Electrochim. Acta*, **7**, 121 (1962).
78. J. C. Bazán and A. J. Arvia, *Electrochim. Acta*, **9**, 17, 667 (1964).
79. T. K. Ross and A. A. Wragg, *Electrochim. Acta*, **10**, 1093 (1965).
80. W. J. Blaedel, C. L. Olson, and L. R. Sharma, *Anal. Chem.*, **35**, 2100 (1963).
81. W. J. Blaedel and L. N. Klatt, *Anal. Chem.*, **38**, 879 (1966).
82. J. Jordan and R. A. Javick, *J. Am. Chem. Soc.*, **80**, 1264 (1958).
83. J. Jordan and R. A. Javick, *Electrochim. Acta*, **6**, 23 (1962).
84. J. Jordan and H. A. Catherino, *J. Phys. Chem.*, **67**, 2241 (1963).
85. O. H. Müller, *J. Am. Chem. Soc.*, **69**, 2992 (1947).
86. F. Štráfelda, *Collection Czech. Chem. Commun.*, **25**, 862 (1960).
87. F. Štráfelda and A. Kimla, *Collection Czech. Chem. Commun.*, **28**, 1516 (1963).
88. F. Štráfelda and A. Kimla, *Collection Czech. Chem. Commun.*, **30**, 3606 (1965), and previous papers.
89. Yu. S. Gurinov and S. V. Gorbachev, *Zh. Fiz. Khim.*, **37**, 1141 (1963).
90. K. F. Blurton and A. C. Riddiford, *J. Electroanal. Chem.*, **10**, 457 (1965).
91. A. C. Riddiford, *Advan. Electrochem. Electrochem. Eng.*, **4**, 47 (1966).
92. K. B. Prater and R. N. Adams, *Anal. Chem.*, **38**, 153 (1966).
93. W. J. Albery and R. P. Bell, *Proc. Chem. Soc.*, **1963**, 169.
94. L. Holleck, B. Kastening, and H. Vogt, *Electrochim. Acta*, **8**, 255 (1963).

95. Yu. B. Ivanov and V. G. Levich, *Dokl. Akad. Nauk SSSR*, **121**, 503 (1958).
96. S. Bruckenstein and G. A. Feldman, *J. Electroanal. Chem.*, **9**, 395 (1965).
97. W. J. Albery, *Trans. Faraday Soc.*, **62**, 1915 (1966).
98. W. J. Albery and S. Bruckenstein, *Trans. Faraday Soc.*, **62**, 1920 (1966).
99. W. J. Albery, S. Bruckenstein, and D. T. Napp, *Trans. Faraday Soc.*, **62**, 1932 (1966).
100. W. J. Albery, S. Bruckenstein, and D. C. Johnson, *Trans. Faraday Soc.*, **62**, 1938 (1966).
101. W. J. Albery and S. Bruckenstein, *Trans. Faraday Soc.*, **62**, 1946 (1966).
102. K. B. Prater and R. N. Adams, unpublished data, 1964.
103. L. N. Nekrasov and N. P. Besezine, *Dokl. Adak. Nauk SSSR*, **142**, 855 (1962).
104. L. N. Nekrasov and L. Mueller, *Dokl. Akad. Nauk SSSR*, **149**, 1107 (1963).
105. L. Mueller and L. N. Nekrasov, *Dokl. Akad. Nauk SSSR*, **154**, 437 (1964); *Electrochim. Acta*, **9**, 1015 (1964).
106. P. A. Malachesky, K. B. Prater, G. Petrie, and R. N. Adams, *J. Electroanal. Chem.*, **16**, 41 (1968).
107. K. E. Heusler and H. Schurig, *Z. Physik. Chem.*, (*Frankfurt*), **47**, 117 (1963).
108. R. P. Buck and H. E. Keller, *Anal. Chem.*, **35**, 400 (1963).
109. J. M. Hale, *J. Electroanal. Chem.*, **6**, 187 (1963); **8**, 332 (1964).
110. Yu. G. Siver, *Zh. Fiz. Khim.*, **34**, 577 (1960).
111. I. Fried and P. J. Elving, *Anal. Chem.*, **37**, 464, 803 (1965).
112. A. J. Arvia, J. S. W. Carrozza, and S. L. Marchiano, *Electrochim. Acta*, **9**, 1483 (1964).
113. J. Newman, *J. Phys. Chem.*, **70**, 1327 (1966).
114. J. Newman, *J. Electrochem. Soc.*, **113**, 501, 1235 (1966); **114**, 239 (1967).
115. J. Newman and L. Hsueh, *Electrochim. Acta*, **12**, 417 (1967).
116. L. Hsueh and J. Newman, *Electrochim. Acta*, **12**, 429 (1967).
117. S. L. Marchiano and A. J. Arvia, *Electrochim. Acta*, **12**, 801 (1967).
118. J. S. W. Carrozza, S. L. Marchiano, J. J. Podesta and A. J. Arvia, *Electrochim. Acta*, **12**, 809 (1967).
119. J. S. Bazán, S. L. Marchiano, and A. J. Arvia, *Electrochim. Acta*, **12**, 821 (1967).

SPECIAL BIBLIOGRAPHY ON ROTATED DISK ELECTRODES

B. Levich, *Acta Physicochim. URSS*, **17**, 257 (1942); **19**, 117, 133 (1944).
B. Levich, *Discussions Faraday Soc.*, **1**, 37 (1947).
Yu. G. Siver and B. N. Kabanov, *J. Phys. Chem. USSR*, **22**, 53 (1948); **23**, 428 (1949).
C. W. Tobias, M. Eisenberg, and C. R. Wilke, *J. Electrochem. Soc.*, **99**, 359C (1952).
L. L. Bircumshaw and A. C. Riddiford, *Quart. Rev.* (*London*), **6**, 157 (1952).
E. A. Aykazyan and A. I. Fedorova, *Dokl. Akad. Nauk SSSR*, **86**, 1137 (1952).
I. A. Bagotskaya, *Dokl. Akad. Nauk SSSR*, **85**, 1057 (1952).
E. A. Hogge and M. B. Kraichman, *J. Am. Chem. Soc.*, **76**, 1431 (1954).
A. N. Frumkin and E. A. Aykazyan, *Dokl. Akad. Nauk SSSR*, **100**, 315 (1955).
V. G. Levich, *Soviet Electrochem.*, *Proc. Conf. Electrochem.* (*English Transl.*) 4th Moscow, *1956*, **1**, 139 (1961).
A. I. Fedorova, G. L. Vidovich, L. I. Boguslavskii, and V. D. Yukhtanova, *Soviet Electrochem.*, *Proc. Conf. Electrochem.* (*English Transl.*) 4th Moscow, *1956*, **1**, 153 (1961).
D. P. Gregory and A. C. Riddiford, *J. Chem. Soc.*, **1956**, 3756.

REFERENCES

Yu. K. Pleskov, *Dokl. Akad. Nauk SSSR*, **117**, 645 (1957).
E. A. Aikazyan and Yu. K. Pleskov, *Zh. Fiz. Khim.*, **31**, 205 (1957).
A. N. Frumkin and G. Tedoradse, *Dokl. Akad. Nauk SSSR*, **118**, 530 (1958); *Z. Elektrochem.*, **62**, 251 (1958).
R. R. Dogonadze, *Zh. Fiz. Khim.*, **32**, 2437 (1958).
Yu. B. Ivanov and V. G. Levich, *Dokl. Akad. Nauk SSSR*, **121**, 503 (1958).
Yu. V. Pleskov and B. N. Kabanov, *Dokl. Akad. Nauk SSSR*, **123**, 884 (1958).
J. Koutecky and V. G. Levich, *Dokl. Akad. Nauk SSSR*, **117**, 441 (1957); *Zh. Fiz. Khim.*, **32**, 1565 (1958).
V. D. Yukhtanova, *Dokl. Akad. Nauk SSSR*, **120**, 137 (1958).
A. N. Frumkin and E. A. Aykazyan, *Izv. Akad. Nauk SSSR Otd. Khim. Nauk*, **2**, 202 (1959).
V. D. Yukhtanova, *Dokl. Akad. Nauk SSSR*, **124**, 377 (1959).
Y. B. Ivanov and V. G. Levich, *Dokl. Akad. Nauk SSSR*, **126**, 1029 (1959).
A. N. Frumkin and L. N. Nekrasov, *Dokl. Akad. Nauk SSSR*, **126**, 115 (1959).
V. S. Temianko, M. B. Bardin, and Y. S. Lyalikov, *Izv. Vysshikh Uch. Zavedenii Khim i Khim Tekhnol.*, **2**, 503 (1959).
G. Tedoradse, *Zh. Fiz. Khim.*, **33**, 129 (1959).
A. N. Frumkin, L. Nekrasov, B. Levich, and Y. Ivanov, *J. Electroanal. Chem.*, **1**, 84 (1959).
G. P. Dezider'ev and S. I. Berezina, *Dokl. Akad. Nauk SSSR*, **130**, 1270 (1960).
D. P. Gregory and A. C. Riddiford, *J. Electrochem. Soc.*, **107**, 950 (1960).
Yu. G. Siver, *Zh. Fiz. Khim.*, **34**, 577 (1960).
E. Budevskii and S. Toshev, *Dokl. Akad. Nauk SSSR*, **130**, 1047 (1960).
J. K. Delimarskii, I. B. Panchenko, and G. V. Silina, *Collection Czech. Chem. Commun.*, **25**, 3061 (1960).
E. A. Aykazyan and R. A. Arakelyan, *Izv. Akad. Nauk Arm. Khim. Nauk*, **113**, 225 (1960).
Yu. K. Pleskov, *Zh. Fiz. Khim.*, **34**, 623 (1960).
M. V. Stackelberg, W. Vielstich, and D. Jahn, *Anales Real. Soc. Espan. Fiz. Quim (Madrid)*, **B56**, 475 (1960).
Z. Zembura, *Roczniki Chem.*, **34**, 1509 (1960).
M. Breiter and K. Hoffman, *Z. Elektrochem.*, **64**, 462 (1960).
W. Vielstich, *Z. Anal. Chem.*, **173**, 84 (1960).
W. Vielstich and D. Jahn, *Z. Elektrochem.*, **64**, 43 (1960); *Advan. Polarog. Proc. Intern. Congr. 2nd Cambridge, Engl. 1959*, **1**, 281 (1960).
M. P. Belyanchikov, Yu. K. Pleskov, and V. G. Pominov, *Zh. Fiz. Khim.*, **34**, 1638 (1960).
R. Landsberg, W. Geissler, and S. Müller, *Z. Chem.*, **1**, 169 (1961).
Yu. K. Pleskov, *Zh. Fiz. Khim.*, **35**, 2540 (1961).
L. P. Kholpanov, *Zh. Fiz. Khim.*, **35**, 1538, 1567, 2759 (1961).
J. D. Newson and A. C. Riddiford, *J. Electrochem. Soc.*, **108**, 695, 699 (1961).
S. V. Gorbachev and V. A. Belyaeva, *Zh. Fiz. Khim.*, **35**, 2158 (1961).
N. G. Chovnyk and V. V. Vashchenko, *Zh. Fiz. Khim.*, **35**, 580 (1961).
R. Landsberg, W. Müller, and J. Hendel, *J. Electroanal. Chem.*, **2**, 400 (1961).
W. Geissler and R. Landsberg, *Z. Chem.*, **1**, 308 (1961).
R. Landsberg, J. Hendel, and W. Müller, *J. Electroanal. Chem.*, **2**, 484 (1961).
O. L. Kabanova, *Zh. Fiz. Khim.*, **35**, 2465 (1961).

V. D. Yokhtanova, *Zh. Fiz. Khim.*, **35**, 2778 (1961).
Z. Zembura, *Bull Acad. Polon. Sci. Ser. Sci. Chim.*, **9**, 531 (1961).
J. K. Delimarskii, I. D. Pantshenko, and G. W. Shilina, *Dopodivi Akad. Nauk Ukr. RSR*, **1961**, 205; *Collection Czech. Chem. Commun.*, **25**, 3061 (1960).
L. P. Kholpanov, *Zh. Fiz. Khim.*, **36**, 214 (1962).
S. Azim and A. C. Riddiford, *Anal. Chem.*, **34**, 1023 (1962).
Z. Galus, C. Olson, H. Y. Lee, and R. N. Adams, *Anal. Chem.*, **34**, 164 (1962).
L. P. Kholpanov and S. V. Gorbachev, *Zh. Fiz. Khim.*, **36**, 855, 859 (1962).
G. R. Johnson and D. R. Turner, *J. Electrochem. Soc.*, **109**, 918 (1962).
S. V. Gorbachev and V. A. Belyaeva, *Zh. Fiz. Khim.*, **36**, 229 (1962).
F. Nagy, Gy. Horanyi, and Gy. Vertes, *Magy. Kem. Folyoirat*, **68**, 198, 202 (1962).
Z. Galus and R. N. Adams, *J. Electroanal. Chem.*, **4**, 248 (1962).
D. Jahn and W. Vielstich, *J. Electrochem. Soc.*, **109**, 849 (1962).
V. A. Belyaeva, *Zh. Fiz. Khim.*, **36**, 1385 (1962).
Z. Galus and R. N. Adams, *J. Am. Chem. Soc.*, **84**, 2061 (1962).
O. L. Kabanova, *Zh. Analit. Khim.*, **17**, 796 (1962).
S. E. Beacom and R. N. Hollyer, *J. Electrochem. Soc.*, **109**, 495 (1962).
L. N. Nekrasov and N. P. Besezine, *Dokl. Akad. Nauk SSSR*, **142**, 855 (1962).
A. Stanienda, *Z. Physik. Chem. (Frankfurt)*, **33**, 170 (1962).
R. Landsberg and R. Thiele, *Z. Physik. Chem. (Leipzig)*, **221**, 211 (1962); *Z. Chem. Tech.*, **10**, 627 (1962).
G. P. Lewis and P. Reutschi, *J. Electrochem. Soc.*, **67**, 65 (1963).
R. P. Buck and H. E. Keller, *Anal. Chem.*, **35**, 400 (1963).
Z. Galus and R. N. Adams, *J. Phys. Chem.*, **67**, 862, 866 (1963).
W. J. Albery and R. P. Bell, *Proc. Chem. Soc.*, **1963**, 169.
G. T. Rogers and K. J. Taylor, *Electrochim. Acta*, **8**, 887 (1963); *Nature*, **200**, 1062 (1963).
I. D. Pantaschenko and G. W. Schilina, *Zh. Analit. Khim.*, **18**, 920 (1963).
A. Stanienda, *Naturwiss.*, **50**, 731 (1963).
K. E. Heusler and H. Schurig, *Z. Physik. Chem. (Frankfurt)*, **47**, 117 (1963).
Z. Zembura, *Bull. Acad. Polon. Sci.*, **9**, 271 (1963).
J. M. Hale, *J. Electroanal. Chem.*, **6**, 187 (1963).
T. Mussini, *Chim. Ind. (Milan)*, **45**, 1075 (1963).
L. Holleck, B. Kastening, and H. Vogt, *Electrochim. Acta*, **8**, 255 (1963).
P. Lewis and P. Ruetschi, *J. Phys. Chem.*, **67**, 65 (1963).
L. N. Nekrasov and L. Mueller, *Dokl. Akad. Nauk SSSR*, **149**, 1107 (1963).
L. P. Kholpanov, *Zh. Fiz. Khim.*, **37**, 890 (1963).
R. Landsberg, R. Nitzsche, and W. Geissler, *Z. Physik. Chem. (Leipzig)*, **222**, 54 (1963).
S. Bruckenstein and D. C. Johnson, *Anal. Chem.*, **36**, 2186 (1964).
J. M. Hale, *J. Electroanal. Chem.*, **8**, 332 (1964).
M. B. Bardin and V. P. Goncharenko, *Zh. Fiz. Khim.*, **38**, 2626 (1964).
L. Mueller and L. N. Nekrasov, *Dokl. Akad. Nauk SSSR*, **154**, 437 (1964).
L. Mueller and L. Nekrasov, *Electrochim. Acta*, **9**, 1015 (1964).
R. Landsberg, K. Kresse, D. Molch, and W. Geissler, *Z. Physik. Chem (Leipzig)*, **227**, 401 (1964).
Z. Zembura and M. Bierowski, *Wiadomosci Chemi.*, **18**, 215 (1964).
Y. Lu-an, Y. B. Vasil'ev, and V. S. Bagotskii, *Zh. Fiz. Khim.*, **38**, 205 (1964); *Elektrokhimiya*, **1**, 170 (1965).

REFERENCES

V. S. Bagotsky and Yu. B. Vasilyev, *Electrochim. Acta*, **9**, 869 (1964).
S. Bruckenstein and G. A. Feldman, *J. Electroanal. Chem.*, **9**, 395 (1965).
I. Fried and P. J. Elving, *Anal. Chem.*, **37**, 464, 803 (1965).
R. E. Meyer, *J. Electrochem. Soc.*, **112**, 684 (1965).
H. E. Hintermann and E. Suter, *Rev. Sci. Inst.*, **36**, 1610 (1965).
K. F. Blurton and A. C. Riddiford, *J. Electroanal. Chem.*, **10**, 457 (1965).
A. Stanienda, *Naturwiss.*, **52**, 105 (1965); *Z. Physik. Chem.*, *(Leipzig)*, **229**, 257 (1965).
W. J. Albery, *Trans. Faraday Soc.*, **61**, 2063 (1965).
Z. Zembura and A. Fulinski, *Electrochim. Acta*, **10**, 859 (1965).
A. N. Frumkin, E. I. Krushcheva, M. R. Tarasevich, and N. A. Shumilova, *Elektrokhimiya*, **1**, 3 (1965).
E. I. Krushcheva, N. A. Shumilova, and M. R. Tarasevich, *Elektrokhimiya*, **1**, 730 (1965).
V. N. Boronenkov, O. A. Esin, and P. M. Shurygin, *Elektrokhimiya*, **1**, 592 (1965).
B. R. Sundheim and W. Sauerwein, *J. Phys. Chem.*, **69**, 4042 (1965).
A. J. Arvia, J. Bazan, and J. S. W. Carrozza, *Electrochim. Acta*, **11**, 881 (1966).
K. B. Prater and R. N. Adams, *Anal. Chem.*, **38**, 153 (1966).
W. J. Albery, *Trans. Faraday Soc.*, **62**, 1915 (1966).
W. J. Albery and S. Bruckenstein, *Trans. Faraday Soc.*, **62**, 1920, 1946, 2584, 2596 (1966).
W. J. Albery, S. Bruckenstein, and D. T. Napp, *Trans. Faraday Soc.*, **62**, 1932 (1966).
W. J. Albery, S. Bruckenstein, and D. C. Johnson, *Trans. Faraday Soc.*, **62**, 1938 (1966).
H. S. Swofford and R. L. Carman, *Anal. Chem.*, **38**, 966 (1966).
S. Azzim and A. C. Riddiford, *J. Polarog. Soc.*, **12**, 20 (1966).
D. E. Rosner, *J. Electrochem. Soc.*, **113**, 624 (1966).
P. A. Malachesky, L. S. Marcoux, and R. N. Adams, *J. Phys. Chem.*, **70**, 4068 (1966).
D. Haberland and R. Landsberg, *Ber. Bunsenges. Physik. Chem.*, **70**, 724 (1966).
W. Geissler, R. Nitzsche, and R. Landsberg, *Electrochim. Acta*, **11**, 389, 495 (1966).
Y. S. Gorodetski, *Elektrokhimiya*, **2**, 122 (1966).
S. Minc, J. Sobkowski, and K. Kijowska, *Roczniki Chem.*, **40**, 1575 (1966).
C. Wyche and F. X. McCawley, *Electrochem. Technol.*, **4**, 447 (1966).
B. Cavalier, C. Dezael, and J. Jacq, *Bull. Soc. Chim. France*, **1966**, 3210.
J. Newman, *J. Electrochem. Soc.*, **113**, 501, 1235 (1966).
J. Newman, *J. Phys. Chem.*, **70**, 1327 (1966).
J. Newman, *Ind. Eng. Chem. Fundamentals*, **5**, 525 (1966).
L. S. Reishakhrit and L. V. Batasova, *Vestn. Leningr. Univ.*, **22**, 150 (1966).
A. I. Oshe, Y. Y. Kulyavik, T. I. Popova, and B. N. Kabanov, *Elektrokhimiya*, **2**, 1485 (1966).
V. V. Malev and J. V. Durdin, *Elektrokhimiya*, **2**, 1354 (1966).
A. C. Riddiford, in *Advan. Electrochem. Electrochem. Eng.*, **4**, 47 (1966).
A. J. Arvia, S. L. Marchiano, and J. J. Podesta, *Electrochim. Acta*, **12**, 259 (1967).
R. E. Davis, G. L. Horvath, and C. W. Tobias, *Electrochim. Acta*, **12**, 287 (1967).
L. K. J. Tong, K. Liang, and W. R. Ruby, *J. Electroanal. Chem.*, **13**, 245 (1967).
J. Wojtowicz and B. E. Conway, *J. Electroanal. Chem.*, **13**, 333 (1967).
O. Schwarzer and R. Landsberg, *J. Electroanal. Chem.*, **14**, 339 (1967).
S. Bruckenstein and S. Prager, *Anal. Chem.*, **39**, 1161 (1967).
J. Newman, *J. Electrochem. Soc.*, **114**, 239 (1967).
J. Newman and L. Hsueh, *Electrochim. Acta*, **12**, 417 (1967).
L. Hsueh and J. Newman, *Electrochim. Acta*, **12**, 429 (1967).

J. Newman, in *Advan. Electrochem. Electrochem. Eng.*, **5,** 87 (1967).
D. T. Napp, D. C. Johnson, and S. Bruckenstein, *Anal. Chem.*, **39,** 481 (1967).
L. S. Marcoux, J. M. Fritsch, and R. N. Adams, *J. Am. Chem. Soc.*, **89,** 5766 (1967).
L. S. Marcoux and R. N. Adams, *Anal. Chem.*, **39,** 1898 (1967).
M. Daguenet and J. Robert, *Compt. Rend.*, **C264,** 161 (1967).
H. Schurig and K. E. Heusler, *Z. Anal. Chem.*, **224,** 45 (1967).
D. Haberland and R. Landsberg, *Ber. Bunsenges. Physik. Chem.*, **71,** 219 (1967).
F. Lohman and W. Mehl, *Ber. Bunsenges. Physik. Chem.*, **71,** 493 (1967).
R. Landsberg, S. Müller, and R. Thiele, *Acta Chim. Acad. Sci. Hung.*, **51,** 85 (1967).
L. P. Kholpanov, *Zh. Fiz. Khim.*, **41,** 2034 (1967).
G. P. Girina, V. Y. Felinovskii, and L. G. Feoktistov, *Elektrokhimiya*, **3,** 941 (1967).
G. V. Shilina and Yu. K. Delimarskii, *Ukr. Khim. Zh.*, **33,** 245 (1967).
R. N. O'Brien, *J. Electrochem. Soc.*, **114,** 710 (1967).
P. A. Malachesky, K. B. Prater, G. Petrie, and R. N. Adams, *J. Electroanal. Chem.*, **16,** 41 (1968).
B. Miller and R. E. Visco, *J. Electrochem. Soc.*, **115,** 251 (1968).

5 CURRENT–POTENTIAL CURVES

5-1. Convective Mass Transport 115
5-2. Quiet Solutions. 118
5-3. Rapid Voltage Sweep Methods At Stationary Electrodes. . . . 122
5-4. Single-Sweep Peak Voltammetry. 124
5-5. Peak Polarograms of Systems With Coupled Chemical Reactions . 139
5-6. Electron Transfer With Follow-Up Chemical Reactions 140
5-7. Cyclic (Triangular Wave) Voltammetry. 143
5-8. Conclusions 159
References. 160

An intuitive, experimental approach to current–potential curves was presented in Chapter 1. It is necessary to inquire further into the nature of the i–E curves. The i–E behavior in quiet solution depends on the rate of change of potential (voltage sweep rate). Current–potential curves with convective mass transport are relatively insensitive to sweep rates.

Only *reversible* systems will be considered initially. These are defined as those to which the Nernst equation can be applied to the electrode reaction under experimental conditions.

5-1. CONVECTIVE MASS TRANSPORT

The theoretical i–E curve for a stirred solution is best examined in terms of the Nernst-diffusion-layer concept, in spite of the inadequacies mentioned in Chapter 4. The following derivation for a complete anodic–cathodic wave is given in some detail, since it illustrates useful relations involved in current–potential curves.

The starting point is Eq. (4-3), based on a linear concentration gradient:

$$i = \frac{nFAD_O}{\delta_O}(C_O^b - C_O^e) \qquad (4\text{-}3)$$

The problem becomes one of finding one or more relationships for C_O^b or C_O^e in terms of potential.

Considering a solution in which both Ox and Red are present initially, one can write that the sum of the fluxes for Ox and Red must be zero at the electrode surface:

$$\frac{D_O}{\delta_O}(C_O^b - C_O^e) = -\frac{D_R}{\delta_R}(C_R^b - C_R^e) \qquad (5\text{-}1)$$

An individual value is written for the diffusion-layer thickness of both Ox and Red, since there is no valid reason for expecting them to be identical. Nevertheless, to simplify the calculations, it is now assumed that

$$D_O = D_R \quad \text{and} \quad \delta_O = \delta_R \qquad (5\text{-}2)$$

Under these conditions, (5-1) can be rearranged to give

$$C_O^e = C_R^b - C_R^e + C_O^b \qquad (5\text{-}3)$$

Rewriting (4-3) in expanded form as

$$i = \frac{nFAD_O C_O^b}{\delta_O} - \frac{nFAD_O C_O^e}{\delta_O} \qquad (5\text{-}4)$$

It is seen that the first term on the right is the limiting cathodic current $i_{L,c}$ [identical with Eq. (4-4)]. Hence it is only necessary to find a value to substitute in Eq. (5-4) for C_O^e. Equation (5-3) is a proper substitution after introducing in it the concentration–potential dependence.

The latter is accomplished by referring to the Nernst equation for the surface concentration:

$$E = E^{\circ\prime} + \frac{RT}{nF} \ln \frac{C_O^e}{C_R^e} \qquad (5\text{-}5)$$

A formal potential $E^{\circ\prime}$ is written rather than the standard potential E° since concentrations rather than activities are used. Solving for C_O^e/C_R^e:

$$\frac{C_O^e}{C_R^e} = e^{(nF/RT)(E-E^{\circ\prime})} \qquad (5\text{-}6)$$

To simplify the writing, Eq. (5-6) is given the abridged form

$$\frac{C_O^e}{C_R^e} = e^z \qquad (5\text{-}7)$$

CONVECTIVE MASS TRANSPORT

Now substituting in Eq. (5-3) for C_R^e from Eq. (5-7),

$$C_O^e = C_R^b - \frac{C_O^e}{e^Z} + C_O^b \tag{5-8}$$

which can be rearranged to give

$$C_O^e = \frac{C_R^b + C_O^b}{(e^Z + 1)/e^Z} \tag{5-9}$$

Inserting this expression for C_O^e in Eq. (5-4) leads to

$$i = i_{L,c} - \frac{nFAD_O}{\delta_O}\left[\frac{C_R^b + C_O^b}{(e^Z + 1)/e^Z}\right] \tag{5-10}$$

The values C_R^b and C_O^b can be given in terms of the corresponding limiting currents as

$$C_O^b = \frac{\delta_O}{nFAD_O}\cdot i_{L,c} \quad \text{and} \quad C_R^b = -\frac{\delta_R}{nFAD_R}\cdot i_{L,a} \tag{5-11}$$

The value for C_R^b is given a negative sign corresponding to the definition of anodic current. In view of the relations in (5-2), C_R^b also can be written as

$$C_R^b = -\frac{\delta_O}{nFAD_O}\cdot i_{L,a} \tag{5-12}$$

Substituting in Eq. (5-10) and simplifying results in

$$i = i_{L,c} - \left[\frac{-i_{L,a} + i_{L,c}}{(e^Z + 1)/e^Z}\right] \tag{5-13}$$

Equation (5-13) readily rearranges to

$$e^Z = \frac{i_{L,c} - i}{i - i_{L,a}} \tag{5-14}$$

By virtue of Eq. (5-6) and (5-7), this is identical with the more familiar form

$$E = E^{\circ\prime} + \frac{RT}{nF}\ln\frac{i_{L,c} - i}{i - i_{L,a}} \tag{5-15}$$

Without the simplifications introduced by the equalities, $\delta_O = \delta_R$ and $D_O = D_R$, $E^{\circ\prime}$ would be replaced by

$$E_{1/2} = E^{\circ\prime} - \frac{RT}{nF}\ln\frac{D_O\delta_R}{D_R\delta_O} \tag{5-16}$$

so that Eq. (5-15) can be written in the common polarographic form

$$E = E_{1/2} + \frac{RT}{nF} \ln \frac{i_{L,c} - i}{i - i_{L,a}} \tag{5-17}$$

The current–potential curve for stirred solution is seen to be almost identical with the familiar "wave equation" in classical polarography. The $E_{1/2}$ usually cannot be given the thermodynamic significance it has at the DME, but from the practical viewpoint it is constant and independent of the concentration of electroactive material.

Tsukamoto et al. arrived at a similar expression for i–E curves at the RPE in terms of their hydrodynamic treatment (1). In their value for $E_{1/2}$, the modifying term is a ratio of hydrodynamic factors rather than $D_O \delta_R / D_R \delta_O$. The difference is negligible for all practical considerations.

If the electrode reaction involves Z moles of hydrogen ion, as is often the case,

$$\text{Ox} + Z\text{H}^+ + ne \rightleftharpoons \text{Red} \tag{5-18}$$

the cathodic i–E curve is given by

$$E = E_{1/2} + \frac{RT}{nF} \ln \frac{i_{L,c} - i}{i - i_{L,a}} + \frac{ZRT}{nF} \ln [\text{H}^+]$$

and the $E_{1/2}$ varies as

$$E_{1/2} = E° - \frac{RT}{nF} \ln \frac{D_O \delta_R}{D_R \delta_O} + \frac{ZRT}{nF} \ln [\text{H}^+] \tag{5-19}$$

Equation (5-19) assumes $[\text{H}^+]^e \cong [\text{H}^+]^b$, i.e., for a well-buffered solution.

Very few studies have been made of rapid voltage sweep methods at rotated electrodes. Shain and Crittenden (2) used linear sweep rates of 0.1–1 V/sec at an RPE, recording the i–E curve on an oscilloscope. Half-wave potentials in general were in accord with data obtained at slower rates. [Provided the mass-transfer rate is considerably greater than the voltage sweep rate, no deviations from Eq. (5-17) are to be expected for a reversible system. This is true until the mass-transfer rate becomes competitive with the electron transfer, when departure from reversible behavior will be observed—see Chapter 4.]

5-2. QUIET SOLUTIONS

As mentioned before, current–potential curves at stationary electrodes, i.e., unstirred solution, depend on the rate of change of applied voltage.

QUIET SOLUTIONS

The case is first considered where E_{app} is either varied manually and the current measured at each voltage increment or where the sweep rate is so slow that the instantaneous current is not affected by the changing voltage. Such processes are called by Delahay (3) voltammetry at *constant* voltage. The term *constant* implies that E_{app} is invariant during the time interval of the current measurement. Such conditions are approached only as a limiting condition in any recording technique, where E_{app} is, in reality, continuously varied. For the discussion below, linear diffusion conditions are assumed to be operative.

It was shown in Chapter 3 that the maximum instantaneous current (corresponding to E_{app} such that $C_O^e = 0$) in linear diffusion was of the form

$$i_L = K_O C_O^b \tag{5-20}$$

Further, the current at any E_{app}, where C_O^e is constant but $\neq 0$, is given by

$$i = K_O(C_O^b - C_O^e) \tag{5-21}$$

These are instantaneous currents and necessarily undergo the $t^{1/2}$ decay with time (if mass transfer is by linear diffusion). K_O is merely a convenient summation of the remaining linear diffusion factors of Eq. (3-11) and (3-12).

In a similar fashion, C_R^e can be related to the current by

$$i = K_R C_R^e \tag{5-22}$$

where the implicit time factor is the same as in the previous case. Eqs. (5-20)–(5-22) may be solved for the values of C_O^e and C_R^e as

$$C_O^e = \frac{i_L - i}{K_O} \quad \text{and} \quad C_R^e = \frac{i}{K_R} \tag{5-23}$$

If electrochemical equilibrium is assumed to operate at the electrode (reversible process), the Nernst equation in terms of surface concentrations is (for 25°C)

$$E = E^{\circ\prime} + \frac{0.059}{n} \log \frac{C_O^e}{C_R^e} \tag{5-24}$$

Substitution of the corresponding values of the surface concentrations from Eq. (5-23) leads to

$$E = E^{\circ\prime} + \frac{0.059}{n} \log \frac{K_R}{K_O} \frac{i_L - i}{i} \tag{5-25}$$

or

$$E = E^{\circ\prime} - \frac{0.059}{n} \log \frac{K_O}{K_R} + \frac{0.059}{n} \log \frac{i_L - i}{i}$$

Equation (5-25) is usually consolidated in the form

$$E = E_{1/2} + \frac{0.059}{n} \log \frac{i_L - i}{i} \qquad (5\text{-}26)$$

with

$$E_{1/2} = E^{\circ\prime} - \frac{0.059}{n} \log \frac{K_O}{K_R} \qquad (5\text{-}27)$$

Equation (5-26) is the familiar expression for a cathodic polarographic wave. The curve is symmetrical about an inflection point which coincides with the value $E_{1/2}$. The latter is defined as the potential at which $i = i_L/2$. At a fixed concentration the slope of the wave depends on n, a two-electron wave rising more sharply than a one-electron wave.

Equation (5-26) is valid for linear diffusion. However, the symbol i_L rather than i_d is used since it is rare that pure diffusion control is operative at a solid electrode. On the other hand, the same i–E characteristics can probably be used for many electrode systems in unstirred solution, provided *some form of semi-steady-state measurements* are made. This implies either a manual determination or recording with a very slow voltage sweep. Julian and Ruby (4) found current–potential curves in substantial agreement with Eq. (5-26) at a platinum-wire electrode in unstirred solution. Voltage sweep rates of not greater than 40 mV/min were recommended. In practice wire electrodes in quiet solution may be the only electrode system which will behave in accord with Eq. (5-26) using a recording technique. Wire electrodes reach a semi-steady-state condition quite rapidly. However, shielded electrodes or even horizontal surfaces without shielding tend to give peak polarograms (described in the next section) with even very slow scan rates. However, such measurements are unsatisfactory from the viewpoint of convection and mechanical disturbances over the long time interval involved. Indeed, the value of carrying out i–E curves in quiet solution with *slow* sweep rates is very questionable.

The approach leading to Eq. (5-26) has been widely used in polarography (5–7). A more rigorous treatment involves solution of the differential equation for both $C_O(x, t)$ and $C_R(x, t)$ in terms of their potential dependence (3). A derivation based on electrochemical rates is given by Milner (8). All lead to an expression similar to Eq. (5-25), where, ignoring activity coefficients, the ratio K_O/K_R can be identified with $(D_O/D_R)^{1/2}$. The $E_{1/2}$ is seen to differ slightly from the value in stirred solution [compare Eq. (5-16)].

Equation (5-26) applies only to a cathodic reduction. The more general expression for a composite anodic–cathodic process is

$$E = E_{1/2} + \frac{0.059}{n} \log \frac{i_{L,c} - i}{i - i_{L,a}} \tag{5-28}$$

where $i_{L,c}$ and $i_{L,a}$ are the limiting cathodic and anodic currents, respectively, and i is the current at any potential E. The curve for a purely anodic reaction is obviously obtained when $C_O^b = 0$ or $i_{L,c} = 0$.

Cathodic, anodic, and composite waves of this form were obtained at stationary platinum-wire electrodes by Laitinen and Kolthoff for soluble redox reactions (9).

The current–potential curve for the deposition of a metal on a solid electrode is a difficult problem from both the experimental and theoretical viewpoints. If it can be assumed that the electrode surface is covered completely with at least a monolayer, the i–E curve is given by

$$E = E_{1/2} + \frac{0.059}{n} \log (i_L - i) \tag{5-29}$$

Since i_L is directly proportional to C_O^b it is evident that $E_{1/2}$ is not independent of concentration but varies by $0.059/n$ V per tenfold change in concentration of metal ion.

The i–E curve has the theoretical shape shown by the solid line of Fig. 5-1. There is no inflection point and the wave rises steeply from the

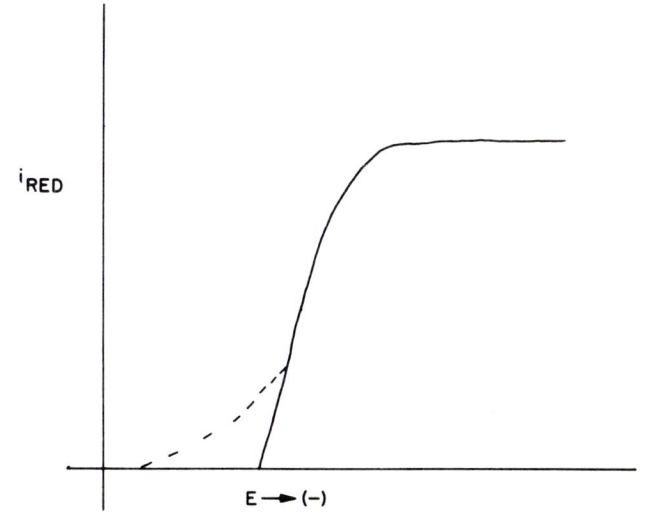

Fig. 5-1. Current–potential curve for metal deposition.

point $i = 0$. In practice, the beginning of the wave follows the dashed line. This is due to variable activity of the surface until the entire electrode is covered with plated metal. Equation (5-29) was derived on the basis of unit activity for the electrode. A precise definition of the variable surface activity during the early stage of electrolysis is very difficult. Rogers and co-workers have carried out extensive investigations of metal deposition on solid electrodes, using tracer techniques for very low-level concentrations (*10–14*). The reader is referred to the original literature and the monograph by Haissinsky for details (*15*). It should be noted that i–E curves for metal deposition from stirred solution are qualitatively the same as for quiet solution. Modifications of the equation for deposition from complexing media have been given by Cohen et al. (*16*).

5-3. RAPID VOLTAGE SWEEP METHODS AT STATIONARY ELECTRODES

It was mentioned previously that the voltage sweep rate has little effect on polarograms in stirred solution. On the other hand, the i–E curve in quiet solution depends to a high degree on the sweep rate. While rapid sweep techniques can thus be classified as "belonging" to quiet solution studies, they are treated separately here since, in themselves, they form a complete and rather unique branch of voltammetry. In addition, rapid sweep techniques are only now being widely applied to solid electrodes and are worthy of being singled out for attention.

The applied voltage waveform has a decided influence on the type of i–E curve obtained. The term waveform is used here in its broadest sense, to include relatively slow signals with durations of a few seconds or even minutes. Some of the waveforms which have been used are shown in Fig. 5-2.

Single-sweep methods are those in which the i–E curve is recorded during a unit sweep or pulse. In *multisweep* techniques the i–E curve is recorded for varying periods while the applied voltage undergoes its periodic variations with time.

The single-sweep sawtooth signal is most commonly used. Here E_{app} is varied linearly with time and then (depending on the "quality" of the sawtooth signal) the potential of the electrode is returned almost instantaneously to its starting value. Actually, the descending portion of the sawtooth is not used; the entire i–E curve is recorded during the linear rise of E_{app}. In this sense a sawtooth single-sweep method is no different

RAPID VOLTAGE SWEEP METHODS AT STATIONARY ELECTRODES 123

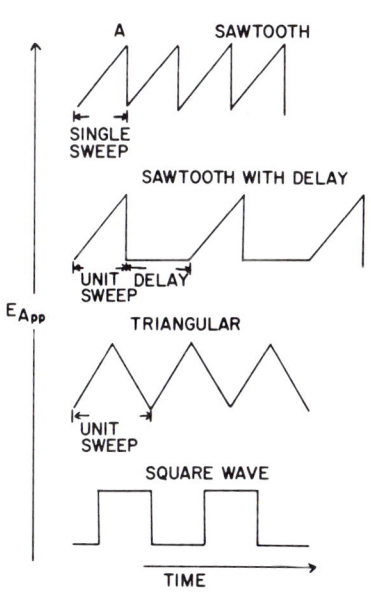

Fig. 5-2. Some applied voltage waveforms used in rapid sweep methods.

in principle from any recording polarographic technique. In the latter, E_{app} is varied with time and the polarogram is recorded during the time interval up to point A of Fig. 5-2. The only difference is that the time interval in ordinary polarography is approximately 3–5 min, while the rapid sweep techniques have durations from less than 1 sec to a few seconds.

Other waveforms which are of value are the sawtooth with variable quiescent periods between sweeps (17,18) and the triangular (isosceles) waveform, which, instead of returning the potential instantaneously to the starting value, provides a slow, linear return. Such a signal gives a combined cathodic–anodic i–E pattern (19), and the technique has become known as cyclic voltammetry. The square wave has very recently achieved considerable importance (8). The use of ordinary sine-wave potential sweeps is, of course, not prohibited. Indeed, the applied voltage may be varied in any desired fashion. In general, linear sweeps are more amenable to simple theoretical evaluations.

Although all the above rapid sweep techniques are of importance, the next discussion is restricted to one method—that employing a single, linear sweep of applied voltage. This is then extended to multisweep operation, i.e., cyclic voltammetry. For further details on the other methods, called oscillographic polarography, the reader is referred to the monographs by Kolthoff and Lingane (7), Delahay (3), and Milner (8).

5-4. SINGLE-SWEEP PEAK VOLTAMMETRY

The current–potential curve is approached in much the same fashion as in previous situations. One considers the reductions of substance Ox in a solution containing an excess of background electrolyte. The solution is unstirred and linear diffusion is maintained. The Nernst equation is assumed applicable to the electrode process.

The sweep voltage can be represented by

$$E = E_i - Vt \tag{5-30}$$

where E_i is the initial potential, V the rate of potential change in volts per second, and t the time interval of electrolysis. E_i is set at a value where Ox is not reduced. Substance Red is assumed absent at time $t = 0$. With the exception of the variable potential term (Vt), conditions are identical with the slow voltage sweep situation (where E_{app} is assumed constant during the i–E measurement interval).

The diffusion problem was first solved and translated into a current–potential curve independently by Randles (20) and Sevcik (19). An excellent account of the details and references to original literature is given by Delahay (3). A valuable graphical representation of the process in terms of a three-dimensional diagram has been presented by Reinmuth (30). The full theory of this technique has evolved during the course of this writing. The development which follows is presented in a somewhat historical manner, since it is illuminating and instructive.

The current–potential curve exhibits a maximum or peak and the peak current is given by

$$i_p = kn^{3/2}AD_O^{1/2}C_O^b V^{1/2} \tag{5-31}$$

where i_p = current, A
      A = area of electrode, cm²
      V = rate of potential change, V/sec
      k = a constant, called the Randles–Sevcik constant

The remaining symbols have their usual significance. The value of i_p is seen to be directly proportional to C^b and $V^{1/2}$. It should be observed that Eq. (5-31) is a current–potential curve of slightly different form than previous equations for polarographic waves. These i–E curves are most commonly called peak polarograms and the techniques has become known as peak voltammetry, peak polarography, or linear sweep chronoamperometry.

The shape of the peak polarogram can be visualized (Fig. 5-3) as the superposition of an *i–t* decay curve on a steady-state polarogram. (This diagram is not intended to be rigorous in any way—it is merely a useful pictorial representation of the overall phenomenon.) The usual steady-state polarogram shows a fixed i_L where electroactive material continues to be replenished at the electrode surface by some means of convective

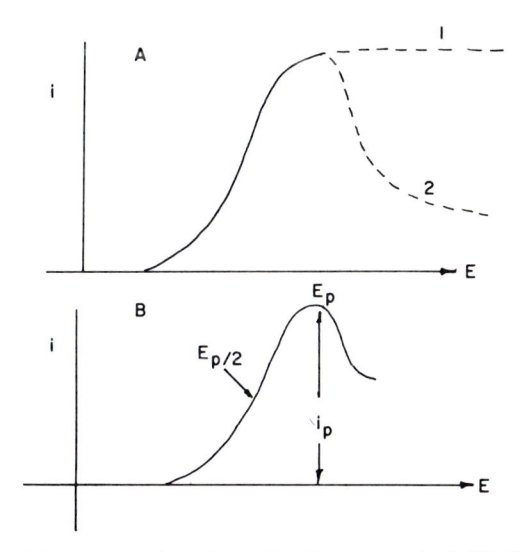

Fig. 5-3. Pictorial representation of a peak polarogram. A, 1. "Ordinary" polarogram with i_L maintained by convection; 2. current–time decay for fixed E_{app} in quiet solution. B, "Resultant peak" polarogram.

mass transport across a layer of constant thickness. On the other hand, with constant E_{app} at a stationary electrode, one obtains a decay of i with t ($i \propto t^{1/2}$ if diffusion is linear as shown in Chapter 3). As the diffusion-layer thickness increases, the current falls. The peak polarogram can be thought of as a composite of these two situations—the depletion effect of C_O^e takes over the process and the peak or maximum in the current results.

The curve may be characterized by the peak potential E_p, or the half-peak potential $E_{p/2}$. The latter is defined as the potential at which $i = i_{p/2}$. Equation (5-31) describes a rapid charge-transfer (reversible) process.

Matsuda and Ayabe addressed themselves to the rather difficult problem of examining the relations in peak voltammetry for reversible, quasi-reversible, and totally irreversible systems, i.e., when one must consider the relative rates of charge transfer and mass transfer (*31*). For the reversible case they found that simple relations exist between the E_p, $E_{p/2}$, and the conventional $E_{1/2}$ of the DME, such as at 25°C:

$$E_p = E_{1/2} - \frac{0.029}{n} \text{ V} \tag{5-32}$$

$$E_{p/2} = E_{1/2} + \frac{0.028}{n} \text{ V} \tag{5-33}$$

$$|E_p - E_{p/2}| = \frac{0.057}{n} \text{ V} \tag{5-34}$$

It can be seen from Eq. (5-34) that the peak polarogram of a reversible system is sharp, spanning a voltage range of roughly 0.12 V for a one-electron system. The peak potential is $0.029/n$ V more cathodic (for a reduction) than the corresponding $E_{1/2}$ at the DME, and, further, the $E_{p/2}$ is $0.028/n$ V more anodic. Thus $E_{1/2}$ of the DME corresponds to *approximately* $E_{3/4}$ on a peak polarogram of the same substance.

The equations for a totally irreversible peak polarogram are considerably more involved (*3,31,61*). No simple relation exists between E_p and $E_{1/2}$—the expression involves both the heterogeneous rate constant for the charge transfer and the transfer coefficient. The product of the transfer coefficient α and the number of electrons in the rate-determining step n_a can be found from the effective slope of the peak polarogram:

$$E_{p/2} - E_p = \frac{0.048}{\alpha n_a} \text{ V} \tag{5-35}$$

It can be observed that as (αn_a) decreases, the peak polarograms become more spread out and the peaks tend to be rounded rather than sharp. This situation is frequently met in organic oxidations. Further discussions of irreversible peak polarograms and evaluation of kinetic parameters are given later. The quasireversible case is involved, and the original literature should be consulted for details (*31*).

Figure 5-4 shows some typical experimental results of anodic peak voltammetry with several solid electrode systems. Generally speaking, satisfactory results are obtained over a wide range of sweep rates. Commercial recording polarographs with fixed sweep rates of 200–300 mV/min usually give peak polarograms. While semiinfinite linear diffusion conditions are assumed in Eq. (5-31), unshielded electrodes and even wires give peak polarograms. The ideal range for pen and ink recording

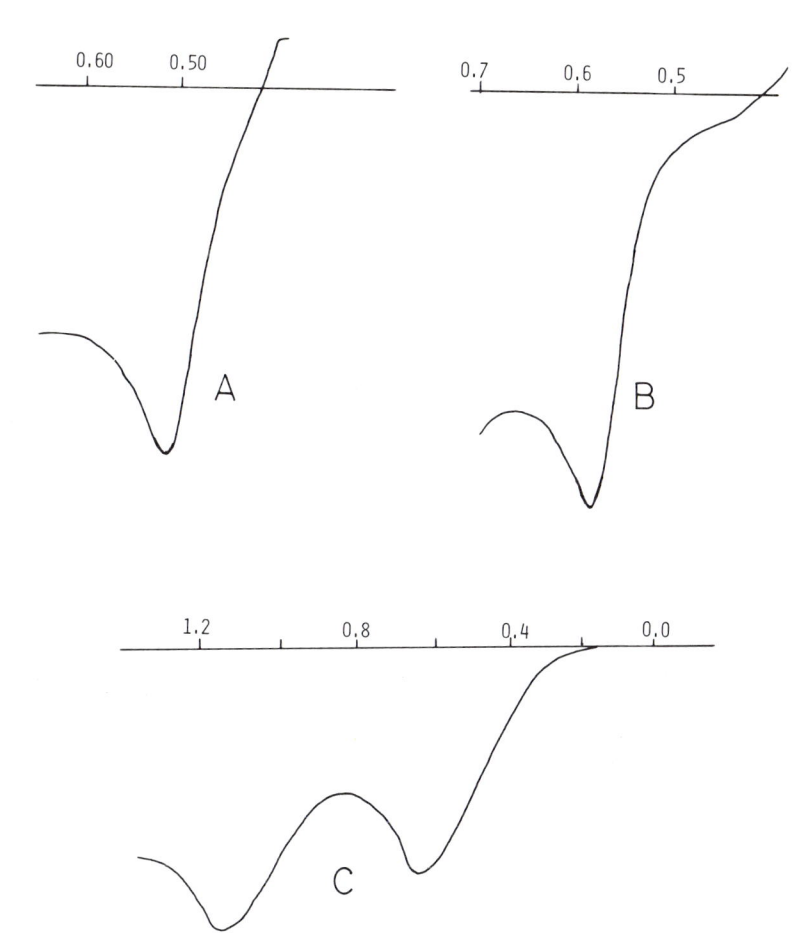

Fig. 5-4. Typical peak polarograms. A, 8 × 10⁻⁵ M N,N′-tetramethylbenzidine, pH 2, Pt-wire electrode. B, Ca. 10⁻⁴ M hydrazobenzene, pH 2, Pt-wire electrode. C, 10⁻³ M *p*-Aminodiphenylamine in benzonitrile, 0.1 M tetraethylammonium perchlorate, Pt-wire electrode. Potentials vs. aqueous SCE in all cases.

is probably about 0.5–3.0 V/min. Rates as high as 10 V/min have been used with 1-sec-pen-response recorders.

Table 5-1 is an example of optimum data that can be obtained at a typical unshielded carbon paste electrode. These results, expressed as the ratio $i_p/C \cdot A$, are for the oxidation of 3,3′-dimethoxybenzidine, commonly called *o*-dianisidine (*o*-DIA) in 2.02 M sulfuric acid. It is to be noted that the carbon paste surface was renewed prior to each run. Thus the data represent polarograms measured at 97 *different* electrodes.

TABLE 5-1
Reproducibility of Peak Polarograms for Oxidation of o-Dianisidine at Carbon Paste Electrodes[a]

Conc. o-DIA, mM	i_p, μA	Runs	$i_p/C \cdot A$[b]
2.270	8.19 ± 0.03	13	131.9
4.560	16.4 ± 0.1	22	131.9
6.430	23.2 ± 0.2	36	132.2
8.212	29.3 ± 0.2	19	130.9
		Mean	131.7 ± 0.4%

[a] Data of Zimmerman (33).
[b] Electrode area 2.73 cm², scan rate 0.500 V/min, unshielded carbon paste (Nujol) electrode.

The value of the Randles–Sevcik constant k in Eq. (5-31) was in some doubt, since Randles originally gave 2.72 ($\times 10^5$) while Sevcik's calculations showed 2.17. The discrepancy arises as a result of numerical or graphical integrations in the diffusion problem. Matsuda and Ayabe calculated a value of $k = 2.69$, and Nicholson's work on peak polarograms at cylindrical (wire) electrodes appears to extrapolate to about 2.72 for the linear diffusion case (27). Delahay indicated 2.72 as the most reliable value (3). Reinmuth suggested that Sevcik's value is too small, owing to an error in calculation (48). A rigorously derived value is discussed later.

The experimental evaluation of k via Eq. (5-31) involves a knowledge of A, the electrode area, and D, the diffusion coefficient of the electroactive species in the particular supporting electrolyte. The latter quantity is often open to question. Mueller reasoned that if peak polarograms and $it^{1/2}$ curves were run on the same system under strict conditions of semiinfinite linear diffusion (all other variables being equivalent), it should be possible to interrelate the corresponding equations and eliminate some of the experimental quantities.

Taking Eq. (5-31) and the corresponding linear diffusion equation (Chapter 3) for the current–time decay,

$$i_p = kn^{3/2} A D_O^{1/2} C_O^b V^{1/2} \tag{5-31}$$

$$it^{1/2} = \frac{nFAC_O^b D^{1/2}}{\pi^{1/2}} \tag{3-16}$$

a simultaneous solution gives

$$i_p = it^{1/2} \frac{(n\pi V)^{1/2}}{F} k \tag{5-36}$$

If i_p is expressed in μA and $it^{1/2}$ in $\mu A\text{-sec}^{1/2}$, the expression can be rearranged to give

$$k \times 10^{-5} = \frac{9.45 i_p}{it^{1/2} n^{1/2}} \tag{5-37}$$

for *scan rates of 200 mV/min*.

Thus it can be seen that a value for the Randles–Sevcik constant can be determined for ordinary polarographic scan rates without a knowledge of the electrode area, D value, or even the bulk concentration. This

TABLE 5-2

Evaluation of Randles–Sevcik Constant at Various Solid Electrodes

Observed k, $\times 10^5$	Electrode configuration
2.52–2.58	Various shielded Pt electrodes
2.56	Shielded carbon paste, $\tfrac{3}{16}$-in. diam.
2.59	Shielded carbon paste, $\tfrac{1}{4}$-in. diam.
2.61	Unshielded carbon paste, SCE auxiliary
2.64	Unshielded carbon paste, Pt-foil auxiliary
2.64	"Bard" electrode, Pt or SCE auxiliary[a]

[a] A shielded Pt-foil electrode provided by Professor A. J. Bard, similar in construction to designs given in Chapter 3.

equation was treated at a shielded boron carbide (Norbide) electrode using the oxidation of ferrocyanide (0.2–4 mM) in KCl as supporting electrolyte. Provided the KCl concentration was $>0.5\ M$ (this ensures that the ferri-ferrocyanide oxidation is reversible), a value of 2.72 ± 0.05 was observed for k (± 0.05 was the standard deviation of the mean at 95% confidence level for ca. 300 determinations of i_p). For the oxidation of o-DIA in 1 M sulfuric acid (two-electron oxidation), the k value between 2.62 and 2.72×10^5. Both two- and three-electrode (controlled-potential polarograph) scans were used, along with several electrodes of different areas and surface roughness. From this data it appeared that the value 2.72×10^5 was a fairly reliable factor to be used for either fast or slow sweep rates in peak voltammetry (*32*).

The Randles–Sevcik constant evaluated at a variety of electrodes under carefully controlled conditions does not agree too well with the previous data at boron carbide electrodes. Table 5-2 summarizes some typical data. It should be emphasized that these measurements were made under conditions comparable to those of Table 5-1; i.e., the reproducibility of

i_p is not responsible for apparent deviations in the k value. Rather, the data of Table 5-2 suggest there may be some relation between the physical configuration of the electrode system (including placement of auxiliary electrode) and the apparent k. This point cannot be demonstrated conclusively at present but the observed data cause some concern in precise correlations of peak voltammetry (*33*).

Experimentally, Mueller and Adams found that a plot of log $(i_p - i)/i$ vs. E for reversible systems at the boron carbide electrode gave straight-line plots with reciprocal slopes close to $0.056/n$ V (*32*). This, however, is inconsistent, since it implies the peak polarogram is symmetrical about $E_{p/2}$, which is not the case.

Several workers contributed materially to the theoretical development of peak voltammetry. Reinmuth proposed equations which have the form of series expansions. It was suggested that plots of log $(i_p - i)^{1/2}/i$ vs. E are linear over most of the rising portion of the wave and yield slopes of RT/nF for rapid charge-transfer processes and $RT/\alpha n_a F$ for irreversible systems (*34*). The dependence of current on $V^{1/2}$ was discussed by Reinmuth (*35*). Buck presented a modified mathematical approach which verified some of these points (*36*). Gokhshtcin and Gokhshtein derived equations for irreversible peak polarograms from which it is possible to calculate the heterogeneous rate constants from the value of i_p and E_p (*37*).

The most definitive treatment of peak voltammetry is that recently given by Nicholson and Shain. Already a classical paper in the area, this study surveys both single-sweep and cyclic voltammetry for simple systems and those with various chemical reactions coupled to reversible and irreversible charge transfers. Tables of numerical functions are presented for calculating theoretical peak polarograms. This work can also be consulted for a complete listing of previous literature on peak voltammetry (*61*).

In the Nicholson and Shain treatment the solution of the boundary-value problem for a reversible system is in the form of an integral equation which may be solved in several ways. The numerical solution gives rise to a current function which, when multiplied by ordinary parameters of the electrode process, gives the current as a function of applied potential. In final form the equation can be written in terms of the current function, χ', as

$$i = 602n^{3/2}AD^{1/2}V^{1/2}C^b(\chi') \qquad (5\text{-}38)$$

Here the current is in amperes, the scan rate V in volts per second, and the bulk concentration in moles/liter. The terminology χ' corresponds to the quantity $[\pi^{1/2}\chi(at)]$ used by Nicholson and Shain (*61*). Table 5-3

TABLE 5-3
Current Function (χ') for Reversible Peak Polarogram at a Planar Electrode[a]

Location on peak polarogram	$E - E_{1/2}$, mV $n = 1$	$E - E_{1/2}$, mV $n = 2$	χ'
i increasing	120	60	0.009
	100	50	0.020
	80	40	0.042
	60	30	0.084
	50	25	0.117
	45	22.5	0.138
	40	20	0.160
	35	17.5	0.185
	30	15	0.211
i at $E_{p/2}$	28	14	0.223
	25	12.5	0.240
	20	10	0.269
	15	7.5	0.298
	10	5	0.328
	5	2.5	0.355
i at $E_{1/2}$	0	0	0.380
	−5	−2.5	0.400
	−10	−5	0.418
	−15	−7.5	0.432
	−20	−10	0.441
	−25	−12.5	0.445
i at peak	−28.50	−14.25	0.4463
i decreasing	−30	−15	0.446
	−35	−17.5	0.443
	−40	−20	0.438
	−50	−25	0.421
	−60	−30	0.399
	−80	−40	0.353
	−100	−50	0.312
	−120	−60	0.280
	−150	−75	0.245

[a] From data of Nicholson and Shain (61), semiinfinite linear diffusion conditions assumed. The original reference also contains current functions for a spherical electrode.

is a simplified version of their current function data for a reversible system. The function χ' is given in terms of the potential increment, $E - E_{1/2}$, here in millivolts. For convenience in calculating, the increments are tabulated for $n = $ one and two electrons. Since the potential increments are spaced from $E_{1/2}$, the minus signs in Table 5-3 correspond to points past $E_{1/2}$. The point 0.0285 V past $E_{1/2}$ (for $n = 1$) is that for which χ' is a maximum and hence corresponds to the peak current. In passing it can be observed that, all other factors being equal, the i_p for a two-electron oxidation is ca. 2.83 times that of a one-electron process, owing to the $n^{3/2}$ factor in Eq. (5-31) or (5-38). The $E_{p/2}$ occurs at a potential $0.028/n$ V before $E_{1/2}$. A comparison of these data with Eq. (5-32)–(5-34) indicates they represent small corrections to previous calculations. Finally, the value of $E_{1/2}$ (which ordinarily is obtained at the dropping mercury electrode) can be ascertained from a stationary electrode polarogram. It is the potential 85.17% up the peak polarogram. It can be noted that multiplication of the constants in Eq. (5-38) by the maximum value of χ' (0.4463) gives a value of 269 for the Randles–Sevcik constant. There seems to be little doubt now that 269 can be taken as a reliable "theoretical" value for the Randles–Sevcik constant. [Since the concentration units of Eq. (5-31) are moles/cc, whereas they are moles/liter in Eq. (5-38), the Randles–Sevcik constant for the latter equation is 2.69×10^2, or 269. The latter seems preferable.] Although the use of Eq. (5-31) in terms of peak currents is firmly established (and useful when electrode response and analytical applications are considered), the form of Eq. (5-38) emphasizes the proper nature of the peak polarogram.

To plot a theoretical peak polarogram it is only necessary to substitute the various values of χ' in Eq. (5-38) and plot the resulting current vs. potential. Alternatively, to check on an experimental system which is supposedly reversible, one determines $E_{1/2}$, chooses convenient potential increments, and compares the theoretical and experimental values. How well this latter procedure works, using the data of Nicholson and Shain, is beautifully illustrated by Fig. 5-5.

The solid line of Fig. 5-5 has been inked over for clarity but is a photograph of the original recording of a peak polarogram for the two-electron oxidation of 5.32×10^{-3} M o-dianisidine in 0.65 M sulfuric acid. The electrode was a carbon paste with $A = 0.238$ cm². The scan rate was 1.75 V/min (0.0292 V/sec). The diffusion coefficient, from other measurements, was taken to be 4.36×10^6 cm²/sec. This polarogram was selected at random from a series of routine measurements made on o-dianisidine in November 1963. Using the above data and the χ' functions of Nicholson

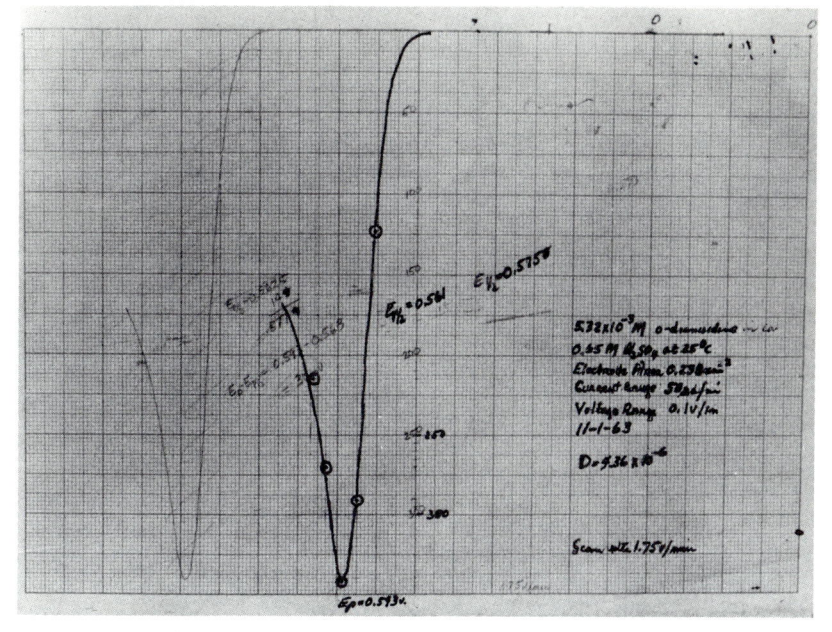

Fig. 5-5. Comparison of calculated and experimental peak polarograms—reversible system.

and Shain, currents at various points were calculated in February 1965. The theoretical peak current was 342.4 μA, and the experimental value read from Fig. 5-5 was 341.2 ± 0.5. The agreement is within 0.4%. Only at points 100 mV or so past the peak does the experimental curve deviate at all significantly from the calculations. The agreement of these data is a slight misrepresentation, since the diffusion coefficient was actually evaluated from an experimental peak current. However, substitution of $D = 4.7 \times 10^{-6}$ cm^2/sec from a totally independent tracer diffusion experiment still gives experimental and theoretical i_p agreement within 4%.

How well the experimental and calculated reversible peak polarograms agree in nonaqueous media is illustrated in Fig. 5-6 for the oxidation of trianisylamine in acetonitrile at a platinum electrode. This again is an exact copy of the chart paper from a routine experiment. Under these conditions trianisylamine undergoes a 1e oxidation to the cation radical.

It is clear that the results of Nicholson and Shain may be used with great confidence for calculating peak polarograms. However, the data of Table 5-2 show that caution must be used in assessing the reliability

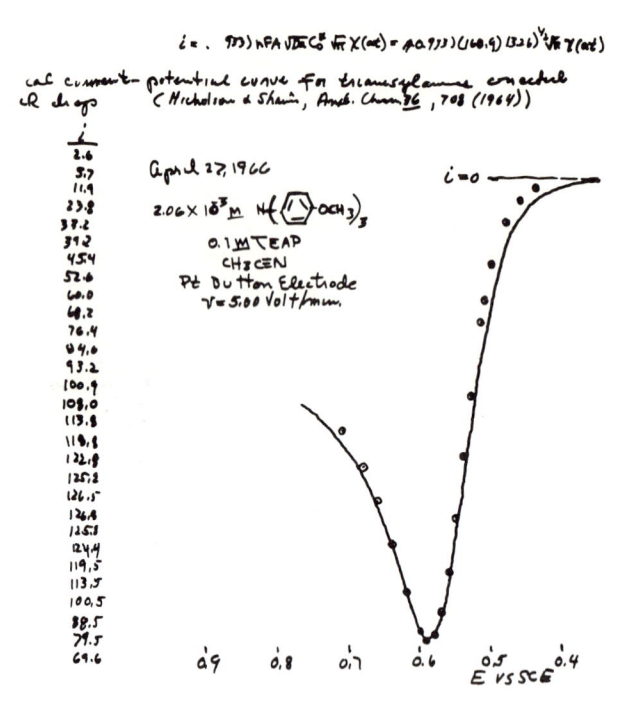

Fig. 5-6. Experimental and calculated peak polarograms—reversible system in a nonaqueous medium.

of experimental peak polarograms. These apparent variations in the Randles–Sevcik constant must be interpreted as unknown deviations of the experimental conditions from those required for the solution of the boundary-value problem.

For totally irreversible electron-transfer processes, the Nicholson and Shain treatment, while it provides a useful and different approach, mainly develops calculational refinements on the original Delahay solution to the problem (3). The current can be expressed in a form analogous to the reversible case as

$$i = 602n(\alpha n_a)^{1/2} A D^{1/2} V^{1/2} C^b(\chi'') \tag{5-39}$$

where (χ'') is a current function for the irreversible case. A table of precise values of χ'' is given by Nicholson and Shain (61). Substitution of the maximum value of χ'' (0.4958) gives an "irreversible Randles–Sevcik constant" of 2.98, compared to 3.01 calculated by Delahay.

With a polarogram of a new system in hand, one of the first questions everyone attempts to answer is: Is it reversible or irreversible? Realizing that the question is an improper statement of the problem, since "reversibility" is a relative evaluation depending on the measuring technique, it nevertheless can be agreed that everyone understands the *practical* question being asked. There are several approaches in peak voltammetry. First, one can compare the experimental i_p with theory for the reversible and irreversible situations. Equations (5-38) and (5-39) can be written with the maximum values of χ' and χ'', respectively, as

$$i_{p(\text{rev})} = 602n^{3/2}AD^{1/2}V^{1/2}C^b(0.4463)$$

and

$$i_{p(\text{irr})} = 602n(\alpha n_a)^{1/2}AD^{1/2}V^{1/2}C^b(0.4958)$$

For an irreversible one-electron transfer, where both n and n_a are necessarily unity, and all other experimental factors remain constant, it is seen that

$$\frac{i_{p(\text{irr})}}{i_{p(\text{rev})}} = \frac{0.4958}{0.4463}\alpha^{1/2} = 1.11\alpha^{1/2}$$

This comparison was first pointed out by Matsuda and Ayabe. If one assumes a reasonable value for the transfer coefficient of $\alpha = 0.5$, the comparison takes the form

$$\frac{i_{p(\text{irr})}}{i_{p(\text{rev})}} = 1.11(0.5)^{1/2} \simeq 0.77$$

In spite of the fact that the maximum χ'' for an irreversible system exceeds that for the reversible case, the actual i_p is only ca. 75–80% that of the reversible i_p. For a one-electron process one can compare the observed i_p with theory. If the system is indeed reversible, its i_p should agree with theory to within $\pm 10\%$. To obtain such agreement a fairly precise knowledge of the diffusion coefficient is required for calculation of the theoretical i_p. This can be derived from an $it^{1/2}$ curve (Chapter 3). The validity of the $it^{1/2}$ curve depends on knowing that the system does not involve follow-up chemical reactions producing products electroactive at the applied potential. The latter limitation can be ascertained quickly by cyclic voltammetry, as shown later.

Instead of calculating a theoretical i_p, one can compare the observed i_p with that of an established one-electron reversible system and assume similar D values. Two processes which adhere to all the reversibility criteria in peak voltammetry are the oxidation at carbon paste electrodes of o-dianisidine in 1 M sulfuric acid (a two-electron process) and the first oxidation peak of N,N-dimethyl-p-anisidine in aqueous media from 1 M

acid to pH 9. The first anodic peak of N,N'-tetramethyl-*p*-phenylenediamine (Wurster's Blue) is also reversible at carbon paste electrodes in pH 7 buffer medium. For nonaqueous comparisons, the *first* oxidation peak of 9,10-dihydro-9,10-dimethylphenazine is reversible at platinum electrodes in acetonitrile, propylene carbonate, and dimethylsulfoxide (tetraethylammonium perchlorate as supporting electrolyte). Notice that none of the comparisons indicated above are unequivocal. There may be combinations of α, n_a, and D which can give fortuitous agreement between irreversible and reversible peak currents.

Instead of examining currents, one can compare the shape of the peak polarogram with theoretical reversible and irreversible plots. The latter is not so simple to construct as its reversible counterpart. The potential axis is in terms of the quantity

$$(E - E^\circ)\alpha n_a + \frac{RT}{F} \ln \frac{\pi^{1/2} D^{1/2} (\alpha n_a FV/RT)^{1/2}}{k_s}$$

where k_s is the heterogeneous rate constant (61). To plot the "theoretical" peak polarogram for a given system one actually needs to know the rate parameters, the existence of which he is trying to establish. Actually, the right-hand term in the expression above only shifts the position of the peak polarogram on the potential axis. It is necessary to have an approximate value of αn_a so that potential increments of $(E - E^\circ)\alpha n_a$ can be selected for corresponding values of χ''. The value of αn_a can be estimated from the relation

$$\alpha n_a = \frac{0.048}{E_p - E_{p/2}}$$

If the system is irreversible, the peak may be quite rounded off and selection of the correct E_p involves quite a bit of uncertainty. The reliability of αn_a measured this way may be questionable.

Another means of obtaining αn_a is to compare again i_p for a suspected irreversible system with $i_{p(\text{rev})}$ for a standard one-electron process. One has then

$$\frac{i_{p(\text{irr})}}{i_{p(\text{rev})}} = \frac{n(\alpha n_a)^{1/2}(1.11)}{n^{3/2}}$$

and if a decision is available for the total n corresponding to the irreversible case, αn_a can be evaluated.

It can be seen that irreversible peak polarograms are more spread out as αn_a decreases. The i_p is significantly less than the reversible situation, with $i_{p(irr)}$ falling rapidly for decreasing αn_a. With possible experimental αn_a values as low as 0.2, it is easy to see how a stationary electrode polarogram may hardly have any peak characteristics and even may be missed at slow sweep rates. E_p and $E_{p/2}$ for irreversible processes vary with potential sweep rate. This differentiates them from reversible processes, but the variation is only about $0.03/\alpha n_a$ V per tenfold change in sweep rate. This does, however, provide one of the best practical ways of evaluating αn_a. By measuring the shift in $E_{p/2}$ at two sweep rates, one eliminates uncertainties in measuring a rounded E_p and comparisons with standards. For an oxidation process one has

$$E_{p/2(2)} - E_{p/2(1)} = \frac{0.0128}{\alpha n_a} \ln \frac{V_2}{V_1}$$

where the subscripts (1) and (2) correspond to the sweep rates V_1 and V_2, respectively. The expression above is a rearranged form of Eq. (53) given by Nicholson and Shain (61). Perone has pointed out potential advantages of derivative peak voltammetry for irreversible systems (85).

The case of successive (n electroactive components) or multistage electron transfers has been considered by Gokhshtein and Gokhshtein (38). If the observed peak currents for two successive processes are $i_{p(1)}$ and $i_{p(2)}$, respectively, then the true $i_{p(2)}$ is obtaining by subtracting a contribution from process 1. The amount to be deducted is the diffusion decay value of i_1 calculated at the E_p of process 2. The equations are involved and the original literature should be consulted for details. Great care should be used in applying this practice to multicomponent organic systems where coupling of intermediate products can give rise to extraneous results (39). A detailed treatment of multistep processes has been given recently by Polcyn and Shain (67).

Several complications exist in rapid sweep methods: the effect of ohmic drop in the electrolytic cell, which distorts the linear potential variation, and the high-capacity current, which is a consequence of charging the electrode double layer. These effects have been considered by Delahay (3) and most recently by R. S. Nicholson (62) and De Vries and Van Dalen (72).

Instead of maintaining strictly linear diffusion conditions in rapid sweep methods, M. M. Nicholson investigated the utility of ordinary wire electrodes (21,22). A glass bead was sealed in the end of the wire and essentially cylindrical diffusion conditions applied. A theoretical treatment involving the curvature of the electrode was achieved and experimental results were compared. Using the ferri-ferrocyanide couple as a test system, very close agreement with theory was obtained. With very rapid sweep rates linear diffusion theory can be applied. This is the case with rates of about 300 mV/sec, where oscillographic recording is used. Nicholson used a pen and ink recorder and a sweep rate of 2.47 mV/sec derived from a commercial polarograph. Under these conditions cylindrical diffusion is operative. This alters Eq. (5-31) to the extent that i_p is no longer exactly proportional to $V^{1/2}$ nor is it exactly proportional to $n^{3/2}$ and $D^{1/2}$. Plots of the experimental dependence of i_p on V are compared to the theoretical values for both plane and cylindrical electrodes in the original paper. Convection effects became evident at sweep rates approaching 1 mV/sec.

The analytical possibilities of this work are indicated by the fact that i_p was reproducible with an average deviation of $\pm 1\%$ (maximum of $\pm 2\%$) for the ferri-ferrocyanide system. The oxidation of alkyl and aryl sulfides was examined also by Nicholson.

Rapid sweep methods involving metal depositions are plagued with the same difficulty met before—variable surface activity of the deposit. Berzins and Delahay derived an expression for the i–E curve based on a surface activity of unity during the entire electrolysis period (23). The form of the i–E curve is identical with that for a soluble redox reaction [Eq. (5-31)] with only a change in numerical constants:

$$i_p = 3.67 \times 10^5 n^{3/2} A D_O^{1/2} C_O^b V^{1/2} \tag{5-40}$$

The value of E_p is shifted toward more cathodic values with decreasing C_O^b. Berzins and Delahay checked the theoretical equation for deposition of cadmium from a potassium chloride background medium. A platinum-wire electrode, sealed on both ends in glass, gave a close approach to linear diffusion conditions. The value of i_p varied with $V^{1/2}$ as predicted, but E_p values were considerably more anodic than was expected. The peak currents were also highly dependent on electrode pretreatment, etc., and the authors concluded that the technique could not be recommended for analytical applications.

Very recently Mamantov and co-workers have reexamined the reversible deposition of metals on solid electrodes using peak voltammetry (68). Using silver ion deposition as a test, they concluded the results were in reasonable agreement with a theoretical derivation which predicts that a plot of log $(i_p - i)$ vs. E is linear with a slope of $2.2nF/RT$. The linear relation holds in the interval 0.5–$0.9 i_p$. These workers once again pointed out the deviations from theoretical behavior caused by uncertainties in the activity of freshly deposited metals.

It would appear that analytical procedures involving metal ion–metal systems are much better handled by the technique of *anodic stripping*. This method was applied initially by Lord et al. (*24*) for small amounts of silver deposited on platinum. Rogers and co-workers also used the technique for the determination of cadmium, zinc, and lead, employing stripping from mercury-plated stationary platinum electrodes (*25,26*). In general, the procedure consists of exhaustive or partial deposition of the metal on the electrode (depending on the concentrations involved) at a constant potential. Controlled potential deposition can be used for selectivity in cases of mixtures. The metal deposit is then anodically stripped using a rapid voltage sweep. A peak-type i–E curve is obtained and the amount of metal can be calculated coulometrically from the area under the peak or by other methods. Nicholson has examined the theory of anodic dissolution of thin metal deposits (*27*).

Since the method inherently involves a concentrating step in the deposition, it has extreme sensitivity. This advantage has been exploited in applications to the hanging drop electrode by De Mars and Shain (*28*), and to small mercury pools by Nikelly and Cooke (*29*). Excellent analytical results at carbon paste and graphite electrodes have been reported by Jacobs (*69*) and Perone and Kretlow (*70,71*).

5-5. PEAK POLAROGRAMS OF SYSTEMS WITH COUPLED CHEMICAL REACTIONS

The previous discussion of peak polarograms involved only reversible and irreversible electron-transfer processes. Especially important in organic electrode reactions are situations where homogeneous chemical reactions are coupled to the electron (charge) transfers. A variety of possibilities exist. Chemical reactions may precede (preceding or precursor reactions) a reversible or irreversible electron transfer. Chemical reactions may follow (follow-up reactions) a reversible or irreversible charge transfer. Here only the first possibility is of interest, since, if the electron transfer is totally irreversible, a follow-up chemical reaction has no direct effect on the primary peak polarogram. The chemical follow-up reactions themselves may be reversible or irreversible. Finally, if the follow-up reactions regenerate starting electroactive material they are called catalytic processes, and these can affect both reversible and irreversible peak polarograms. In addition, an important, more complex, situation is frequently met where a chemical reaction is coupled between two successive electron transfers, i.e., electron transfer–chemical reaction–electron transfer—commonly referred to as an ECE mechanism. Combinations of the above sequences are presumably present in multistage organic electrode reactions. The common term *kinetic processes* has been applied to all the systems described above. This is sometimes confusing, for one reads of the kinetics of slow electron transfer processes (without chemical complications). The designation kinetic process should be used only for an electrode reaction with coupled chemical involvement.

Quantitative treatments of peak polarograms with coupled chemical reactions are now available (*35,57,58,61–66*). Nicholson and Shain give tables of current functions from which the peak currents can be calculated for various coupled chemical reaction cases (*61*). Saveant and Vianello recently presented a general approach to coupled chemical reactions in peak voltammetry (*83*) and Saveant has considered new two-stage sweep programs for ECE reactions (*84*). It is first fruitful to consider an intuitive approach which emphasizes the most important feature of stationary electrode voltammetry. This feature, of particular significance when either single-sweep or cyclic methods are applied to kinetic processes, is the experimental control one has over the time interval of electrolysis. The variation of sweep rate can be used like a time interval gating circuit and, in some instances, the chemical complications can be "tuned in or out." In all cases, there are direct relationships between the sweep rate and the rates of the chemical reactions. This is well illustrated by considering the following kinetic process.

5-6. ELECTRON TRANSFER WITH FOLLOW-UP CHEMICAL REACTIONS

This very common organic electrode process in which the chemical reaction is irreversible can be generalized as

$$Ox + ne \rightleftharpoons Red$$

$$Red \xrightarrow{k_f} Z$$

where k_f is usually a pseudo-first-order rate constant for interaction with solvent or other material in solution. The product of the chemical reaction Z, is assumed to be electroinactive. By this one means Z is electroinactive in the potential region of the primary process. It is often the case that Z is oxidizable or reducible at some potential considerably removed from that of the primary electrode reaction. In such instances it does not contribute faradaic current in the potential region where Ox is reduced. The symbol Z is usually reserved for substances which are electroinactive in this context.

Now it is quite clear that for a relatively small value of k_f, there can be applied a sweep rate such that practically no chemical reaction can occur in the short time interval of the sweep. Under these conditions one would expect the peak polarogram to be simply that of a reversible charge transfer, i.e., the sampling gate is too short for the slow chemical reaction to affect the response. This is indeed the result; the peak current can be calculated directly from Table 5-3, and the peak polarogram shows all the characteristics previously mentioned for a reversible system.

In the other limiting situation when k_f is very large or the sweep interval long (the latter is limited by the onset of natural convection which will disturb quantitative results), the chemical reaction will predominate. Since the net effect of the chemical reaction is to remove Red from the oxidation–reduction equilibrium, one might expect the peak polarogram to have irreversible character in this case. This happens and the quantitative treatment by Nicholson and Shain shows the current function to be of the *same form* as that for an irreversible charge transfer. The value of E_p (or $E_{p/2}$) is a function of k_f and the sweep rate. Since, for instance, either a very small k_f or a very rapid sweep rate produces the same general effect, most of the quantitative data is evaluated in terms of the ratio of these quantities or a closely related function (i.e., k_f/a, where $a = nFV/RT$). Tables of the current function vs. potential for varying values of k_f/a are available to calculate theoretical peak polarograms (*61*). The maximum value of the current function is about 0.491 at $k_f/a = 10$ [Table X in Nicholson and Shain (*61*)] and the equation for this current function is of the same form as for an irreversible charge transfer. Notice, however, that to calculate the peak current for the follow-up reaction case, the value 0.491 is multiplied not by the factors for an irreversible charge transfer but those for a reversible case. Thus, with the absence of the multiplier (αn_a), the actual i_p (as well as the current function) will exceed that for a reversible charge transfer, in contrast to the previous distinction between i_p for reversible and irreversible charge transfers. Hence the effect of a fast follow-up reaction on a reversible electron transfer is actually to slightly increase the i_p or "sharpen" the peak polarogram.

The most useful experimental approach is to observe the variation of i_p and $E_{p/2}$ (or E_p) with increasing sweep rate. It should be noted that an increasing value of V corresponds to a decreasing ratio of k_f/a, since each system has a fixed value of k_f. Experimental plots can be made of $i_p/V^{1/2}$ vs. V or $V^{1/2}$. For reversible and irreversible charge transfers with no chemical complications these plots are horizontal straight lines. The irreversible follow-up reaction case approaches the reversible line at fast sweep rates and the irreversible line for slow sweeps (observe the distinction between current function and i_p noted above.) Actually, the peak current changes only very slightly with a several order of magnitude change in V and is difficult to evaluate experimentally. The shift in E_p or $E_{p/2}$ is more significant. At low values of V, E_p or $E_{p/2}$ shifts ca. $30/n$ mV cathodic for each tenfold increase in V, which is typical behavior for an irreversible charge transfer. As V increases, the shifts of E_p or $E_{p/2}$ with V tend toward zero, which is characteristic of a reversible charge transfer.

Caution must be exercised in applying variation of sweep rate techniques. First, to obtain the full range of response normally requires varying the sweep rate over several orders of magnitude. This often implies changing from pen and ink to oscillographic recording. Difficulties sometimes attend correlating data from two recording devices. Further, departure from linear diffusion conditions at slow sweep rates, as well as charging current effects at high rates, may make quantitative evaluations difficult. Finally, in the oxidation of organic substances at solid electrodes, a wide choice of experimental variables is simply not always available, owing to electrode filming, adsorption, etc. A more detailed discussion of practical applications of these techniques is given in Chapter 8.

If the follow-up chemical reaction is reversible, then k_f, k_b, and K, the equilibrium constant of the chemical reaction, enter into the response. Nicholson and Shain (61) examined the results in terms of the kinetic parameter $K(a/l^{1/2})$, where $l = (k_f + k_b)$. Several limiting situations, where the response becomes similar to that for a reversible or irreversible charge transfer, can be deduced according to the magnitude of the quantities l/a and $K(a/l)^{1/2}$. This system can be differentiated from that of an irreversible follow-up reaction, particularly by observing the shift of E_p with sweep rate (61). The reversible follow-up reaction would seem to be unusual for organic electrode processes where the initial electron transfer is an anodic oxidation, but future work may reveal examples. On the other hand, consideration of the previous reaction in a backward sense, i.e., the reduction of an oxidized organic compound followed by reversible protonation, provides numerous examples of the reversible follow-up situation.

It should be noted again that if the initial electron transfer is irreversible, then follow-up chemical reactions have no effect on the primary wave.

This is so because the net effect of a follow-up reaction is to render the conjugate redox product electroinactive in the potential region of the primary wave. If the charge transfer is irreversible, this effect already is reflected in the overpotential of the system, and the chemical effect cannot be discerned in the primary wave. The chemical reaction may still be detected and its kinetics determined because Z is often electroactive in some other potential region.

In summary, single-sweep peak voltammetry is an excellent technique for analytical purposes. In solid electrode applications there is no doubt that one should use sweep rates rapid enough to achieve peak polarograms as opposed to steady-state (very slow) sweep measurements. When coupled with anodic stripping procedures it provides high sensitivity.

In spite of the apparent attractiveness for studying electrode processes, single-sweep peak voltammetry, *by itself, and especially applied to solid electrodes*, is severely limited in capabilities. First, the most useful diagnostic criteria for distinguishing various types of electrode reactions depend on variations of i_p or $E_p(E_{p/2})$ vs. sweep rate. Typically the variations of i_p are at most only about $\pm 20\%$ and require a sweep range of three or more orders of magnitude. It is not at all simple to achieve such a wide variation of sweep rate in actual practice. Superimposed on an expected $\pm 20\%$ variation in i_p will often be a ± 5–10% experimental precision in measuring i_p at solid electrodes. With only two orders of magnitude of sweep range, experimentally accessible interpretations are quite difficult. The point to be emphasized is that one should not limit himself to single-sweep methods. The technique of cyclic voltammetry discussed next provides much more information with less experimental restrictions on the nature of electrode processes. If a choice is to be made, one can derive much more information *initially* by the use of cyclic voltammetry. This can be followed up with single-sweep techniques and, in fact, all other available electrochemical methods, to more precisely define individual points about the electrode process.

5-7. CYCLIC (TRIANGULAR WAVE) VOLTAMMETRY

Cyclic voltammetry (CV) employs the isosceles-triangle wave of Fig. 5-2. The technique was apparently first practiced by Sevcik (*19*). An excellent series of papers by Kemula and co-workers employing the hanging mercury drop electrode pointed the way to studies of electrode mechanisms (*40–44*).

In cyclic voltammetry the current (both cathodic and anodic segments) is followed during the complete excursions of the applied triangular voltage sweep. The sweep rates can be about the same as in single-sweep peak voltammetry, but values of 10–100 V/min are frequently of interest. Large-amplitude signals of 0–3 V are used, depending upon the potential range of interest. The latter condition distinguishes CV from small-amplitude triangular wave methods in which much higher frequencies are used. The data are best presented on an X-Y recorder (1-sec pen response), although conventional X-time or strip-chart recording can be used. Oscillographic recording is used for high sweep rates.

Since cyclic polarograms are observed at quiet electrodes and the time interval between reverse sweeps is relatively short, products of say, a cathodic reduction, are available at and near the electrode surface for reoxidation on the anodic-going segment of the cycle. Ideally, CV should employ strictly linear diffusion conditions. However, after several sweeps the concentration profiles near the electrode surface are rather involved. For qualitative examinations of electrode processes, little is gained by using special shielded electrodes. For instance, vertical wire electrodes with sweep rates of 2–3 V/min often give quite a satisfactory picture of the overall electrode process (45). Since, however, qualitative CV is almost always followed up by refined measurements, it is wise always to use an electrode configuration which gives satisfactory correlation with linear diffusion theory.

The ideal apparatus for CV employs a controlled-potential polarograph (operational-amplifier type with a triangular wave scanner) for a three-electrode system (46). CV can be carried out, of course, using only two electrodes. In this case the triangular wave generator is a low-resistance slide wire driven in a cyclic fashion and connected directly to two electrodes as with the usual polarographic bridge circuit. A suitable design of this type was given by Juliard and Shalit (47).

Intuitively it is easy to see that a rapid charge-transfer process (reversible system) will show cyclic polarograms as in Fig. 5-7a, which represents several sweeps taken at a carbon paste electrode of a solution 10^{-3} M in Fe(III), 1 M in hydrochloric acid, and 5 M in calcium chloride. At this high chloride concentration the Fe(III)–Fe(II) system behaves almost reversibly at carbon paste with the scan rate of 2 V/min. Actually, for a reversible system, the anodic and cathodic peaks are not at exactly the same potential, but a separation is predicted. This is verified by Eq. (5-32), which shows that $E_{p,c}$ for a reduction is $0.029/n$ V more *cathodic* than $E_{1/2}$. Correspondingly, $E_{p,a}$ for the oxidation of the same system will be $0.029/n$ V more *anodic* than $E_{1/2}$. The potential increment between the

CYCLIC (TRIANGULAR WAVE) VOLTAMMETRY 145

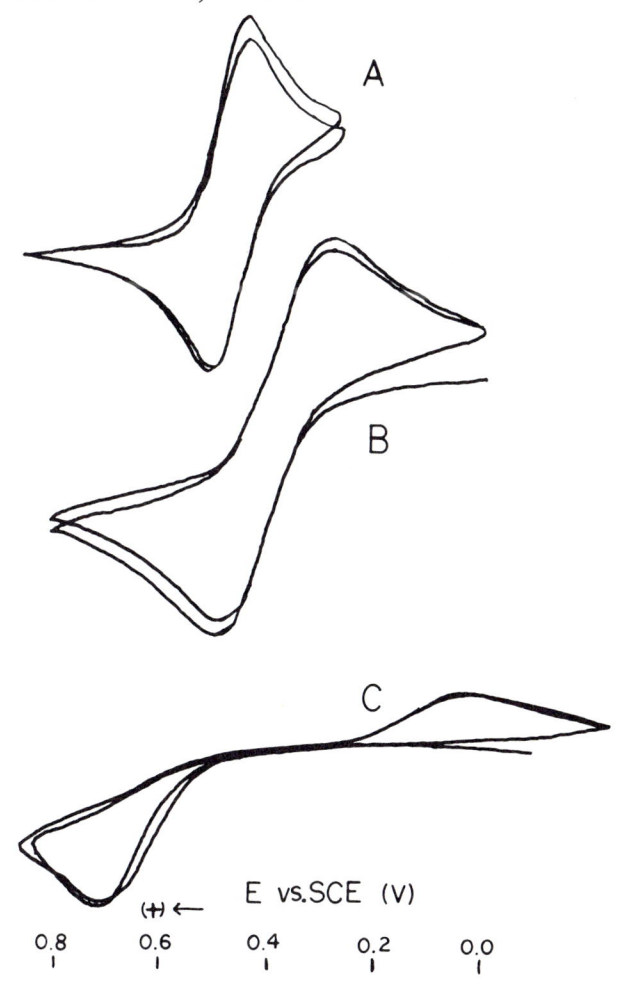

Fig. 5-7. Cyclic polarograms as function of charge-transfer rate. A, 10^{-3} M Fe(III) in 1 M HCl, 5 M CaCl$_2$, carbon paste, sweep rate 2 V/min. B, 10^{-3} M Fe(III) in 1 M H$_2$SO$_4$, Pt wire, sweep rate 1 V/min. C, Same as B, carbon paste electrode.

peaks for a reversible system will be $(3,31)$

$$E_{p,a} - E_{p,c} \simeq 2(0.029/n) = \frac{0.058}{n} \text{ V} \qquad (5\text{-}41)$$

Equation (5-41) holds for single-sweep polarograms of individual solutions of Ox and Red. Conditions in the actual cyclic polarogram are slightly different. Consider the reduction scan of Fig. 5-7A. The concentration of Fe(III) at the electrode surface is not zero at the peak potential and only becomes zero at potentials considerably beyond $E_{p,c}$. [At first glance this seems surprising, but it is helpful to remember that even at 85.17% $E_{p,c}$, the value of C_O has decreased only to the extent that $(C_O/C_R)_{x=0} \simeq 1$.] Since all Fe(II) for reoxidation comes from the cathodic sweep, if the switching (cyclic reversal) potential is too "close" to $E_{p,c}$, the value of $[Fe(II)]_{x=0}$ will be slightly less than the initial bulk concentration of Fe(III). Thus the initial condition for the anodic reversal is not the same as for the starting cathodic sweep. The result is a small anodic shift of the oxidation wave, and the separation of peak potentials will be slightly greater than that predicted by Eq. (5-41).

Nicholson and Shain have calculated the shift of E_p for the reverse sweep in terms of the switching potential E_λ [E_λ is the potential increment between the initial potential and the point of the triangular wave reversal (61)]. Provided the sweep is carried far enough such that $(E_{1/2} - E_\lambda)n$ is of the order of 2–300 mV, the position of the reverse wave is relatively unaffected for sweep rates as slow as 3 V/min. This represents a time interval of only 4–6 sec past the peak.

A distinction should be made between the first two or three cycles and continuing cycles. Many of the quantitative data discussed later are based on the first cycle (by cycle one means the complete potential excursion, i.e., from 0.0 to +1.0 V and back to 0.0 V). In the *absence of coupled chemical reactions*, continuing cycles merely gradually alter the concentration profiles near the electrode surface. The anodic and cathodic peaks slightly change shape and decrease until a steady state is achieved. With slow sweep rates this steady state appears after about 5–10 cycles. For a reversible system Matsuda showed that the difference between initial and steady-state E_p's was only a few millivolts and relatively insignificant. A thorough discussion of cyclic polarograms at steady-state conditions is given in the original paper by Matsuda (59). The slight changes observed in cycling to steady-state conditions are very different from the gross alterations that may occur with coupled chemical reactions. The latter are most marked during the first few cycles. For most cases the most revealing information is contained within the first five cycles and continuing on to the steady-state level is uninteresting.

A quasi-reversible system shows a greater separation in E_p's. The polarograms are more drawn out and the peaks more rounded. This behavior is seen in Fig. 5-7B, which is of 10^{-3} M Fe(II) in 1 M sulfuric acid at a platinum electrode, using a sweep rate of 1 V/min in this case.

For a very slow charge-transfer process, one sees the complete separation

CYCLIC (TRIANGULAR WAVE) VOLTAMMETRY   147

of anodic and cathodic processes as presented in Fig. 5-7C. This represents the same system as in B, but run on a carbon paste electrode, where the reaction is markedly more irreversible. Obviously the artificial distinctions between reversible, quasi-reversible, and irreversible reactions are only gradations of the charge-transfer rate, but these terms are common and will undoubtedly remain in use. The examples above were used to illustrate the profound effect of both background electrolyte and electrode material on charge-transfer rates as reflected in the CV.

R. S. Nicholson has used the separation in anodic and cathodic peak potentials to measure rate constants for electron transfer (73). Application of absolute rate theory to the electrode process and numerical solution of an integral equation provides a correlation of the separation in peak potentials ΔE_p with a function ψ given by

$$\psi = \gamma^\alpha \frac{k_s}{\pi^{1/2} D_O (nF/RT)^{1/2} V^{1/2}}$$

where γ is the square root of the ratio of the diffusion coefficients D_O and D_R, α the transfer coefficient, and k_s the standard rate constant at $E = E°$. The other terms have their usual significance. Since the difference in D values is ordinarily small, γ^α is unity for many practical situations. With certain restrictions the relation between i_p and ψ is essentially independent of α.

Using the approximation $\gamma^\alpha = 1$, $D_O = 1 \times 10^{-5}$ cm²/sec, and $F/RT = 39.2$ V^{-1} gives a practical relation between ψ, k_s, and V, which is useful for estimating the sweep rates required to evaluate a given k_s. The sweep rate V is in volts per second:

$$\psi \simeq 28.8 \frac{k_s}{V^{1/2}}$$

The variation of ΔE_p with ψ, either in tabular form or as a working curve (Table 1 or Fig. 3 of Ref. 73) shows that a ψ value of 0.5 is about optimum for measuring purposes. Thus

$$0.5 = 28.8 \frac{k_s}{V^{1/2}}$$

and it is readily seen that for $k_s = 1 \times 10^{-3}$, the sweep rate needs to be of the order of 3–4 V/min. Obviously, for fairly rapid reactions the required sweep rate is very high and oscillographic presentation is used.

One can either "tune in the rate constant" by increasing the sweep rate or, if the system already shows peak separation in excess of reversible behavior, vary the sweep rate to realize ΔE_p values which fall in the calculation range for ψ.

The above treatment is very approximate and the original paper by Nicholson should be consulted for proper evaluation of k_s. For instance, at high sweep rates one must be particularly cautious of uncompensated iR drop, which gives experimental results almost identical to the k_s effect (73). However, even in the crude form illustrated above, the technique

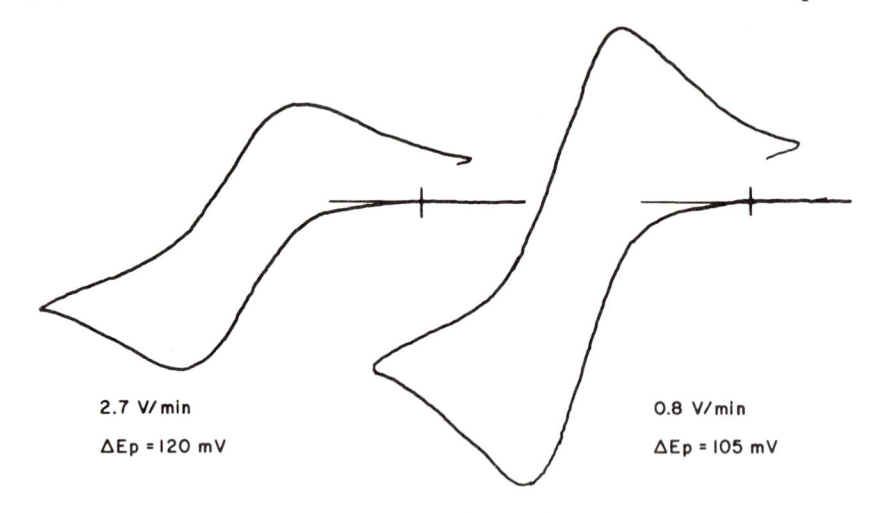

2.7 V/min
$\Delta E_p = 120$ mV

0.8 V/min
$\Delta E_p = 105$ mV

Fig. 5-8. Measurement of charge-transfer rate via cyclic voltammetry.

is probably the most attractive and useful method devised for approximating k_s values. For instance, k_s values of 10^{-3} cm/sec are common in solid electrode studies and hence can be evaluated at sweep rates of a few volts per minute without serious problems of iR drop. Just how well the technique works is illustrated by the following data.

Previous rotated disk studies had shown that k_s for ferri-ferrocyanide in 0.1 M KCl at carbon paste electrodes was $ca.$ 1.5 × 10^{-3} cm/sec. A 5 × 10^{-4} M solution of ferrocyanide in 0.1 M KCl was subjected to cyclic voltammetry at a stationary carbon paste electrode with sweep rates between 0.27 and 2.7 V/min. Figure 5-8 shows two of the cyclic polarograms (first cycle only) at sweep rates of 0.81 and 2.7 V/min with the corresponding ΔE_p values of 105 and 120 mV, respectively. The k_s values were calculated from the approximation $k_s = V^{1/2}\psi/28.8$ given above. The data for four such determinations are summarized in Table 5-4.

Obviously there is some experimental error between the last two entries in the table since both have the same E_p for sweep rates varying by about 2. Nevertheless, the values are all in acceptable agreement with the rotated disk work. Furthermore, the data above represent about ½hr of experimental and calculation time. For order of magnitude evaluation, or trends in k_s for different compounds, the Nicholson technique is extremely useful. (It should be emphasized that the calculations and experimental illustration used here are purposely crude. The refinements given in the original paper should be employed in practice.)

Although CV has advantages for studying charge-transfer rates, it is most ideally suited for investigating the *overall* processes, both chemical

TABLE 5-4

Sweep Rate				
V/min	V/sec	ΔE_p, mV	ψ^a	k_s, cm/sec
0.27	0.0045	85	1	2.2×10^{-3}
0.81	0.0135	105	0.5	2.0×10^{-3}
1.35	0.0225	120	0.35	1.6×10^{-3}
2.7	0.045	120	0.35	2.4×10^{-3}

^{a} ψ corresponding to closest ΔE_p value in Table 1, Ref. 73.

and electrochemical, which may occur in a complex electrode reaction. CV is a relatively simple technique experimentally, and a maximum of information can be gained using it in a short period of time. The advantages are best illustrated by examples of organic oxidations at solid electrodes.

Often one of the most striking features of a series of cyclic polarograms is a complete change between the first and all subsequent sweeps in a given direction. This is well illustrated by the series of cyclic polarograms in Fig. 5-9. This represents the oxidation of the triphenylmethane dyestuff, crystal violet, in 1 M sulfuric acid at a carbon paste electrode. The sweep was initiated in an anodic direction and the first anodic sweep is marked as 1F (forward) on the diagram. It can be readily seen that essentially no oxidation occurs on sweep 1F until ca. +0.7 V, when the primary oxidation of the dye gives rise to two peaks marked A and B. The sweep is reversed at about +1.1 V and on the return cathodic cycle there is very little evidence of any reversible character to the dyestuff oxidation (i.e., no reductions corresponding to A and B). There is however, a marked cathodic peak C' at about +0.55 V. On the second and all other anodic cycles there is now an anodic peak at a potential less anodic than that of

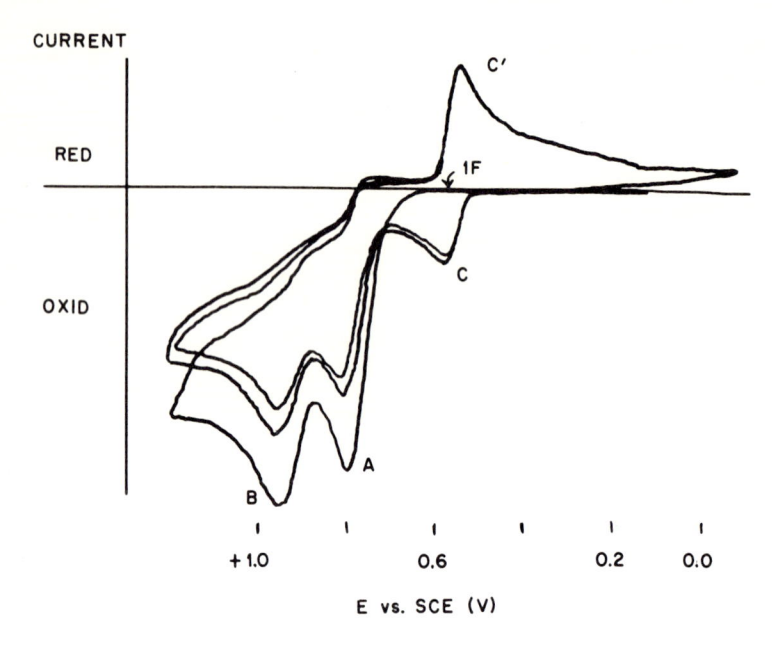

Fig. 5-9. Cyclic voltammetry of crystal violet in 1 M sulfuric acid.

the dyestuff, i.e., peak C at +0.55 V. This anodic peak C, which appears only *after* initial oxidation of the dye, immediately marks the system as one in which some chemical follow-up reaction has occurred after the initial charge transfer. In this case the almost reversible redox system C,C′ at +0.55 V can positively be identified with the oxidation–reduction of N,N′-tetramethylbenzidine. This material results from decomposition of the triphenylmethane dye following a two-electron oxidation.

Tentative identification of new redox systems formed in the fashion indicated above often can be made by running the cyclic voltammetry of suspected products and matching potentials. This is hardly unequivocal with complex organic compounds, and further physical evidence is necessary for positive identification. There are also quite a few variations of CV which are useful in identifying either the intermediates or their source. For instance, in the crystal violet case, by stopping the anodic sweep at +0.85 V it can be shown that only peak A contributes to the formation of the tetramethylbenzidine. This can be done by examining the peak height of the reduction wave for the oxidized form of tetramethylbenzidine (peak C′) as a function of the limit of the anodic sweep and the time spent at this potential. Hence one of the useful variables in CV is the sweep interval. Similarly, by increasing the sweep frequency, one may

CYCLIC (TRIANGULAR WAVE) VOLTAMMETRY 151

observe transient intermediates whose lifetimes are too short to be seen with longer time cycles.

Another striking example of the effect of a follow-up chemical reaction is seen in Fig. 5-10 for the oxidation of 4,4'-oxydiphenol in 1 M perchloric acid at a carbon paste electrode. The first anodic sweep (1F) shows the initial oxidation (peak A) at ca. +0.7 V. On the reverse cathodic sweep

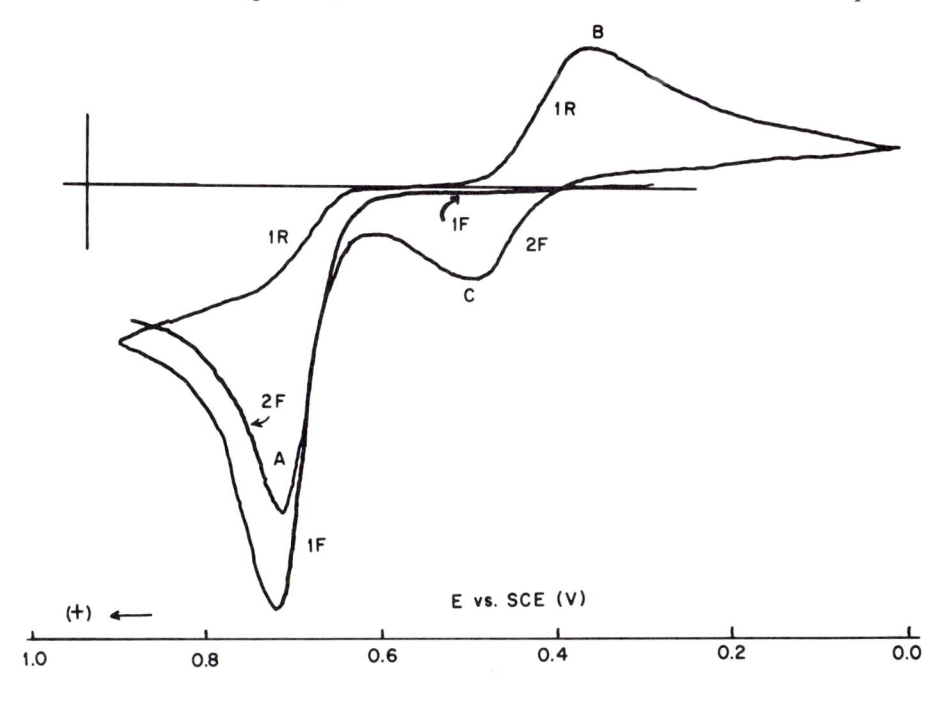

Fig. 5-10. Cyclic voltammetry of 4,4'-oxydiphenol in 1 M perchloric acid.

(1R) essentially no current appears until about +0.4 V (peak B). On the second anodic sweep (2F) peak C now appears, corresponding to reoxidation of reductant generated during peak B. The semireversible system corresponding to peaks B and C can be unequivocally identified as hydroquinone–benzoquinone. Thus, following initial charge transfer, oxydiphenol splits into 2 moles of benzoquinone. More detailed studies of this reaction are given later.

Will and Knorr employed a much faster type of CV using oscillographic recording exclusively. They utilized it to great advantage in studying the sorption of hydrogen and oxygen on platinum and other metal electrode surfaces (49). Böld and Breiter have also used CV in similar studies of electrode surfaces (50,51). Breiter described the use of CV with a small, superimposed a-c component. In this way one can make impedance measurements during the course of the periodic triangular wave voltage excursions and study adsorption on the electrode surface. A series of papers in this area by Breiter and Gilman has described adsorption and charge-transfer processes on solid electrodes (52–54). Buck and Griffith presented a thorough study of the oxidation of methanol, formaldehyde, and formic acid at platinum employing CV at moderate sweep rates (55). Vielstich and Vogel have employed CV to study the oxidation of alcohols and acids (79,80). A recent review of CV and single sweep methods by Srinivasan and Gileadi is limited in interest to adsorption-controlled processes (81). Kublik has discussed the applicability of CV to platinum electrodes and compared results for selected systems with the hanging mercury drop electrode (56).

If CV is to be useful for quantitative purposes it is first required that one be able to measure true i_p's from the cyclic polarograms. There is no problem with the first peak of the initial sweep—it is identical to a single-sweep peak polarogram. Its base line, from which i_p is measured, usually is the zero current axis or at least is a constant, flat portion at the foot of the peak. The reverse cycle, or any other peaks, will not have any flat base lines from which to measure. Essentially the entire problem of quantitative i_p measurement in CV is to establish the proper base line. The simplest situation, that of an uncomplicated electron transfer, is considered first.

Lee investigated the ratio of reverse to forward peak currents for a few reversible and quasireversible systems at an unshielded carbon paste electrode using relatively slow sweep rates (60). She was able to show that provided the sweep was carried a certain minimum time past E_p, the ratio $i_{p(r)}/i_{p(f)}$ for the first cycle was very close to unity. Table 5-5 summarizes the results for the oxidation of N,N'-tetramethylbenzidine in pH 2 buffer. The requirement that the forward sweep (in this case forward is in the direction of oxidation) be carried past E_p is expressed at Δt, both in seconds and millivolts past E_p.

The ratio of peak currents is very close to unity. This ratio is less than unity whenever Δt is too small. Note a sweep rate of 1 V/min and Δt of 20 sec correspond to a potential increment of about 340 mV past E_p.

Identical results were obtained for the oxidation of o-dianisidine in 1 M sulfuric acid. The results for ferrocyanide oxidation and ferricyanide reduction showed greater deviations than the two organic systems. Starting with ferrocyanide oxidation (2 × 10^{-4} M in 3 M KCl) and a sweep rate of 3 V/min, the ratio $i_{p(r)}/i_{p(f)}$ was 1.010. If the reduction of ferricyanide in the same medium was the forward process, the ratio of i_p's decreased to 0.932. The density gradients in this particular electrolysis may account for the differences observed when the initial species is interchanged.

TABLE 5-5

Experimental Determination of $i_{p(r)}/i_{p(f)}$ for N,N'-Tetramethylbenzidine[a]

Sweep rate		$i_{p(f)}$,	$i_{p(r)}$,	Δt		
V/min	V/sec	uA	uA	sec	mV	$i_{p(r)}/i_{p(f)}$
3.0	0.050	7.05	7.22	7.5	375	1.02
		7.13	7.28	6.6	330	1.02
		7.28	7.20	5.6	280	0.99
		7.50	7.22	4.6	230	0.96
2.0	0.033	6.45	6.57	14.5	483	1.02
		6.54	6.62	13.0	432	1.01
		6.45	6.58	8.5	283	1.02
		6.50	6.50	7.1	236	1.00
		6.48	6.35	4.4	147	0.98
1.0	0.0167	4.85	4.90	23.6	401	1.01
		4.88	4.87	20.3	345	1.00
		4.85	4.93	17.4	296	1.02
0.667	0.0111	3.88	3.88	34.6	381	1.00
		3.83	3.88	30.1	331	1.01
		3.80	3.89	21.1	232	1.02
		3.81	3.70	12.1	133	0.97
					Av.	1.00₄

[a] 1×10^{-4} M tetramethylbenzidine in Britton and Robinson buffer pH 2.0, unshielded carbon paste electrode, upward diffusion.

The systems chosen by Lee were fortunate in that they provided approximately correct base lines for the measurement of the reverse i_p at values of Δt used. In any event, it is clear that with proper measuring conditions the ratio of anodic/cathodic peak currents for simple electron transfer is unity.

The correct measuring conditions are embodied in the quantitative treatment of Nicholson and Shain (61). It was shown that, provided the switching potential E_λ is not too close to E_p, the reverse current ratio is unity and independent of E_λ. To establish the proper base line for measuring

$i_{p(r)}$ one uses the extension of the forward peak polarogram which would have been obtained if the potential sweep had not been reversed. Nicholson and Shain present their cyclic polarograms for this purpose using X-time recording. A typical example for o-dianisidine oxidation is shown in Fig. 5-11.

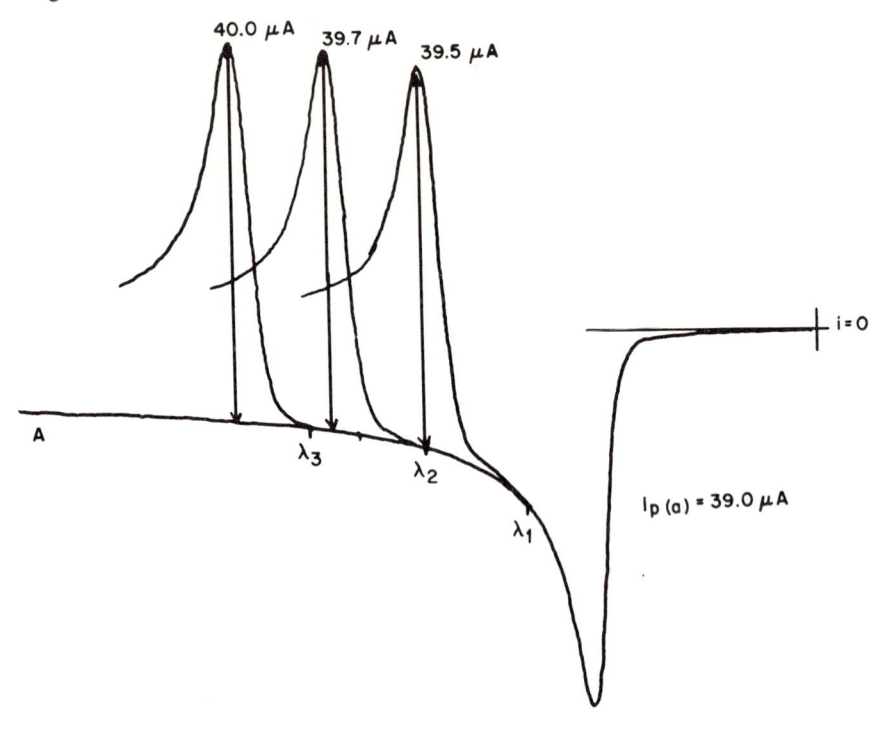

Fig. 5-11. Cyclic polarograms with varied switching potentials using X-time recording.

Line A in Fig. 5-11 is the extended forward (anodic) sweep and was obtained experimentally by continuing the sweep once to a potential beyond the intended switching points. For subsequent switching potentials, λ_1, λ_2, and λ_3, the corresponding $i_{p,c}$ was measured vertically to the extended anodic curve. It is clear that the ratio $i_{p,c}/i_{p,a}$ is essentially unity and independent of E_λ for each of the reversals shown in Fig. 5-11. If the forward sweep cannot be extended because of interference of background or other reactions, the "decay" line can be obtained by stopping the sweep beyond E_p and recording the usual $it^{1/2}$ decay curve. The use of X-time recording has the advantage that it gives a clear picture of the base line

CYCLIC (TRIANGULAR WAVE) VOLTAMMETRY

for the reverse current measurement. For most other purposes it is a cumbersome way to record CV. Olmstead and Nicholson have presented diagrams of this type for the ratio of anodic/cathodic peak currents for a reversible system (74).

It is far more practical to develop a useful scheme for measuring peak currents based on X-Y recording. One very satisfactory method was used by Hawley and Adams (75,76). The principle is illustrated in Fig. 5-12, where the forward sweep is again in the anodic sense. In Fig. 5-12A, the anodic sweep starts at 0.0 V and is reversed at E_λ. The true $i_{p,c}$ is obviously not just the current above the zero current line designated as i'_c. To i'_c must be added whatever value of i_a exists at the time when the

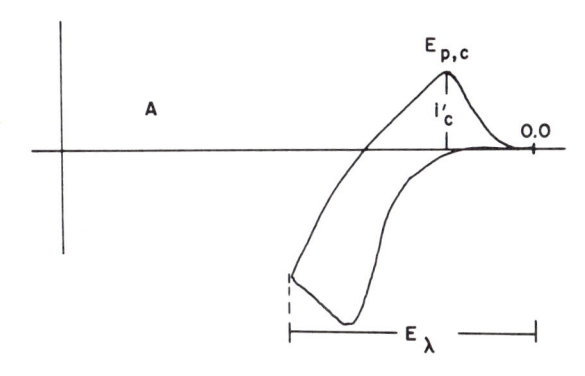

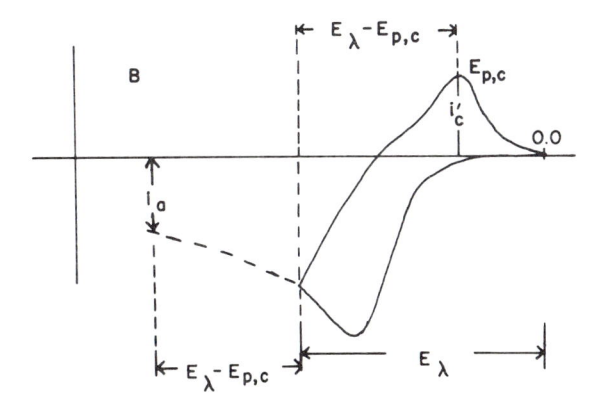

Fig. 5-12. Measurement of true reverse peak current with X-Y recording.

reverse sweep has returned to $E_{p,c}$. Hence the decayed level of i_a must be determined at this time.

The "time" it takes to get from E_λ back to $E_{p,c}$ (with the constant sweep rate) can be expressed as the potential increment $E_\lambda - E_{p,c}$. Thus one anodic sweep can be extended past E_λ by *at least* the increment $E_\lambda - E_{p,c}$, as indicated in Fig. 5-12B. The value of i_a observed at this point is to be

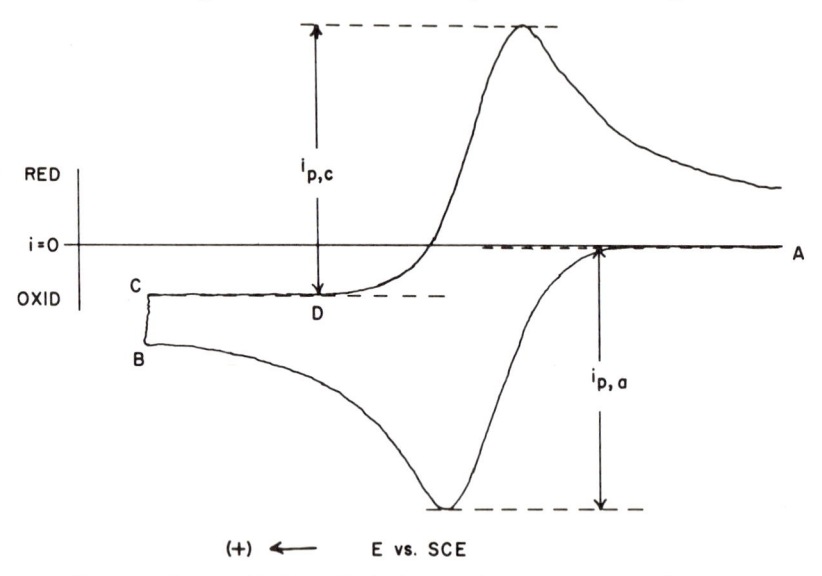

Fig. 5-13. Potential hold method of measuring true reverse peak current.

added to i'_c, or, as is clear from the diagram,

$$(i_{p,c})_{\text{true}} = i'_c + i_a$$

This procedure is entirely equivalent to the Nicholson and Shain method of finding the decay line. The only difference is that the "geometry" is arranged for X-Y recording. The reverse i_p is still being measured from the extension of the forward current as it decays.

Rather than establish the base line at a predetermined potential as above, it is possible to sweep, hold the potential constant, allow the $it^{1/2}$ decay to occur, and then after a proper holding interval initiate the reverse sweep. This results in a flat base line for the reverse peak. The technique is illustrated in Fig. 5-13. The forward direction is the one-electron oxidation of 5,10-dihydro-5,10-dimethylphenazine in acetonitrile to the cation radical (77).

CYCLIC (TRIANGULAR WAVE) VOLTAMMETRY

The forward sweep is initiated at A in Fig. 5-13 and continued anodic to B. The sweep is stopped and held constant at B, during which time the recorded anodic current decays to C. The time interval for this "hold" is about 15–30 sec. Then the reverse sweep is initiated and results in the flat base line CD, from which $i_{p,c}$ is measured. Here $i_{p,a}$ is measured in the usual sense and the ratio $i_{p,a}/i_{p,c}$ is 1.00 ± 0.01. (This method is very sensitive to chemical follow-up reactions, since a relatively long time interval is available for the reaction. If these are present, the current ratio will be very different from unity.)

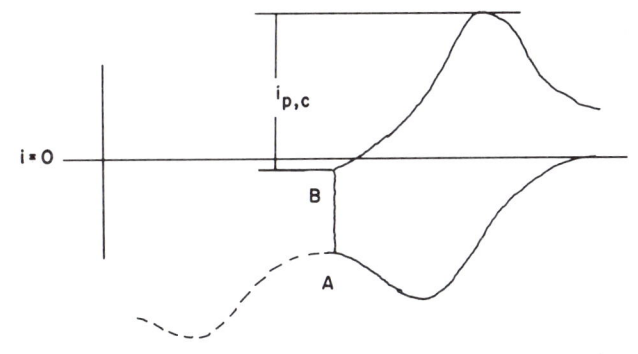

Fig. 5-14. Potential hold method for case of successive reactions.

Figure 5-13 represents, in a sense, two single-sweep polarograms, since the constant-potential electrolysis at point B amounts to generating a considerable quantity of oxidized product. Actually, generation of Ox occurs during the anodic sweep, the holding interval, and also all the way back to about 85% up the cathodic peak (i.e., 85% of the cathodic peak is $E°$ for the reversible system, hence only then does appreciable reduction begin). A calculation of the concentration–distance profile for Ox generated only during the holding period shows it is *close* to C^b (i.e., (i.e., $C_O/C_R^b \simeq 1$) for some distance out from $x = 0$. It should be emphasized that the holding operation is not for this purpose—a 1:1 ratio of peak currents would be obtained with a fast sweep reversal *if one could find a base line from which to measure $i_{p,c}$ under such conditions.* The holding and current decay provides such a base line, and this is the purpose.

If background or other reactions interfere, again one can hold at potentials just past the forward sweep, as illustrated in Fig. 5-14, where the dotted curve represents a complicating second step in the oxidation. Here the sweep is stopped at A. Notice that, regardless of the holding time, in this case the reverse sweep will start up the cathodic wave almost immediately, since the holding potential is practically at the foot of the peak. There is no way to obtain a base line as such. However, by holding until the $it^{1/2}$ decay is virtually constant (30–60 sec), point B becomes a pseudo-steady-state current level. The $i_{p,c}$ can be measured from point B as illustrated.

The methods above where the potential is held for fixed periods before the reversing of the sweep are analogous in principle to the potential step-linear scan method for investigating chemical follow-up reactions reported by Schwarz and Shain (78). They are used here in CV as empirical procedures for establishing base lines to measure the reverse peak currents. The criticism can be made that the holding times are such that natural convection can develop. Also, the concentration of product formed during the forward sweep and hold can begin to diffuse out to the bulk of solution (since its concentration *in the bulk* is zero). Certainly with holding times of 30–60 sec, natural convection develops at unshielded electrodes. Nevertheless, the procedures work admirably well on quite a few systems which have been tested. The situation is much like that encountered in measuring transition times in chronopotentiometry—one must have practical geometrical procedures for evaluating transition times and peak currents. In the present case, the theoretical justification is better than in chronopotentiometry. The primary difficulty is that one is overextending the time interval during which the theory holds. The procedures can be used safely if one keeps these limitations and potential difficulties in mind.

A very convenient semiempirical procedure for measuring ratios of anodic–cathodic peak currents, specifically for cases of chemical follow-up reactions, has been described very recently by Nicholson. The graphical procedure employs currents measured directly from a complete cyclic polarogram and does not involve holding times (82).

It should be emphasized further that all the peak-current measuring techniques were discussed for fairly idealized systems. *Regardless of theory*, if two or more peaks are considerably overlapped, true peak-current measurements become very questionable in practice. Other complications, such as adsorption and filming on solid electrode surfaces, present problems which have not been investigated properly to date. Since the full exploitation of CV depends very much on the measurement of true peak currents and current ratios, the practical problem of measuring these quantities is of the greatest importance in CV. The theory of CV has been well developed. The practice, on real systems at solid electrodes, requires much further study. Further discussions and illustrations of quantitative CV applied to organic systems will be given in Chapters 8 and 10.

5-8. CONCLUSIONS

Current–potential curves usually represent the overall recorded measurement of solid electrodes, and one might assess the relative merits and potentialities of various electroanalytical techniques in terms of the material discussed in this chapter.

First, slow voltage sweeps applied to quiet solutions appear to have little to recommend their continued use. There is no particular advantage to the slow sweep and, unless considerable precautions are taken, natural convection destroys the linear diffusion conditions one is trying to preserve. It would appear that this technique is employed in the belief that strict adherance to linear diffusion conditions ("more like the dropping electrode") leads to more reliable data for half-wave potentials. Only in rare instances can $E_{1/2}$ values at solid electrodes be identified justifiably with thermodynamic $E°$'s. $E_{p/2}$ values, for instance, are every bit as significant for certain purposes as $E_{1/2}$'s. It is unfortunate that commerical polarographs generally available in the United States are slow sweep types.

Rotated electrodes represent very practical systems for current–potential curve determinations. Despite the difficulties attendant with turbulent flow conditions, rotated electrodes give accurate and reproducible results if the experimental conditions are held constant. Many of the solid electrode polarographic data have been obtained using rotated electrodes. The effect of rate of change of applied voltage is a minimum here, which allows a choice of sweep rates as the occasion dictates. The rotated disk electrode, with precise limiting current behavior, probably is the most promising of all solid electrode systems. It should be kept in mind, however, that with quasi-reversible systems, the rotated disk electrode may supply high-enough mass-transport rates that the electrochemical system shows mixed charge and mass-transfer control. Hence $E_{1/2}$ values may shift due to increasing irreversibility as rotation rate increases with systems which are on the borderline of reversibility.

Rapid voltage scanning (single sweep) at stationary electrodes appears very attractive from the analytical viewpoint. This method has inherent sensitivity, results are reproducible, and the advantage of rapidity of analysis is not to be overlooked. The recent theoretical developments make the method very useful in electrode mechanism studies.

Finally, cyclic voltammetry shows much promise in electrochemical mechanism studies. When coupled with other techniques, such as single-sweep voltammetry and rotated disk studies, it is an extremely powerful tool. In fact, it seems logical to discontinue designing single-sweep polarographs, since so much can be gained by having single plus repetitive sweep operations in the same instrument. It should also be mentioned that for any precise work with fast sweep techniques, the use of two-electrode polarography is to be discouraged. It is hoped that a wider range of commercial instruments having controlled potential characteristics and rapid and multisweep operations will be available soon. In the meantime, however, with the availability of operational amplifier components and the excellent circuits in the literature, one can construct these instruments without much difficulty.

REFERENCES

1. T. Tsukamoto, T. Kambara, and I. Tachi, *Proc. 1st Intern. Polarog. Congr., Prague, 1951*, p. 525.
2. I. Shain and A. L. Crittenden, *Anal. Chem.*, **26,** 281 (1954).
3. P. Delahay, *New Instrumental Methods in Electrochemistry*, Wiley (Interscience), New York, 1954, Chap. 3.
4. D. B. Julian and W. R. Ruby, *J. Am. Chem. Soc.*, **72,** 4719 (1950).
5. J. Heyrovsky and D. Ilkovic, *Collection Czech. Chem. Commun.*, **7,** 198 (1935).
6. J. J. Lingane, *J. Am. Chem. Soc.*, **61,** 2099 (1939).
7. I. M. Kolthoff and J. J. Lingane, *Polarography*, Wiley (Interscience), New York, 1952, Chap. 11.
8. G. W. C. Milner, *The Principles and Applications of Polarography and Other Electroanalytical Processes*, Longmans, London, 1957, Chap. 4.
9. H. A. Laitinen and I. M. Kolthoff, *J. Phys. Chem.*, **45,** 1061 (1941).
10. L. B. Rogers and A. F. Stehney, *J. Electrochem. Soc.*, **95,** 25 (1949).
11. L. B. Rogers, D. P. Kraus, J. C. Griess, and D. B. Erlinger, *J. Electrochem. Soc.*, **95,** 33 (1949).
12. J. C. Griess and L. B. Rogers, *J. Electrochem. Soc.*, **95,** 129 (1949).
13. J. C. Griess, J. T. Byrne, and L. B. Rogers, *J. Electrochem. Soc.*, **98,** 447 (1951).
14. J. T. Byrne and L. B. Rogers, *J. Electrochem. Soc.*, **98,** 452, 457 (1951).
15. M. Haissinsky, *Electrochemie des Substances Radioactives et des Solutions Extrement Diluees*, Hermann, Paris, 1946.
16. S. H. Cohen, R. T. Iwamoto, and J. Kleinberg, *J. Am. Chem. Soc.*, **82,** 1844 (1960).
17. P. Delahay, *J. Phys. Colloid Chem.*, **54,** 402 (1950).
18. P. Delahay and G. L. Stiehl, *J. Phys. Colloid Sci.*, **55,** 570 (1951).
19. A. Sevcik, *Collection Czech. Chem. Commun.*, **13,** 349 (1948).
20. J. E. B. Randles, *Trans. Faraday Soc.*, **44,** 327 (1948).
21. M. M. Nicholson, *J. Am. Chem. Soc.*, **76,** 2539 (1954).
22. See Ref. 3, pp. 120–123.
23. T. Berzins and P. Delahay, *J. Am. Chem. Soc.*, **75,** 555 (1953).
24. S. S. Lord, R. C. O'Neil, and L. B. Rogers, *Anal. Chem.*, **24,** 209 (1952).
25. K. W. Gardiner and L. B. Rogers, *Anal. Chem.*, **25,** 1393 (1953).
26. T. L. Marple and L. B. Rogers, *Anal. Chim. Acta*, **11,** 574 (1954).
27. M. M. Nicholson, *J. Am. Chem. Soc.*, **79,** 7 (1957).
28. R. D. DeMars and I. Shain, *Anal. Chem.*, **29,** 1825 (1957).
29. J. G. Nikelly and W. D. Cooke, *Anal. Chem.*, **29,** 933 (1957).

REFERENCES

30. W. H. Reinmuth, *Anal. Chem.*, **32**, 1509 (1960).
31. H. Matsuda and Y. Ayabe, *Z. Elektrochem.*, **59**, 494 (1955).
32. T. R. Mueller and R. N. Adams, *Anal. Chim. Acta*, **25**, 482 (1961).
33. J. Zimmerman, Ph.D. thesis, Univ. Kansas, Lawrence, 1963.
34. W. H. Reinmuth, *Anal. Chem.*, **33**, 1793 (1961); **34**, 1446 (1962).
35. W. H. Reinmuth, *Anal. Chem.*, **32**, 1891 (1960).
36. R. P. Buck, *Anal. Chem.*, **36**, 947 (1964).
37. A. Y. Gokhshtein and Y. P. Gokhshtein, *Dokl. Akad. Nauk SSSR*, **131**, 601 (1960).
38. Y. P. Gokhshtein and A. Y. Gokhshtein, *Advan. Polarog. Proc. Intern. Polarog. Congr., 2nd, Cambridge, Engl., 1959*, **2**, 465 (1960).
39. H. Y. Lee and R. N. Adams, *Anal. Chem.*, **34**, 1587 (1962).
40. W. Kemula and Z. Kublik, *Nature*, **182**, 793 (1958).
41. W. Kemula and Z. Kublik, *Roczniki Chem.*, **32**, 941 (1958).
42. W. Kemula, *Advan. Polarog. Proc. Intern. Polarog. Congr., 2nd, Cambridge, Engl., 1959*, **1**, 105 (1960).
43. W. Kemula, Z. R. Grabowski, and M. K. Kalinowski, *Naturwiss.*, **22**, 1 (1960).
44. W. Kemula, Z. Kublik, and R. Cyranski, *Roczniki Chem.*, **36**, 1349 (1962).
45. Z. Galus, H. Y. Lee, and R. N. Adams, *J. Electroanal. Chem.*, **5**, 17 (1962).
46. J. R. Alden, J. Q. Chambers, and R. N. Adams, *J. Electroanal. Chem.*, **5**, 152 (1962).
47. A. L. Juliard and H. Shalit, *J. Electrochem. Soc.*, **110**, 1002 (1963).
48. W. H. Reinmuth, *J. Am. Chem. Soc.*, **79**, 6358 (1957).
49. F. G. Will and C. A. Knorr, *Z. Elektrochem.*, **64**, 258, 270 (1962).
50. W. Böld and M. Breiter, *Electrochim. Acta*, **5**, 145, 169 (1961).
51. M. Breiter, *Electrochim. Acta*, **7**, 25 (1962).
52. M. W. Breiter and S. Gilman, *J. Electrochem. Soc.*, **109**, 622 (1962).
53. M. W. Breiter, *Electrochim. Acta*, **7**, 533 (1962).
54. S. Gilman and M. W. Breiter, *J. Electrochem. Soc.*, **109**, 1099 (1962).
55. R. P. Buck and L. P. Griffith, *J. Electrochem. Soc.*, **109**, 1005 (1962).
56. Z. Kublik, *J. Electroanal. Chem.*, **5**, 450 (1963).
57. W. M. Schwarz, Ph.D. thesis, Univ. Wisconsin, Madison, 1960.
58. D. E. Smith, Ph.D. thesis, Columbia Univ., New York, 1960.
59. H. Matsuda, *Z. Elektrochem.*, **61**, 489 (1957).
60. H. Y. Lee, M.S. thesis, Univ. Kansas, Lawrence, 1962.
61. R. S. Nicholson and I. Shain, *Anal. Chem.*, **36**, 706 (1964).
62. R. S. Nicholson, *Anal. Chem.*, **37**, 667 (1965).
63. J. M. Saveant and E. Vianello, *Advan. Polarog. Proc. Intern. Congr., 2nd, Cambridge, Engl., 1959*, **1**, 367 (1960).
64. J. M. Saveant and E. Vianello, *Compt. Rend.*, **256**, 2597 (1963).
65. J. M. Saveant and E. Vianello, *Anal. Chim. Acta*, **8**, 905 (1963).
66. R. S. Nicholson and I. Shain, *Anal. Chem.*, **37**, 178 (1965).
67. D. S. Polcyn and I. Shain, *Anal. Chem.*, **38**, 370, 376 (1966).
68. G. Mamantov, D. L. Manning, and J. M. Dale, *J. Electroanal. Chem.*, **9**, 253 (1965).
69. E. S. Jacobs, *Anal. Chem.*, **35**, 2112 (1963).
70. S. P. Perone, *Anal. Chem.*, **35**, 2091 (1963).
71. S. P. Perone and W. J. Kretlow, *Anal. Chem.*, **37**, 968 (1965).
72. W. T. De Vries and E. Van Dalen, *J. Electroanal. Chem.*, **10**, 183 (1965).
73. R. S. Nicholson, *Anal. Chem.*, **37**, 1351 (1965).
74. M. L. Olmstead and R. S. Nicholson, *Anal. Chem.*, **38**, 150 (1966).

75. D. Hawley, Ph.D. thesis, Univ. Kansas, Lawrence, 1965.
76. D. Hawley and R. N. Adams, *J. Electroanal. Chem.*, **10,** 376 (1965).
77. R. Nelson, E. T. Seo, D. Leedy, and R. N. Adams, *Z. Anal. Chem.*, **224,** 184 (1967).
78. W. M. Schwarz and I. Shain, *J. Phys. Chem.*, **70,** 845 (1966).
79. W. Vielstich, *Chem. Ingr.-Tech.*, **35,** 262 (1963); *Z. Instrumententenk.*, **71,** 29 (1963).
80. W. Vielstich and V. Vogel, *Ber. Bunsenges. Physik. Chem.*, **68,** 688 (1964).
81. S. Srinivasan and E. Gileadi, *Electrochim. Acta*, **11,** 321 (1966).
82. R. S. Nicholson, *Anal. Chem.*, **38,** 1406 (1960).
83. J. M. Saveant and E. Vianello, *Electrochim. Acta*, **12,** 629 (1967).
84. J. M. Saveant, *Electrochim. Acta*, **12,** 753 (1967).
85. S. P. Perone, *Anal. Chem.*, **37,** 1061 (1965).

6 ELECTROCHEMICAL METHODS EMPLOYING CONTROLLED CURRENT

6-1. Current Sweep Voltammetry 164
6-2. Chronopotentiometry 165
6-3. Chronopotentiometric Study of Electrode Processes: Methods Employing $i_0\tau^{1/2}$ Variation 172
6-4. Chronopotentiometric Study of Electrode Processes: Application of Current Programs 177
6-5. Practical Measurement of Transition Times 183
References 184

All the polarographic techniques discussed so far have involved application of a selected voltage to an electrolytic cell and measurement of the resulting current. If the current through the cell is varied and the corresponding potential of the working electrode is measured under steady-state conditions (convective mass transport), the technique may be called *current scanning* or *current sweep polarography*. One can also use a constant current and measure transient potential-time behavior of the working electrode during a period of exhaustive electrolysis of the interface layer under linear diffusion control. The latter method is known as *chronopotentiometry*. Neither of these techniques is new. Some degree of control of the current density at solid electrodes was apparently the method of choice in early studies of *i–E* curves (*1*). The technique has persisted in corrosion and electrode kinetics studies, but only recently has it been widely used in electroanalytical work.

The theory of chronopotentiometry was developed some 30 years before polarography. Early chronopotentiometric measurements were used to evaluate diffusion coefficients (*2–4*). The method remained essentially buried in the literature until the already classical work of Gierst and Juliard (*5*) pointed out its potentialities as an analytical tool.

The relationships between voltammetric processes employing constant potential and constant current are very well illustrated by the three-dimensional model involving time dependence presented by Reinmuth (6). Current-sweep voltammetry is not covered in this representation since it is time invarient as a result of the convective mass transport. However potential–time responses upon application of various current impulses to stationary electrodes are clearly depicted.

6-1. CURRENT SWEEP VOLTAMMETRY

Consider the typical polarogram depicted in Fig. 6-1, which represents the anodic oxidation of ca. 10^{-3} M iodide ion at a RPE. The background is 1 M sulfuric acid. The solid line is the i–E curve obtained by a manual plot or the voltage sweep technique described in Chapter 5. Suppose the leads from the voltage divider are now disconnected and replaced by a source of small, constant current. The arrangement is such that the RPE is still an anode. If a 2-μA increment of anodic current is now passed through the RPE, the potential with respect to the SCE is found to be at point A. If the current is raised stepwise to, say, 4, 8, and 10 μA, the potential corresponds to points B, C, and D, respectively. In other words, controlling either the applied voltage or the current results in an identical rising portion of the polarogram. With 14 μA of applied current in this hypothetical case, one is approaching the limiting current plateau. There is always some degree of fluctuation in this region at an RPE due to stirring irregularities, and, if the current is maintained for a period of time at 14 μA, some variability in potential would usually be observed, as indicated at "point" E. With a very slight further increase in current, say to 14.5 μA, there is a rapid shift in potential to point F, indicating the limiting current region. Continued increase of current results in potentials following the solid line into the next electrode process—in this case evolution of oxygen from the background electrolyte. Thus by varying the applied current and measuring the resulting potential one can trace an i–E identical with that obtained by the reverse experimental technique (7).

The equipment for such manual current sweep polarography is quite simple. The method only requires that one be able to vary the applied current relatively independent of any influence of the polarographic cell. A battery with a large, variable series resistance will supply currents in the microampere range, and, if the source voltage (batteries) is large

CHRONOPOTENTIOMETRY

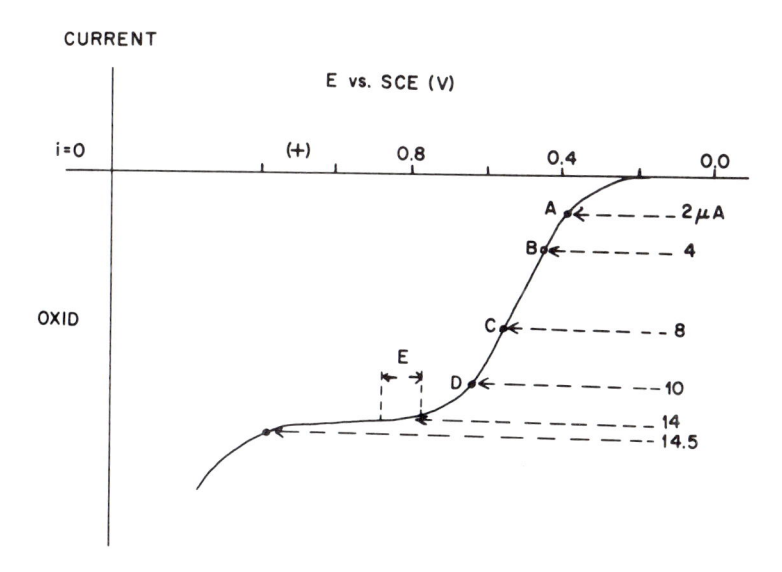

Fig. 6-1. Comparison of voltage and current sweep polarography.

enough, the current is relatively independent of the iR drop across the cell. Equipment for manual or automatic recording of current sweep voltammetry has been given and several applications to anodic oxidations described (8–10). One distinct advantage is that the current sweep technique is directly adaptable to automatic recording of iR corrected polarograms. This can be done without resorting to X-Y recording or iR compensation techniques (9). As a consequence, the current sweep method is well suited to high resistance (nonaqueous) systems. Badoz-Lambling has recommended current sweep techniques for certain solid electrode applications (11). It should be emphasized that the current sweep method as described above applies only to electrodes with convective mass transport. When a linear current sweep is applied to a stationary electrode, the technique becomes a form of chronopotentiometry.

6-2. CHRONOPOTENTIOMETRY

Chronopotentiometry involves the measurement of the potential–time variations at a working electrode during a short period of exhaustive electrolysis carried out at constant current. The electrolysis is accomplished as far as possible under linear diffusion conditions (treatments for cylindrical or other modes of diffusion are possible). The constant-current level is chosen so that the electrolysis may be terminated in less than about 60 sec to prevent natural convection from developing.

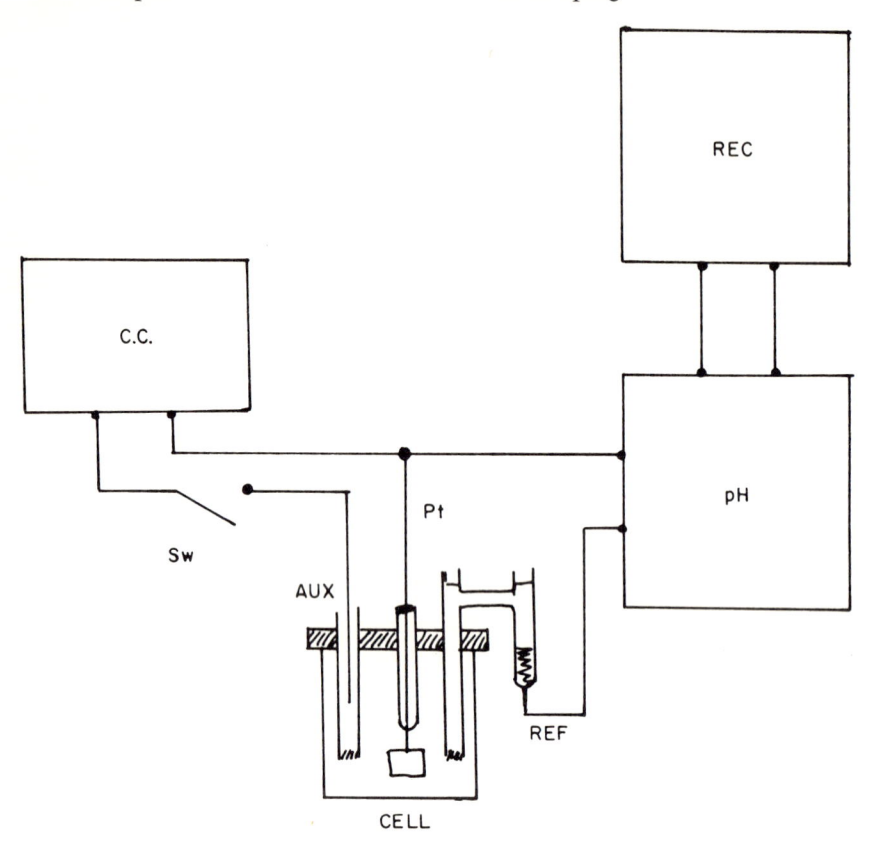

Fig. 6-2. Simple equipment for chronopotentiometry. C.C., constant-current source; S_w, current on–off switch; AUX, auxiliary electrode; Pt, platinum working electrode; REF, reference electrode; pH, pH meter or vacuum-tube voltmeter; REC, recorder.

The equipment for chronopotentiometry is seen in Fig. 6-2. The constant-current source can be a simple battery and series resistor combination. Currents commonly used in solid electrode work range from a few milliamperes down to 50–100 μA. The current source is connected to the working electrode (Pt foil) and an auxiliary electrode. The potential of the WE is monitored with respect to the reference electrode (SCE) by a line-operated pH meter or VTVM. The output of the pH meter is fed

CHRONOPOTENTIOMETRY 167

to a pen and ink recorder having a 1-sec pen response. The chart is normally operated at about 8 in./min. (An oscilloscope is used for potential–time response of ca. <2 sec.)

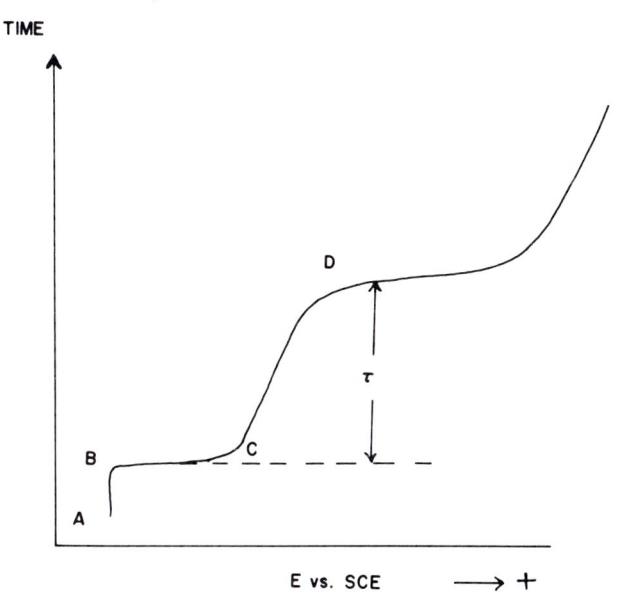

Fig. 6-3. Typical chronopotentiogram for oxidation of iodide ion. Background, 1 M sulfuric acid.

Figure 6-3 shows a typical potential–time pattern or *chronopotentiogram* for the oxidation of iodide ion in air-saturated 1 M sulfuric acid as the background electrolyte. The course of the potential–time curve can be explained in the following qualitative manner.

The potential of a platinum electrode dipping in the air-saturated sulfuric acid containing the small amount of iodide ion has no particular significance. After a few minutes, however, the potential of the platinum electrode reaches some fairly steady value and, if the recorder is turned on, the line AB represents this steady potential with no current flow. At point B the electrolysis current is initiated and the potential shifts rapidly out to C, which marks the beginning of the iodide oxidation. The potential now undergoes only a slight change from C to D, during which time the concentration of iodide next to the electrode surface is being progressively depleted through the oxidation reaction

$$2I^- \rightleftharpoons I_2 + 2e \qquad (6\text{-}1)$$

At D, the concentration of iodide next to the electrode is completely exhausted, and the potential of the electrode shifts rapidly to the next most easily oxidized component in solution—in this case the background electrolysis. The interval from C to D then represents the time required to oxidize completely all iodide ion in the immediate vicinity of the electrode and is called the *transition time*, τ. It is this transition time which is of primary analytical importance, for $\tau^{1/2}$ is directly proportional to bulk concentration.

A derivation of the theoretical E–t curve in chronopotentiometry involves calculations of the functions $C_O(x, t)$ and $C_R(x, t)$ under the conditions of constant-current electrolysis (constant flux at electrode surface). From these values one deduces C_O^e and C_R^e in terms of t. For a reversible process, the values of C_O^e and C_R^e are substituted in the Nernst expression to give the potential–time curve. Details of the derivation are to be found in Delahay's monograph and the original literature (2–5,12,13). The final form of interest here can be expressed as

$$E = E_{1/2} + \frac{0.059}{n} \log \frac{\tau^{1/2} - t^{1/2}}{t^{1/2}} \qquad (6\text{-}2)$$

where E is the potential at any time t, $E_{1/2}$ the familiar polarographic half-wave potential, and τ the transition time. Equation (6-2) is identical with the polarographic "wave equation" (5-25), where i_L and i are replaced by $\tau^{1/2}$ and $t^{1/2}$, respectively. Just as $E_{1/2}$ in the polarographic equation is the value of E when $i = i_L/2$, evidently $E_{1/2}$ in chronopotentiometry is obtained when $t = \tau/4$. Hence Eq. (6-2) is often written as

$$E = E_{1/4} + \frac{0.059}{n} \log \frac{\tau^{1/2} - t^{1/2}}{t^{1/2}} \qquad (6\text{-}3)$$

where $E_{1/4}$ is called the quarter-wave potential (13,15). The quarter-wave potential is identical with the polarographic $E_{1/2}$ if a mercury electrode is used and, in turn, is given by the equation

$$E_{1/4} = E_{1/2} = E^\circ - \frac{RT}{nF} \ln \frac{D_O^{1/2}}{D_R^{1/2}} \qquad (6\text{-}4)$$

The $E_{1/4}$ is characteristic of the electroactive substance and independent of its concentration.

CHRONOPOTENTIOMETRY

The quantitative aspects of chronopotentiometry are embodied in the expression

$$\tau^{1/2} = \frac{\pi^{1/2} n F D_O^{1/2} C_O^b}{2 i_0} \quad (6\text{-}5)$$

where τ = transition time, sec
i_0 = current density, A/cm²
C_O^b = concentration, moles/cm³

and the other symbols have their usual meaning. Equation (6-5) is the counterpart of the Ilkovic equation in classical polarography and is often referred to as the Sand equation. A plot of $\tau^{1/2}$ vs. C^b gives a straight-line calibration curve which serves as the basis of quantitative chronopotentiometry. When working with solid electrodes, the electrode area is constant, so that i and i_0 are interchangeable.

An alternative form of Eq. (6-5) which employs current densities and concentrations in units more appropriate to solid electrode work is (30)

$$\tau^{1/2} = \frac{\pi^{1/2} n F D^{1/2} C^b}{2 i_0} = \frac{8.55 \times 10^4 n D^{1/2} C^b}{i_0} \quad (6\text{-}6)$$

where now i_0 is the current density in microamperes per square centimeter, C^b is given in millimoles per liter, and the other symbols have their previous significance. The function $i_0 \tau^{1/2}/C^b$ is frequently called the transition-time constant, or simply the chronopotentiometric constant, and is analogous to the "diffusion current constant" of dropping mercury polarography. If Eq. (6-5) or (6-6) is valid, the product of $i_0 \tau^{1/2}$ should be independent of i_0 and one can establish a linear calibration plot $\tau^{1/2}$ vs. C which involves perhaps several different current levels. While such a procedure is valid for several reactions at solid electrodes (15,16), it has generally not been found to hold for organic oxidations at platinum electrodes (16). In this case one must be satisfied with calibration data over limited ranges using a single current level. This is no real disadvantage, since several calibration plots can be made to overlap. Since $\tau^{1/2}$ is inversely proportional to i or i_0, a wide range of transition times can be selected very simply by varying i_0. In normal work, current density (or concentration) should be adjusted to transition times not exceeding 60 sec.

All the current in chronopotentiometry is not consumed directly in electrochemical reaction. Since the potential varies markedly during the electrolysis, some current is consumed in charging the electrode double layer. This results in a distortion of all chronopotentiograms. Transition times are therefore measured by empirical procedures largely inspired by polarographic practice in the measurement of limiting currents. Once the shape of the *E–t* curve has been recorded for a new system, *routine* measurements of apparent τ values can be made by timing the interval between two preselected potentials using a stopwatch. Reilley and coworkers have demonstrated that this method, which dispenses with the recorder, is quite precise (15). It can be recommended only for analytical applications. Further details on the measurement of τ will be given later.

A study of the chronopotentiometric oxidation of oxalic acid at platinum and gold wires (17) led Lingane and co-workers to a thorough evaluation of chronopotentiometry at cylindrical electrodes. Peters and Lingane deduced a modification to the Sand equation for cylindrical electrodes, which is (18)

$$\frac{i\tau^{1/2}}{C^b} = \frac{nFA\pi^{1/2}D^{1/2}/2}{\left[1 - \frac{\pi^{1/2}D^{1/2}\tau^{1/2}}{4r_0} + \frac{D\tau}{4r_0^2} + \frac{3\pi^{1/2}D^{1/2}\tau^{1/2}}{32r_0^3} + \cdots\right]} \tag{6-7}$$

where A is the area of cylindrical wire electrode, in square centimeters, r_0 is the radius of the wire, in centimeters, and the other symbols have their previous significance. Coupling the numerator and the left side of Eq. (6-7) is seen to give the familiar Sand equation for semiinfinite linear diffusion. The denominator then accounts for the enhancement of τ due to the cylindricity of the wire. With a cylindrical diffusion field the effective area increases with increasing distance from the electrode surface. More electroactive material can get to the electrode surface, and a longer time τ, is required for exhaustive electrolysis to the interface region.

As τ decreases, the denominator of Eq. (6-7) approaches the first term (unity), and the results approximate the linear diffusion case. Even with τ in the range of 1 sec, however, the cylindrical effect increases the product $(i\tau^{1/2}/C^b)$ by 5%, according to Peters and Lingane. This effect increases as the wire radius decreases or as the diffusion coefficient of the electroactive species increases (i.e., the diffusion layer thickness increases with D). Lingane (19) showed that the increase of $i\tau^{1/2}/C^b$ for hydrogen ion reduction ($D \simeq 9 \times 10^{-5}$ cm²/sec) far exceeded that for ferric ion reduction ($D \simeq 0.5 \times 10^{-5}$ cm²/sec). To account properly for a species with a large D value like hydrogen ion requires further terms in the expansion in Eq. (6-7). The complexities of the case of successive chronopotentiometric reactions were examined by Peters and Lingane (20).

To facilitate using chronopotentiometry at wire electrodes Evans and Price tabulated values of the correction factor in terms of practical values of the function $D^{1/2}/r_0$. They also pointed out the correction can be neglected (<1% deviation from the linear diffusion calculations) with a careful selection of conditions. Thus, for τ about 25 sec (with $D \sim 10^{-5}$ cm^2/sec), r_0 should be 0.7 cm or greater (*21*). The preparation of wire electrodes with radii of this size, having negligible end effects, is not particularly simple.

As Lingane has pointed out, wire electrodes are useful for practical measurements since they can be constructed more easily than plane electrodes designed for strict control by semiifinite linear diffusion. However, the deviations from ideality with a metal foil are probably no more troublesome that the cylindrical diffusion corrections for a wire surface. Foil electrodes of platinum or gold are relatively easy to prepare. Indeed, it seems a matter of individual choice whether the advantages of wire electrodes outweight the complexities attendant with cylindrical diffusion in chronopotentiometry.

The potentialities of chronopotentiometry as an analytical technique were pointed out by Reilley et al. (*22*). Excellent results have been reported for mercury pools by a variety of workers, and solid electrode applications have been promising (*23–25*). In the first study of the oxidation of organic compounds at platinum foils results were shown to be quite satisfactory provided, in general, calibration plots of $\tau^{1/2}$ vs. concentration were not extended over too wide a range (*16*). One of the obvious advantages, the speed of the actual chronopotentiometric measurement, is partially nullified by the usual need to clean the solid electrode surface after each run. This difficulty is often lessened in nonaqueous media, and Ward has shown excellent results for the quantitative determination of a variety of antioxidants by anodic chronopotentiometry in acetonitrile and ethanol at graphite electrodes (*26*). Other studies of chronopotentiometry, particularly in acetonitrile, have proved valuable for nonanalytical applications (*27–29*).

Solid electrode chronopotentiometry can be utilized for routine analytical applications with confidence provided one takes care of the usual problems encountered with solid surfaces. On the other hand, in the writer's opinion, single-sweep peak voltammetry is preferable for analysis. Both methods require restricted diffusion control, but peak-current measurements are subject to less uncertainty than transition times. Further, for quite a few organic oxidations, peak polarograms are far better defined than the corresponding chronopotentiograms. In both techniques, the speed of any analytical determination is predominately determined by electrode cleaning procedures. Indeed, while there is not reason to deny the analytical possibilities of chronopotentiometry, the applications of real utility lie in the study of electrode processes which are discussed below. The treatment is not detailed, since excellent discussions, specifically concerning solid electrode applications, are to be found in the original literature.

6-3. CHRONOPOTENTIOMETRIC STUDY OF ELECTRODE PROCESSES: METHODS EMPLOYING $i_0\tau^{1/2}$ VARIATION

Before attempting to discuss the utility of chronopotentiometry in the study of electrode processes, it is necessary to point out several disconcerting facts. First, although chronopotentiometry is probably one of the most versatile and powerful of the electroanalytical techniques to be applied to the study of electrode mechanisms, its very versatility requires great care in interpretation of the results. Specifically, the wide range of current densities which can be employed actually provide complications not present in the simpler voltammetric methods. Second, raw experimental data in chronopotentiometry at solid electrodes are frequently unusable.

The latter complication was demonstrated forcibly by Bard in his study of electrode configurations and their effect on transition times (*30*). The transition-time constant ($i_0\tau^{1/2}/C^b$) should be completely independent of i_0 (or τ) for a straightforward electrode reaction (no complications such as preceding chemical reactions). Yet Fig. 6-4, taken from Bard's data, shows that this is not at all the case for the oxidation of iodide ion at a platinum electrode. (To cover a wide range of τ, the value of $i_0\tau^{1/2}/C^b$ is plotted vs. log τ.) The value of $i_0\tau^{1/2}/C^b$ increases very much at both short and long transition times, contrary to theory.

The increase at long times can be traced easily to both departure from linear diffusion conditions and, even worse, if present, to natural convection resulting from density gradients. This is well illustrated by Fig. 6-5, also from Bard's work, which represents the reduction of silver ion on a well-shielded, linear diffusion electrode. Here one expects little contribution from partial spherical diffusion and, indeed, this is verified by curve 1. However, if the electrode orientation is such that the density gradient resulting from electrolysis develops stirring, then this natural convection is extremely important at values of τ around the usually quoted 60-sec minimum. This is seen in curves 2 and 3 of Fig. 6-5. Hence, for a given electrolysis, proper attention to electrode design and orientation will eliminate or certainly minimize the increase of $i_0\tau^{1/2}/C^b$ at long transition times.

METHODS EMPLOYING $i_0\tau^{1/2}$ VARIATION

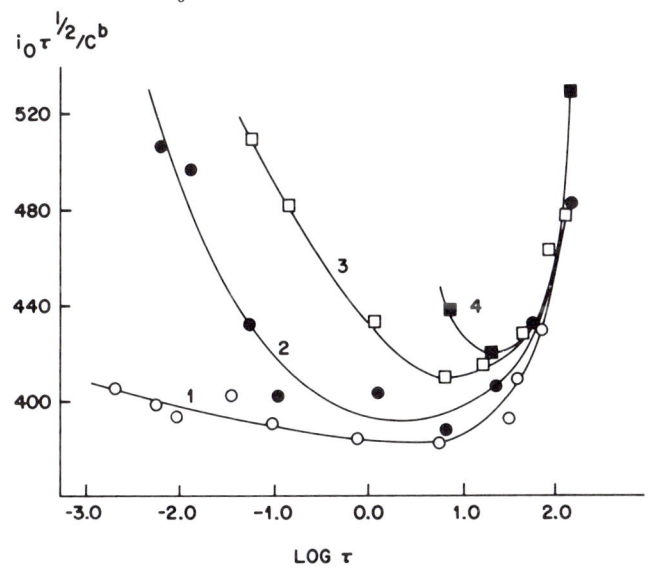

Fig. 6-4. Dependence of $i_0\tau^{1/2}/C^b$ on τ. Concn. iodide ion: 1, 10 mM; 2, 5 mM; 3, 1 mM; 4, 0.1 mM. All in 1.0 M sulfuric acid. [After A. J. Bard, *Anal. Chem.*, **33**, 11 (1961).]

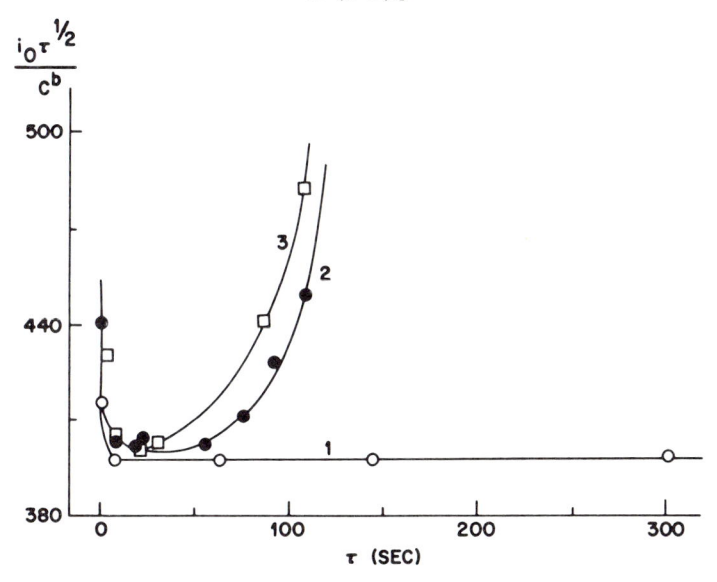

Fig. 6-5. Dependence of $i_0\tau^{1/2}/C^b$ on τ for reduction of AG$^+$ at shielded electrodes. 1. Horizontal, upward diffusion; 2. horizontal, downward diffusion; 3. vertical. All 5 mM Ag$^+$ in 0.2 M KNO$_3$ and 0.01 M HNO$_3$. [After A. J. Bard, *Anal. Chem.*, **33**, 11 (1961).]

The increase at short electrolysis times can hardly be caused by disturbances to the mass transport and, indeed, is a bit more subtle. In general, this effective increase in τ is caused by the current being distributed between the electrode reaction and other side processes. For the diffusion-controlled electrode reaction we have $\tau^{1/2}$ varying with $1/i_0$, as predicted by the Sand equation. For the side processes, generally τ itself varies with $1/i_0$. At shorter transition times this distribution of the current to the side processes becomes more important. These side processes which consume current can be (1) charging of the double layer, (2) formation or reduction of oxide (or other) films especially on noble metal electrodes, and (3) oxidation or reduction of species adsorbed on the electrode surface. (Surface roughness may also be a problem when the diffusion-layer thickness is so small as to be comparable in magnitude to the microscopic size of surface crevices, etc.).

Bard has derived a generalized equation to account for such side processes, assuming that the total current density is divided into fractional *constant* currents for each of the side processes (*31*). The general equation is

$$\frac{i_0 \tau^{1/2}}{C^b} = \begin{pmatrix} \text{Sand equation} \\ \text{component} \end{pmatrix} + \begin{pmatrix} \text{double-layer} \\ \text{charging} \end{pmatrix} + \begin{pmatrix} \text{oxide} \\ \text{effects} \end{pmatrix} + \begin{pmatrix} \text{adsorption} \\ \text{effects} \end{pmatrix}$$

$$= \frac{10^{-3} \pi^{1/2} nFD^{1/2}}{2} + \frac{10^{-3}(C_{dl})_{av}\Delta E}{C^b \tau^{1/2}} + \frac{Q_{ox}}{C^b \tau^{1/2}} + \frac{10^3 nF\Gamma}{C^b \tau^{1/2}} \quad (6\text{-}8)$$

Here current density has been put in milliamperes per square centimeter and C^b is in millimoles per liter. The double-layer capacity $(C_{dl})_{av}$ in microfarads per square centimeter is the average value over the potential interval ΔE in volts which the chronopotentiogram spans. This is usually obtainable in approximate form from experimental measurements on the particular electrode. Q_{ox} represents the millicoulombs required for formation or dissolution of an oxide film on the particular electrode surface. The quantity Γ is the moles per square centimeter of electroactive material adsorbed on the electrode surface.

Equation (6-8) is, of necessity, approximate. This is principally seen in the correction factor for charging the double layer. Thus, in an actual chronopotentiogram, there is normally a rapid shift of potential (from the "rest" potential to the foot of the chronopotentiometric wave) and another rapid shift at the transition time, where the potential usually switches out to the background process. The charging current in these regions is quite high. However, during the rising portion of the chronopotentiogram, where the charge of E with respect to time is very low, the charging component is quite negligible. Hence the current component of i_0 which goes to double-layer charging cannot be constant. As a matter of fact, however, the approximation can be shown to be quite satisfactory. This same approximation with regard to constant currents is inherent in the correction for adsorption.

All the corrective factors in Eq. (6-8) are of the form $(\text{constant}/C^b\tau^{1/2})$ so one can write, as did Bard, a simplified version in terms of the true chronopotentiometric constant (Sand component) and a summed correction:

$$\frac{i_0\tau^{1/2}}{C^b} = 10^{-3}\frac{\pi^{1/2}nFD^{1/2}}{2} + \frac{B}{C^b\tau^{1/2}} \tag{6-9}$$

where the units are the same as previously given. For a given electrode and electroactive system, one or more of the individual components of the B term may be missing. For instance, adsorption may play no role, or, for a platinum electrode whose potential excursion does not exceed say, 0.4 V vs. SCE, the oxide effect is negligible. In general, however, the effects are far from nil. As pointed out by Bard, only 2.5 μC of oxide film, or 1/100 of a monolayer of adsorbed material, can cause a 1% error with millimolar solutions and τ's of ca. 1 sec. Applications of Eq. (6-9) to the study of adsorption are considered later. From the form of Eq. (6-8) or, better, (6-9), it is seen that the error can be minimized by working at high bulk concentrations or at long transition times. As a working rule concentrations of at least 5 and preferably 10 M should be used in chronopotentiometry, where the corrective terms become minimal. The use of longer transition times must be balanced against the natural convection effects which may develop with practical electrode designs. If nothing else, the use of longer values of τ (i.e., small current densities) minimizes the double-layer correction, as was initially pointed out by Gierst (32) and Delahay (33).

The most irritating aspect of the corrections discussed above is that they render very difficult some applications in electrode mechanism studies for which chronopotentiometry is ideally suited. One of these is the detection of homogeneous chemical reactions (kinetic processes) prior to the electron transfer process (often called precursor reaction). Assuming the complications absent, this type of reaction is readily examined in chronopotentiometry from consideration of the variation of $i_0\tau^{1/2}$ with increasing i_0.

For a rapid charge-transfer process, with no kinetic complications, a plot of $i_0\tau^{1/2}$ vs. i_0 will be a horizontal straight line. In fact, if the time t is measured at various E's up the chronopotentiometric wave, a plot of $i_0 t_E^{1/2}$ vs. i_0 will give a series of horizontal straight lines for a reversible reaction (i.e., at all points on the chronopotentiogram, the product $i_0 t^{1/2}$ is independent of i_0) (34). For a slow charge-transfer reaction ("irreversible"), again the product $i_0\tau^{1/2}$ vs. i_0 is a horizontal straight line. In this case, if $i_0 t_E^{1/2}$ is plotted (for various potentials along the wave) vs. i_0, the function decreases with increasing i_0, since the rate constant for the electron transfer is an exponential function of E and only at potentials approaching τ is it at a maximum such that $i_0\tau^{1/2}$ will be independent of i_0 (34). Hence in all simple electron-transfer processes, $i_0\tau^{1/2}$ is independent of i_0.

Consider now a preceding chemical reaction, such as an acid–base dissociation, which is usually formulated as follows:

$$Y \underset{k_b}{\overset{k_f}{\rightleftarrows}} Ox + ne \rightarrow Red$$

Substance Y is nonelectroactive in the potential region considered. The rate of reduction of Ox, the electroactive species, is now determined not only by the mass-transfer process of Ox to the electrode surface but also by the rate of the chemical transformation $Y \rightarrow Ox$. This problem was first treated by Gierst and Juliard (5) and more rigorously by Delahay and Berzins (14). Reinmuth has given a thorough analysis of these processes for other than first-order kinetic complications (35).

To describe this process the Sand equation must be modified with a correction P as

$$i_0\tau^{1/2} = \frac{\pi^{1/2} n F D^{1/2} C^b}{2} - P i_0 \qquad (6\text{-}10)$$

where

$$P = \frac{\pi^{1/2}}{2K(k_f + k_b)^{1/2}}$$

or

$$(i_0\tau^{1/2}) = \frac{\pi^{1/2} n F D^{1/2} C^b}{2} - \frac{\pi^{1/2}}{2K(k_f + k_b)^{1/2}} \cdot i_0 \qquad (6\text{-}11)$$

Here K is the equilibrium constant for the chemical transformation. Another form of Eq. (6-11) which is helpful to visualize the effects of the chemical reaction is obtained by dividing (6-11) by i_0:

$$\tau_K^{1/2} = \tau_D^{1/2} - \frac{\pi^{1/2}}{2K(k_f + k_b)^{1/2}} \qquad (6\text{-}12)$$

The "kinetic" transition time τ_K is less than that for pure diffusion control (τ_D), provided the denominator of the second term on the right of Eq. (6-12) is not too large. Detailed discussions of the detectability of such reactions in terms of K and the rate constants are given by Delahay and Berzins (*14,36*) and Reinmuth (*35*).

It is evident from Eq. (6-11) that a plot of $i_0\tau^{1/2}$ vs. i_0 will have a slope of $-\pi^{1/2}/2K(k_f + k_b)^{1/2}$. Knowledge of K allows k_f and k_b to be determined. The difficulty of this type of kinetic analysis at practical solid electrode systems is easy to see. The corrective term P in Eq. (6-10) or its equivalent (6-11) is of the same general form as the B factors mentioned previously. Double-layer, oxide, and adsorption corrections (B factors) lead to positive deviations of a plot of $i_0\tau^{1/2}$ vs. i_0 (note in Fig. 6-5 that if the data were plotted vs. increasing i_0, the short time corrections would appear as a positive deviation at increasing current densities—neglecting any consideration of convection in this case). Hence B-type corrections tend to cancel the P factor in the kinetic case. Working with high concentrations of electroactive material and long transition times one can minimize the B factors. But again, with practical solid electrodes (especially if unshielded) τ's much greater than 5–10 sec show contributions from convective mass transport. Further, at high concentrations of electroactive species, films produced during the course of the electrode reaction may give false information (*16*). The situation for obtaining unequivocal data for preceding chemical reactions at practical solid electrodes is beset with many difficulties. Despite this pessimistic attitude, the dissociation rates of protonated *p*-phenylenediamines and related amines was recently carried out elegantly in this fashion by Mark and Anson (*37*).

6-4. CHRONOPOTENTIOMETRIC STUDY OF ELECTRODE PROCESSES: APPLICATION OF CURRENT PROGRAMS

The techniques discussed in Section 6-3 all involved examination of the variation of ($i_0\tau^{1/2}$) with one or another of the experimentally accessible variables such as i_0 or C^b. This approach, which has been called the kinetic analysis method (*38*), employs constant and unidirectional current during the entire course of a complete chronopotentiogram. Another large segment of applications can be grouped conveniently as programmed current methods. Here the current is varied during the course of the chronopotentiogram so as to obtain a particular potential–time response, or, in the most widely used method, the current is reversed at or before the transition time and the potential–time signals of the electrolysis products are examined. This classification is merely convenient and certainly does not include all applications. A third branch of mechanism studies employing chronopotentiometry is one in which the shape of the E–t curve is examined rigorously (40). This approach, with diagnostic criteria for distinguishing various reaction schemes, has been shown by Reinmuth to be a very powerful tool. It is probably the most difficult to apply to solid electrodes, since the reproducibility of shapes of i–E and E–t curves are often involved with surface-history effects. These methods are not discussed further here.

A wide variety of current program techniques is available and a thorough discussion of their particular advantages is given in the original literature (35,41,42,50). Only the simplest of these, in which the current is reversed at or before the transition time, called reverse-current chronopotentiometry, is discussed below.

Reverse-current chronopotentiometry (RCC) for a simple mass-transfer-controlled process with no complexities was examined in detail initially by Berzins and Delahay (12,43). In RCC the first part of the measurement is identical with that discussed previously. The forward direction of the measurement records τ_f and the corresponding E–t curve. At any time t_f up to but not exceeding τ_f, the current is reversed abruptly and one records the corresponding τ_r and E–t curve.

Considering the usual reduction as the forward process, as electrolysis proceeds, $C_O(x, t)$ decreases and $C_R(x, t)$ builds up in the vicinity of the electrode and also diffuses away from the electrode surface. These concentrations are readily calculated from the standard relations for constant-current electrolysis (see Ref. 12, pp. 193–195). The concentration $C_R(x, t_f)$, where t_f is the time of reversal, becomes the initial condition for calculation of the concentration profile of C_R during the reoxidation cycle. The concentration profiles are discussed in detail in the original literature (43).

For the simple electron transfer

$$\text{Ox} + ne \rightleftharpoons \text{Red}$$

with current reversal at t_f, Berzins and Delahay give the general relation for the ratio (τ_r/t_f) as (43)

$$\frac{\tau_r}{t_f} = \frac{\theta^2}{(\theta + \lambda')^2 - \theta^2} \qquad (6\text{-}13)$$

APPLICATION OF CURRENT PROGRAMS

where

$$\theta = \frac{i_f}{nFD_r}$$

$$\lambda' = \frac{i_r}{nFD_r}$$

The general treatment considers any set of values for i_f and i_r, but, for the most widely used case of equal currents in both directions, the relation (6-13) gives rise to the delightfully simple expression

$$\tau_r = \frac{t_f}{3} \tag{6-14}$$

It should be noted that the relations do not depend on the differences in diffusion coefficients of Ox and Red, nor is the ratio τ_r/t_f a function of the reversibility of the electrode reaction. On the other hand, if the process is reversible, a simple relation exists between the value of the forward quarter-wave potential (E at $t_f = 0.25\tau_f$) and its counterpart on the reverse cycle. This is

$$E_{1/4(f)} = E_{0.22(r)} \tag{6-15}$$

That is, the potential at a point $0.22\tau_r$ is equivalent to the potential at $0.25\tau_f$, the latter being the "ordinary" $E_{1/4}$ for a single-direction chronopotentiogram. This relation may be used to check the reversibility of a process via chronopotentiometry.

Now, the system above, in which the number of electrons transferred in both directions is identical and in which nothing but mass transfer occurs, is not particularly interesting. But two other cases are extremely intriguing. First, suppose the product of electrolysis undergoes rapid, irreversible reaction to give X and Z which are electrochemically active on current reversal. Then Eq. (6-14) must take into account the possible differences for n_f and n_r [i.e., Eq. (6-13) should, for a general scheme be written with n_f and n_r in the functions θ and λ'].

Geske first used this type of analysis to great advantage in elucidating the oxidation mechanism of tropilidine (cycloheptatriene) in acetonitrile medium. According to Geske the oxidation proceeds to tropylium ion and a proton as (44)

Proof of this reaction was obtained by RCC when two distinct rereductions were obtained, corresponding to the reactions:

$$H^+ + e \longrightarrow \tfrac{1}{2} H_2$$

$$2\,[C_7H_7]^+ + 2e \longrightarrow \text{Bitropyl}$$

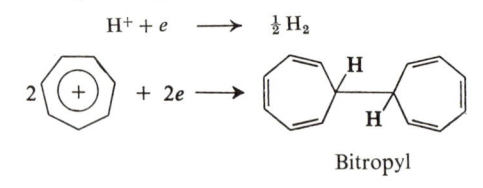

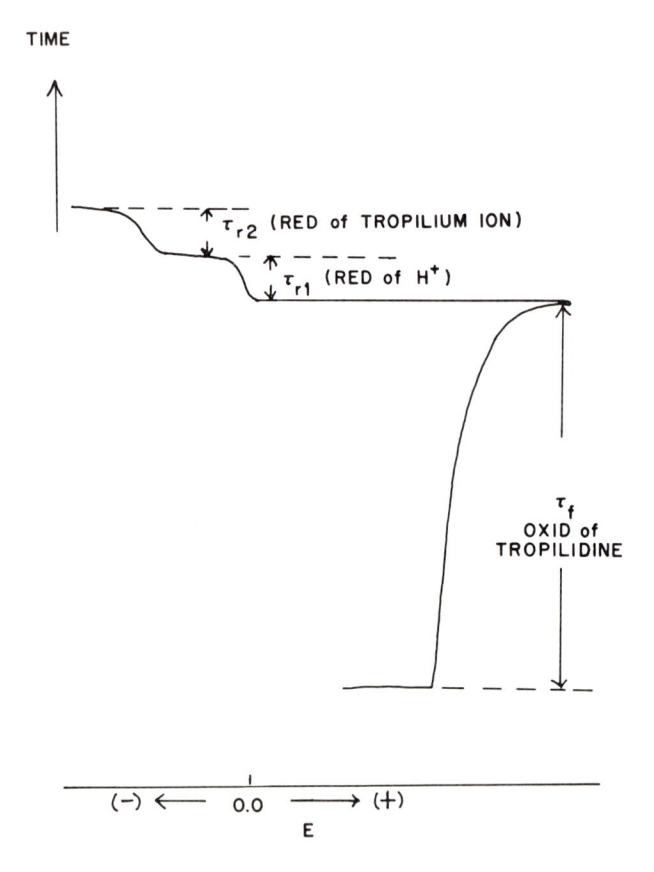

Fig. 6-6. Reverse-current chronopotentiometry of tropilicine.

The RCC of this system is shown in Fig. 6-6, which is reconstructed from Geske's data. Note that upon current reversal, the first reduction process is not connected with a reversible tropilidine system but is displaced

APPLICATION OF CURRENT PROGRAMS   181

strongly to negative potentials. Independently this was identified as due to hydrogen ion reduction. For $n_f = 2$ and $n_r = 1$, Eq. (6-13) predicts $\tau_f/\tau_{r_1} = 8.0$, and Geske found 7.6 experimentally. If the two reverse transition times represent all the products of electrolysis, $\tau_f/\tau_{r_1} + \tau_{r_2}$ should be 3.0, since the same total electrons are involved in forward plus both reverse processes. Solution of the two relations above also predicts $\tau_{r_2}/\tau_{r_1} = 1.67$. A general expression for such systems was presented by Furlani and Morpurgo (45) and King and Reilley (46). Testa and Reinmuth have also considered this type of reaction (47).

Another electrode reaction scheme for which RCC gives valuable information is a reversible charge transfer followed by a homogeneous chemical reaction. A typical example is the hydrolysis of an organic species following electron transfer. Specifically, the oxidation of *p*-aminophenol to *p*-quinoneimine followed by hydrolysis to quinone has been studied by several workers. Since the examples at solid electrodes are predominately oxidations, one switches the usual formulation and writes

$$\text{Red}_1 \rightleftharpoons \text{Ox}_1 + ne$$
$$\text{Ox}_1 + \text{S} \rightarrow \text{Ox}_2 + \text{P}$$

With the forward direction of the chronopotentiogram an oxidation, it is evident that upon current reversal, the τ_r will depend on the rate of the chemical reaction which consumes Ox_1 (in this case hydrolysis by the solvent S). A knowledge of the relation between t_f and τ_r enables one to obtain data for the rate of the chemical reaction. This type of reaction was considered independently by a variety of workers at almost the same time (45,47,48). The fundamental expression which relates the rate constant of the chemical reaction and the values of t_f and τ_r is

$$2 \operatorname{erf}(k\tau_r)^{1/2} = \operatorname{erf}[k(t_f + \tau_r)]^{1/2} \tag{6-16}$$

[It should be noted that the paper of Testa and Reinmuth (47) contains a missing bracket in their equation (1). Equation (6-16) is correct.] Dracka's work simplifies the treatment by assuming that the error function on the right in Eq. (6-16) is unity. The treatment of Furlani and Morpurgo plots the variation of the ratio of reverse to forward transition times as a function of the forward time for selected values of the rate constant. The value of k can be obtained by interpolation (45). Probably the most satisfactory treatment is that of Testa and Reinmuth, in which a working curve of τ_r/t_f vs. kt_f is calculated from which kt_f can be read directly. A further plot of kt_f vs. t_f gives the value of k directly as the slope of a straight line. Since the details of the construction of the working curve are not immediately apparent in the original paper, they are given below. (The quantity $t_f \equiv t_a$ and $\tau_r \equiv \tau_1$ in the paper by Testa and Reinmuth.)

In Eq. (6-16), if one assumes values for the quantity $k\tau_r$, then $2\,\text{erf}(k\tau_r)^{1/2}$ can be determined from standard tables of the error function. Now one has $\text{erf}(kt_f + k\tau_r)^{1/2}$ equal to a numerical value. It is again possible to find from error-function tables the value of $(kt_f + k\tau_r)^{1/2}$. The square of this quantity gives the sum $kt_f + k\tau_r$. Now, arbitrarily let

then
$$k\tau_r = a \quad \text{and} \quad kt_f + k\tau_r = b$$

$$kt_f = b - a$$

$$a = (k\tau_r) = 0.01 \qquad (k\tau_r)^{1/2} = 0.1$$

$$\text{erf}(k\tau_r)^{1/2} = 0.1125 \qquad 2\,\text{erf}(k\tau_r)^{1/2} = \text{erf}(kt_f + k\tau_r)^{1/2} = 0.2250$$

$$(kt_f + k\tau_r)^{1/2} = 0.202 \qquad (kt_f + k\tau_r) = 0.041 = b$$

$$b - a = 0.031 = kt_f$$

Therefore,
$$\frac{\tau_r}{t_f} = \frac{a}{b-a} = \frac{0.01}{0.031} \simeq 0.33 \quad \text{with } kt_f = 0.031$$

A similar calculation with $k\tau_r = 0.225$ gives $\tau_r/t_f \simeq 0.056$ at $kt_f = 4.01$. In this manner one constructs the working curve of τ_r/t_f vs. increasing values of kt_f. As can be seen, the properties of the curve are such that it approaches the theoretical $\tau_r/t_f = 1/3$ as $kt_f \to 0$, i.e., as k becomes negligible or as t_f, the time during which the chemical reaction can occur, decreases. At the other end of the curve, small decreases of τ_r/t_f cause a large change in kt_f, and this is not a particularly advantageous region for measurements. Experimentally, one varies t_f and measures τ_r/t_f. From this value the working curve gives kt_f directly. A linear plot of kt_f vs. the experimental t_f gives k as the slope.

The number of systems which have been subjected to this type of analysis are indeed few. The hydrolysis of p-benzoquinoneimine is, in fact, the only one which has been studied in detail (47). Snead and Remick also studied this hydrolysis by analyzing the shape of the chronopotentiograms as affected by the follow-up chemical reaction (49). Discrepancies between the two sets of results have been described and the RCC studies seem to be substantiated by independent studies (51).

Other treatments of current programmed chronopotentiometry have been given by various workers (*52,53*). The use of RCC in studying adsorption is discussed elsewhere. A recent innovation is cyclic chronopotentiometry, developed by Herman and Bard (*54,60,61*). In this modification, the current is alternately reversed at forward and reverse transition times and resembles cyclic voltammetry. While the same qualitative information is available with both techniques, the cyclic chronopotentiometry allows quantitative treatments. A recent chronopotentiometry review by Paunovic summarizes developments in the study of electrode processes (*62*).

6-5. PRACTICAL MEASUREMENT OF TRANSITION TIMES

It was observed early in the practice of chronopotentiometry that only a very few chronopotentiograms had such idealized shapes that measurement of τ was unequivocal. Even for reversible system the E–t pattern is distorted by charging current complications, and for irreversible systems the chronopotentiogram is often just "drawn out" and τ is quite difficult to evaluate. Empirical methods much akin to those used in polarography must be used in all instances to measure τ. Geometric procedures were suggested by Delahay and co-workers (*55,56*), Reinmuth (*57*), and others. A thorough paper by Russel and Peterson on the evaluation of τ for irreversible systems is very much to the point (*58*). These workers examined several of the accepted graphical procedures and compared the results with a least-squares analysis of the chronopotentiogram. Algebraic methods for calculating τ and αn_a for irreversible chronopotentiograms were developed.

It is interesting to note that the least-squares analysis, deemed by Russel and Peterson to be the most reliable method, agreed best with an empirical procedure developed by Kuwana. This method, which has no "official status" since it was not described per se in the literature, was first used by Kuwana in the writer's laboratory (*59*). It consists of merely determining that point on the transition-time plateau where the E–t curve deviates from a straight line. τ is measured from this point down to the base line, as seen in Fig. 6-7. As pointed out by Russel and Peterson, theory does not predict that the plateau will be linear, but experimentally it almost always occurs. Disregarding the lack of theoretical justification for this method of evaluating τ, it is true that it agrees best with the results

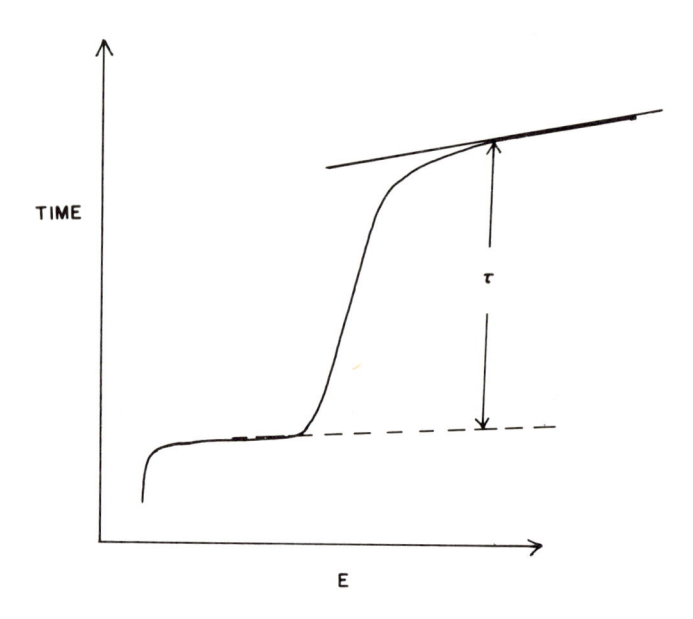

Fig. 6-7. Determination of transition times—method of Kuwana.

of Russel and Peterson and also is in agreement with the derivative techniques presented by Iwamoto (39). Even more convincing evidence of the practicality of Kuwana's method is that it gives the closest correlation of the graphical methods when results of chronopotentiometry, constant-potential electrolysis (i–$t^{1/2}$ curves), and linear potential sweep voltammetry are carefully compared at electrodes with strict linear diffusion control (38). Although the more mathematical approach of Russel and Peterson has advantages in precision, of the simpler graphical methods, that of Kuwana is certainly to be most recommended. The comments in this section, as throughout this section on chronopotentiometry, are oriented mainly toward E–t curves recorded on pen and ink equipment. The special problems associated with oscillographic measurements are treated in the original literature.

REFERENCES

1. S. Glasstone and A. Hickling, *Electrolytic Oxidation and Reduction*, Chapman and Hall, London, 1935, see, for example, Chap. 3.
2. H. F. Weber, *Wied Ann.*, **1**, 536 (1879).
3. H. J. S. Sand, *Phil. Mag.*, **1**, 45 (1901).
4. Z. Karaoglanoff, *Z. Elektrochem.*, **12**, 5 (1906).

REFERENCES

5. L. Gierst and A. Juliard, *J. Phys. Chem.*, **57**, 701 (1953).
6. W. Reinmuth, *Anal. Chem.*, **32**, 1509 (1960).
7. R. N. Adams, C. N. Reilley, and N. H. Furman, *Anal. Chem.*, **25**, 1160 (1953).
8. R. E. Parker and R. N. Adams, *Anal. Chem.*, **28**, 828 (1956).
9. R. N. Adams and J. D. Voorhies, *Anal. Chem.*, **29**, 1690 (1957).
10. J. D. Voorhies and R. N. Adams, *Anal. Chem.*, **30**, 346 (1958).
11. J. Badoz-Lambling, *Anal. Chim. Acta*, **16**, 285 (1957).
12. P. Delahay, *New Instrumental Methods in Electrochemistry*, Wiley (Interscience), New York, 1954, Chap. 8.
13. P. Delahay and G. Mamantov, *Anal. Chem.*, **27**, 478 (1955).
14. P. Delahay and T. Berzins, *J. Am. Chem. Soc.*, **75**, 2486 (1953).
15. C. N. Reilley, G. W. Everett, and R. H. Johns, *Anal. Chem.*, **27**, 483 (1955).
16. R. N. Adams, J. H. McClure, and J. B. Morris, *Anal. Chem.*, **30**, 471 (1958).
17. J. J. Lingane, *J. Electroanal. Chem.*, **1**, 379 (1959).
18. D. G. Peters and J. J. Lingane, *J. Electroanal. Chem.*, **2**, 1 (1961).
19. J. J. Lingane, *J. Electroanal. Chem.*, **2**, 46 (1961).
20. D. G. Peters and J. J. Lingane, *J. Electroanal. Chem.*, **2**, 249 (1961).
21. D. H. Evans and J. E. Price, *J. Electroanal. Chem.*, **5**, 77 (1963).
22. C. N. Reilley, G. W. Everett, and R. H. Johns, *Anal. Chem.*, **27**, 483 (1955).
23. J. D. Voorhies and N. H. Furman, *Anal. Chem.*, **30**, 1656 (1958); **31**, 381 (1959).
24. J. D. Voorhies and J. S. Parsons, *Anal. Chem.*, **31**, 516 (1959).
25. D. G. Davis, *Anal. Chem.*, **33**, 1839 (1961).
26. G. A. Ward, *Talanta*, **10**, 261 (1963).
27. D. E. Bublitz, G. Hoh, and T. Kuwana, *Chem. Ind. (London)*, **1959**, 635.
28. T. Kuwana, D. E. Bublitz, and G. Hoh, *J. Am. Chem. Soc.*, **82**, 5811 (1960).
29. G. K. L. Hoh, W. E. McEwen, and J. Kleinberg, *J. Am. Chem. Soc.*, **83**, 3949 (1961).
30. A. J. Bard, *Anal. Chem.*, **33**, 11 (1961).
31. A. J. Bard, *Anal. Chem.*, **35**, 340 (1963).
32. L. Gierst, thesis, Univ. Brussels, 1952.
33. Ref. *12*, p. 207.
34. L. E. Gierst, *Z. Elektrochem.*, **59**, 784 (1955).
35. W. H. Reinmuth, *Anal. Chem.*, **33**, 322 (1961).
36. Ref. *12*, p. 199; P. Delahay, *Discussions Faraday Soc.*, **17**, 205 (1954).
37. H. B. Mark and F. C. Anson, *Anal. Chem.*, **35**, 722 (1963).
38. J. Zimmerman, Ph.D. thesis, Univ. Kansas, Lawrence, 1964.
39. R. T. Iwamoto, *Anal. Chem.*, **31**, 1062 (1959).
40. W. H. Reinmuth, *Anal. Chem.*, **32**, 1514 (1960).
41. A. C. Testa and W. H. Reinmuth, *Anal. Chem.*, **33**, 1320 (1961).
42. R. W. Murray and C. N. Reilley, *J. Electroanal. Chem.*, **3**, 64, 182 (1962).
43. T. Berzins and P. Delahay, *J. Am. Chem. Soc.*, **75**, 4205 (1953).
44. D. H. Geske, *J. Am. Chem. Soc.*, **81**, 4145 (1959).
45. C. Furlani and G. Morpurgo, *J. Electroanal. Chem.*, **1**, 351 (1960).
46. R. M. King and C. N. Reilley, *J. Electroanal. Chem.*, **1**, 434 (1960).
47. A. C. Testa and W. H. Reinmuth, *Anal. Chem.*, **33**, 1324 (1961).
48. O. Dracka, *Collection Czech. Chem. Commun.*, **25**, 338 (1960).
49. W. K. Snead and A. E. Remick, *J. Am. Chem. Soc.*, **79**, 6121 (1957).
50. A. C. Testa and W. H. Reinmuth, *Anal. Chem.*, **32**, 1518 (1960).
51. D. Hawley and R. N. Adams, *J. Electroanal. Chem.*, **10**, 376 (1965).

52. T. O. Rouse, Ph.D. thesis, Univ. Minnesota, Minneapolis, 1961.
53. W. E. Palke, C. D. Russell, and F. C. Anson, *Anal. Chem.*, **34,** 1171 (1962).
54. H. B. Herman and A. J. Bard, *Anal. Chem.*, **35,** 1121 (1963).
55. P. Delahay and C. C. Mattax, *J. Am. Chem. Soc.*, **76,** 874 (1954).
56. P. Delahay and G. Mamantov, *Anal. Chem.*, **27,** 478 (1955).
57. W. H. Reinmuth, *Anal. Chem.*, **33,** 485 (1961).
58. C. D. Russel and J. M. Peterson, *J. Electroanal. Chem.*, **5,** 467 (1963).
59. T. Kuwana, Ph.D. thesis, Univ. Kansas, Lawrence, 1959.
60. H. B. Herman and A. J. Bard, *J. Phys. Chem.*, **70,** 396 (1966).
61. A. J. Bard and H. B. Herman, in *Polarography 1964* (G. J. Hills, ed.), Macmillan, New York, 1966, p. 373.
62. M. Paunovic, *J. Electroanal. Chem.*, **14,** 447 (1967).

7 ELECTRODE SURFACE CONDITIONS

7-1. Introduction 187
7-2. Adsorbed Hydrogen Films on Platinum and Gold 189
7-3. Oxidation of Platinum and Gold Electrodes 191
7-4. Effect of Electrode History 205
7-5. Operating Procedures for Platinum Electrodes 206
References 208

7-1. INTRODUCTION

All factors which affect the surface prior to, during, or even after the electrode reaction are of concern when one discusses electrode surface conditions. The DME has achieved its prominence due to the fact that the surface effects can be ignored (at least with regard to the past history of the electrode). One "throws away" the old surface with each falling drop. Probably the most serious problem in solid electrode methodology is associated with understanding the true electrode surface conditions and their possible effect on the electrode process. Separating those parts of the observable electrochemistry which belong only to the electroactive system from those determined by the electrode surface is often difficult. At the same time these surface problems are fascinating.

While one often draws schematic diagrams of solid electrodes with smooth, planar edges, it is well known that the actual picture is more like that represented by Fig. 7-1. The true surface is very rough in microscopic detail and consists of a maze of cracks, holes, and projections. Only rarely have polishing techniques been used which might smooth out this roughness. Such a surface will naturally tend to occlude various foreign materials from solution. Once incorporated into the obscure crevices and holes, these substances may be very difficult to remove.

Fig. 7-1. Schematic representation of actual solid electrode surface.

Depending on their electrochemical behavior, such inclusions may or may not be detrimental to the polarographic response of the working electrode. Adsorption, as well as occlusion, may be enhanced by increases in microscopic area.

In many cases following an electrolysis, the electrode is found to be coated with a more or less completely insulating film. These are sometimes easy to remove by simple chemical treatment. In other cases the electrode must be heated to high temperatures to destroy the impermeable film. Such difficulties are often encountered in the oxidation of organic compounds.

Adsorption of solution components onto solid surfaces may serve as the pathway for later incorporation of extraneous substances in the electrode matrix. Adsorption phenomena are undoubtedly present in most every application.

Platinum electrodes which have been used at highly cathodic potentials are found to be covered partially with a sorbed layer of hydrogen atoms. If these electrodes are now used without further treatment, in an anodic sense, an $i–E$ curve can be obtained for the anodic dissolution of the hydrogen-gas film. According to the type of measurement, the dissolution current may be superimposed on the anodic polarogram of the solution species and give false results. Kolthoff and Tanaka investigated this effect, which give rise to anomolous residual currents (1).

Another surface factor which so far we have happily ignored is any possible chemical reaction of the electrode material. Being "inert" metals, it is sometimes assumed that platinum and gold electrodes merely serve as electron sinks and enter into the electron-transfer reaction merely as conductors. Although suspicion has existed for years as to their truly inert character, only recently has unequivocable evidence been presented that these two noble metals are quite readily oxidized when used in voltammetry. This can occur by electrolytic oxidation during the course of an anodic polarogram or may be brought about by chemical oxidants. In both cases the oxidation may, if not taken into account, produce serious errors in the observed electrochemical response. Of all the surface effects, the electrode oxidation together with sorbed hydrogen films have been most completely examined from the viewpoint of their influence on actual polarograms. We shall examine these two problems first in some detail.

7-2. ADSORBED HYDROGEN FILMS ON PLATINUM AND GOLD

Apparently the first evidence for the presence of sorbed hydrogen films on platinum electrodes was contained in the report of Rogers et al., which pioneered the recording of solid electrode polarograms (2). These workers found that the recorded residual current curves were different depending on the applied voltage at which the electrolysis was started. In cases where the recording was initiated with cathodic potentials, the residual current was of opposite sign to that obtained by starting at anodic potentials (3). Similar effects of considerably less magnitude were seen at rotated wire electrodes.

Other investigators observed that polarograms recorded in the "forward and backward" sense were rarely identical (4–8). It remained for Kolthoff

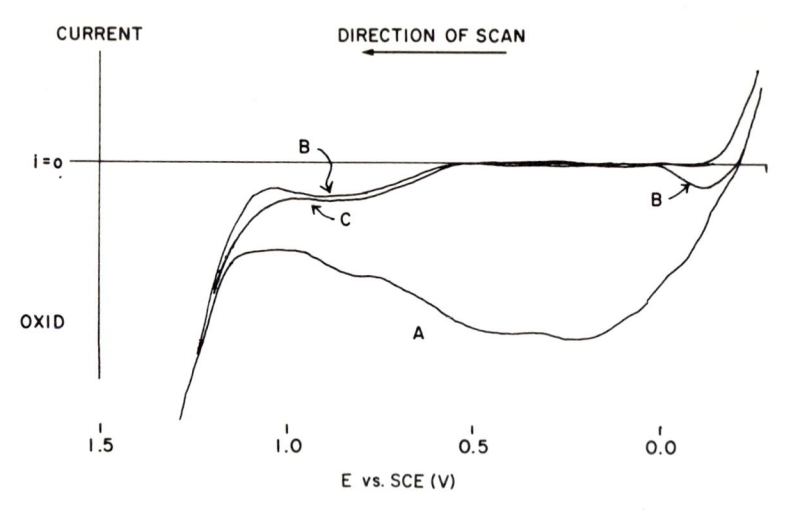

Fig. 7-2. Anodic polarograms of a platinum electrode following cathodic pretreatment. A, Pretreatment for 10 min at −2.0 V; B, same pretreatment, washed with air-free water, polarogram started at −0.3 V; C, same pretreatment followed by treatment with oxidant, polarogram started at −0.3 V. electrolyte: oxygen-free 0.1 M perchloric acid. [After Kolthoff and Tanaka, *Anal. Chem.*, **26**, 632 (1954).]

and Tanaka to give a clear-cut explanation of these effects in terms of a sorbed layer of hydrogen [and oxidation of the electrode surface, which is dealt with in the next section (1)].

Kolthoff and Tanaka showed that in air-free 0.1 M perchloric acid, a platinum electrode which had been subjected to strong cathodic pretreatment (potential held at −2.0 V vs. SCE for 10 min, during which time hydrogen was vigorously evolved at the surface) recorded a large anodic current when the polarogram was started at −0.3 V vs. SCE and continued toward more positive potentials. If, after the same cathodic pretreatment, the electrode was washed well with air-free distilled water, only a small anodic current was obtained. The anodic current could be eliminated completely by following the cathodic pretreatment with immersion of the electrode for 10 min in an acid solution containing an oxidant (ferric or ceric sulfate or bromine water). The results are summarized in Fig. 7-2. It can be seen that the anodic current between about 0.0 and 0.5 V is completely eliminated by treating the electrode with an oxidant.

The anodic current is not found after only a brief pretreatment at cathodic potentials. It is also rarely seen if a manual, point by point polarogram is measured. The potential at which the anodic current begins is quite close to the hydrogen evolution potential. With a palladium electrode which adsorbs and dissolves large amounts of hydrogen, the current level is larger than on platinum. These experimental facts are all in accord with attributing the anodic current to dissolution of a sorbed hydrogen film on the surface of the electrode.

Bauman and Shain have shown that gold electrodes behave distinctly different from platinum following cathodic pretreatment (9). A gold electrode was pretreated at −1.0 V vs. SCE for 15 sec. There was no evidence of the hydrogen dissolution current. The only "anomolous" behavior here is the oxidation of the gold surface beginning at about +0.9 V. According to Baumann and Shain, the lack of a hydrogen dissolution current is in keeping with the lack of absorption of hydrogen by gold.

The large residual currents encountered with untreated graphite electrodes can be removed by proper wax-impregnation techniques. There apparently have been no studies of residual current behavior at wax-impregnated graphite which indicate adsorbed hydrogen films.

7-3. OXIDATION OF PLATINUM AND GOLD ELECTRODES

It is remarkable that chemists over the years have held so strongly to the idea that platinum and gold are truly inert metals and do not undergo oxidation when used as indicator electrodes in potentiometry or voltammetry. The thermodynamic metal–metal oxide potentials indicate that oxidation of either of these metals is certainly possible. Indirect electrochemical evidence for such behavior has existed in the literature for years (10–16). These older data contain valuable background information. The following discussion concentrates on more recent investigations which are primarily concerned with the effects of surface oxidation on the polarographic behavior of the electrodes.

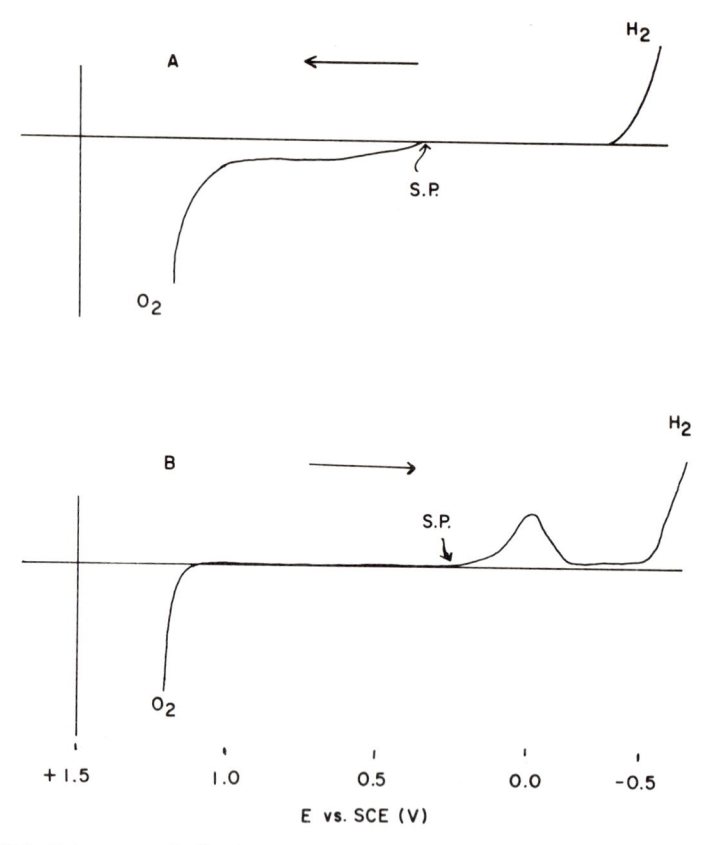

Fig. 7-3. Polarograms indicating oxide formation on platinum electrodes. A, Oxygen-free acetate buffer, pH 5, voltage sweep as indicated by arrow; B, same background, start 1.35 V, voltage sweep as indicated by arrow. [After Kolthoff and Tanaka, *Anal. Chem.*, **26**, 632 (1954).]

A platinum electrode having no sorbed hydrogen film will still show some small residual current effects when the polarogram is recorded from negative to positive potentials (backward). Kolthoff and Tanaka studied this situation at a so-called "clean electrode" (*1*). Their "clean electrode" was normally stored in 10 M nitric acid, then washed well with distilled water, and short-circuited against the reference SCE in air-free 0.1 M perchloric acid until the current decreased to nearly zero. A background polarogram of an acetate buffer solution run with the clean electrode is shown in curve A of Fig. 7-3. The following points are significant in curve A. First, while the recording was made in the backward direction and thus the electrode was initially cathodic, there is not evidence of the anodic dissolution pattern for hydrogen films. A more severe cathodic pretreatment is necessary to produce this effect. A small, well-developed anodic wave, starting at about 0.5 V, precedes the oxygen-discharge current. The half-wave potential is difficult to measure but the starting potential (S.P.) (where the foot of the wave begins) varies regularly with pH. These anodic prewaves, as they have been called, are regularly seen in recorded polarograms but are usually absent in manual determinations. These

TABLE 7-1
Potentials of Cathodic Dissolution Patterns in
Various Background Electrolytes

Background solution	pH	Starting potential[a]	Potential[a] of Pt–Pt(OH)$_2$
0.1 M HClO$_4$	1	0.67	0.67
Acetate buffer	5	0.39	0.43
Borax buffer	9	0.17	0.19
0.1 M NaOH	13	−0.03	−0.04

[a] All values in volts vs. SCE.

prewaves are due to oxidation of the platinum surface. The effect is more strikingly illustrated by curve B of Fig. 7-3. Here the polarograms were started at positive potentials, where oxygen was evolved at the surface of the electrode, and recorded in the forward sense. Curve B is for the same acetate buffer system, but similar results were obtained with a variety of background electrolytes of varying pH. All polarograms showed cathodic dissolution patterns whose starting potentials again varied approximately by 60 MV/pH unit. These cathodic-current peaks are due to reduction of the oxide film formed on the surface.

In Table 7-1, taken from the work of Kolthoff and Tanaka (*1*), are shown the starting potentials of the cathodic dissolution patterns, compared to the standard potentials of the couple Pt–Pt(OH)$_2$ given by Latimer (*17*). Although there may well be some question on the accuracy of the thermodynamic values, there is a surprising agreement which suggests that oxides are involved.

Further support of the contention that oxide (or hydrous oxide) films are formed is given by studies of the current–time relation at constant potential for the anodic prewave. The current–time curves were almost identical whether the electrode was stationary or rotated, indicating that the electrochemical reaction is neither diffusion- nor convection-controlled. This is consistent with oxide-film formation.

It was further found that after treating a clean platinum electrode with various oxidants (dichromate, ceric, permanganate, etc.), cathodic dissolution patterns could be obtained which were identical with those developed by the electrolytic oxidation. The areas under the dissolution peaks depended on the oxidant used and the time of contact between electrode and oxidizing solution. Finally, the oxide films could be removed by treatment with reducing agents. Ferrous sulfate rapidly removed oxide films, as indicated by the disappearance of the cathodic dissolution

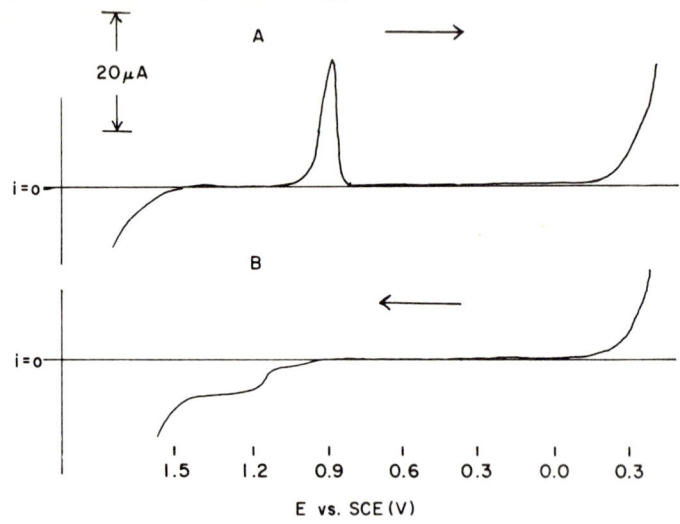

Fig. 7-4. Polarograms indicating oxide formation on gold electrodes. A, Anodic pretreatment; B, cathodic pretreatment. Direction of voltage sweep indicated by arrows; background solution, air-free 1 M perchloric acid. [After Bauman and Shain, *Anal. Chem.*, **29**, 303 (1957).]

pattern. As(III) reduced the film slowly. The oxide film can also be removed, of course, by cathodic reduction.

Bauman and Shain (9) showed that the anodic behavior of gold is very similar to platinum. Thus a small anodic wave precedes oxygen evolution when a polarogram is recorded in the backward direction in air-free 1 M perchloric acid. After anodic pretreatment, a cathodic dissolution pattern is obtained whose starting potential differs only slightly from that of the gold–gold hydroxide couple. The anodic prewave and the cathodic dissolution patterns were independent of rate of stirring but depended on the rate of voltage sweep. Figure 7-4 shows typical results obtained with a gold electrode.

When the gold electrode surface has become oxidized, normal electrode reactions may be hindered. This is well illustrated by the polarogram of Fig. 7-5, representing the oxidation of Fe(II) in 1 M perchloric acid using an oxidized gold electrode. Sweeping from positive to negative potentials, no limiting current level for Fe(II) oxidation is evident until after the cathodic dissolution peak (C) of the oxide film. The current then shifts rapidly to the anodic limiting value (point O) for Fe(II). The remainder of the polarogram is normal.

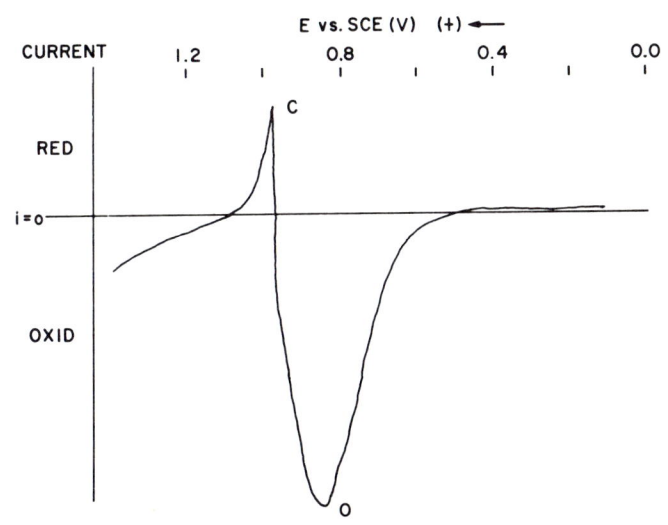

Fig. 7-5. Effect of gold oxide film on anodic polarogram of ferrous ion. Solution, air-free 1 M HClO$_4$, 2 × 10^{-4} M Fe(II). [After Bauman and Shain, *Anal. Chem.*, **29**, 303 (1957).]

Ross and Shain (*18*) examined drifting potentials with platinum indicator electrodes during potentiometric redox titrations. The platinum electrodes were first cleaned of oxide films by immersion in 0.1 M ferrous solution for 10 min. They were then inserted into solutions of various redox couples and the potential–time behavior recorded. Drifting was considered to take place if the potential was changing detectably 1 min after placing in the oxidizing solution.

It was found that drifting was not encountered until the redox potential of the solution approached 0.72–0.74 V vs. SCE. This potential, of course, is close to the platinum–platinous hydroxide potential in 1 M acid. Ross and Shain concluded that drifting potentials were caused by slow formation of oxide films. The chemical reduction of oxide films (for instance, by ferrous ion) was found to be a more rapid process than the formation.

The problem of chemical oxidation of platinum, gold, and palladium electrodes was investigated in a somewhat different fashion by Lee (*19*). It was desired to obtain a complete picture of the response of a noble metal indicator electrode, beginning with an oxide-free surface and ending with the attainment of the solution redox potential. An experimental method was set up so that the electrode could be treated cathodically in a

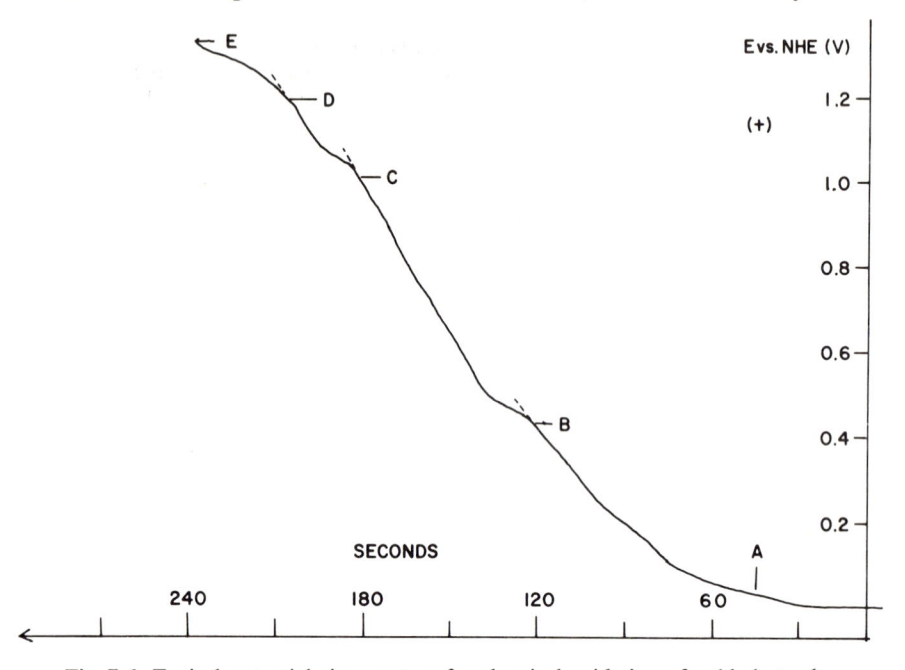

Fig. 7-6. Typical potential–time pattern for chemical oxidation of gold electrode.

solution containing the selected oxidizing solution. After stripping the electrode of oxide, the current was interrupted when the electrode was subject to chemical oxidation by the soluble redox species. The potential of the indicator electrode was measured with a pH meter and the output fed to a fast-pen-speed recorder. The electrical circuitry was practically identical with that used in chronopotentiometry (see Chapter 6). The recorded *E–t* curve showed all changes at the electrode during the period desired.

A typical potential–time pattern for the oxidation of a smooth gold-foil electrode in 1 *N* sulfuric acid is seen in Fig. 7-6. Prior to starting the recording, the electrode was cathodically reduced to the extent that strong evolution of hydrogen occurred on the surface. (The cathodic current level for this cleaning process is adjusted slightly greater than the limiting current of the soluble oxidant in solution. A recording of the stripping pattern thus shows a reduction step for the oxidant followed by oxide stripping and finally hydrogen evolution. These prereduction steps are not shown in Fig. 7-6.)

TABLE 7-2

Starting Potentials of Steps in E–t Curves of a Gold Electrode Immersed in Various Redox Systems

Redox system[a]	Potentials, V vs. NHE				
	Step B	Au$_2$O/Au	Step C	AuO/Au	Step D	Au$_2$O$_3$/Au

Redox system[a]	Step B	Au$_2$O/Au	Step C	AuO/Au	Step D	Au$_2$O$_3$/Au
In 1 N H$_2$SO$_4$, pH = 0						
4.7 × 10^{-3} M Ce^{4+}/Ce^{3+}	—	+0.42	—	+1.04	+1.33	+1.36
2.4 × 10^{-3} M	—		—		1.28	
2.4 × 10^{-4} M	0.47		—		1.23	
1.9 × 10^{-4} M	0.47		1.09		1.22	
9.5 × 10^{-5} M	0.45		1.06		1.21	
2.8 × 10^{-3} M MnO$_4^-$/Mn^{2+}	—		—		1.30	
2.8 × 10^{-4} M	0.45		1.09		1.29	
1.7 × 10^{-4} M	0.47		1.09		1.27	
2.3 × 10^{-4} M Cr$_2$O$_7^{2-}$	0.47		0.98		—	
2.6 × 10^{-4} M BrO$_3^-$	0.47		0.98		—	
In 0.1 N Na$_2$CO$_3$, pH = 11.2						
2.2 × 10^{-4} M MnO$_4^-$/Mn^{2+}	−0.29	−0.24	—	+0.38	—	+0.70

[a] The redox system usually contained equimolar quantities of the conjugate redox pair to simulate conditions midway through a potentiometric titration. No significance of "poised" systems is intended for such mixtures.

Upon interruption of the current, the potential remains relatively constant for a time (A) until chemical oxidation begins to take place (B). The pronounced potential holdups at (B) and (C) as well as the change of slope at (D) are highly persistent and reproducible. Furthermore, they are *completely independent of the chemical oxidant used.* Since they do not depend on the solution species, the potential holdups must be due to reactions on the electrode surface.

Table 7-2 summarizes the starting potentials of the holdups obtained at the gold electrode with four different oxidants in varying concentrations. For comparison, beside the columns marked steps B, C, and D, are given the potentials of the various gold–gold oxide couples, calculated to the same pH conditions.

It can be seen that starting potentials vary with concentration of the oxidant. If too concentrated, the E–t curve rises so rapidly that some of the steps are missed with a pen and ink recorder. The formal redox potentials of dichromate and bromate are not positive enough to reach step D. Considering the difficulty of measuring proper and reproducible starting potentials, the agreement between these values and the metal–metal oxide potentials was taken as a valid indication that stepwise formation of hydrous gold oxides occurs on the surface of the electrode. Further studies showed that the electrolytic oxidation of gold gave identical results.

Chemical and electrolytic oxidation of platinum electrodes was similar in nature to gold. Variations in starting potentials with concentration (and with current density for electrolytic oxidation) were much more pronounced than with gold. The stepwise formation of PtO and PtO_2 (or hydrous oxides) by chemical or electrical oxidation of platinum was postulated by Lee et al. For further details and an account of the oxidation of palladium the reader is referred to the original literature (20).

That $PtO_2 \cdot nH_2O$ is formed by electrolytic oxidation was shown beyond doubt by Altman and Busch, although their electrolysis conditions were considerably different than normally employed in voltammetry (21). Platinum electrodes were oxidized by combined direct and alternating current. Using optimum conditions of anode area equal to 4 cm² available surface, a dc current of 1 A and a total current of 2 A, the hydrated PtO_2 was deposited in an anode compartment at a rate of about 0.025 g/hr. Dilute sulfuric acid served as the electrolyte and the cell was immersed in an ice bath for cooling.

Anson and Lingane employed chronopotentiometry to demonstrate electrolytic oxidation of platinum surfaces (22). The electrode was stripped free of oxides by either cathodic pretreatment or chemical reductants and the electrode oxidized at a current density of 200 $\mu A/cm^2$. If this oxidized electrode was then made a cathode at the same current density, the resulting cathodic chronopotentiogram showed the electrolytic reduction of the oxidized surface. The cathodic E–t patterns were more well defined than the anodic, and τ cathodic was believed to be a more nearly correct measure of the quantity of oxidized platinum. The transition times for both cathodic and anodic chronopotentiograms were examined over a wide range of pH. The number of milliequivalents of oxide formed is given by

$$\frac{(i_{mA})(\tau_{sec})}{96{,}493} = \text{meq} \qquad (7\text{-}1)$$

Hence, at constant current, the transition time τ is a direct indication of the quantity of oxide. The values of both τ_{anodic} and $\tau_{cathodic}$ were found to be essentially independent of pH. The total quantities of oxide present were of the order of 0.5 mg for a 60-cm² foil electrode.

The most significant contribution of the work of Anson and Lingane is that it provided the first direct *chemical* evidence for the formation of different types of surface oxides under conditions normally employed in voltammetry. Previous work had supplied only electrochemical information or, in the investigation by Altman and Busch, had used current densities outside the province of polarography. The chemical identification of PtO and PtO₂ was carried out by stripping, or dissolving, the surface films from a previously oxidized electrode into a solution of 0.2 M hydrochloric acid and 0.1 M sodium chloride. The oxides were slowly soluble in this medium and, from the resulting solution, Pt(II) and Pt(IV) could be identified from the absorption spectra of their chloro complexes, $PtCl_4^{2-}$ and $PtCl_6^{2-}$. Standard spectrophotometric techniques allowed quantitative determination of Pt(II) and Pt(IV). Details are given in the original paper (22).

A typical determination consisted of oxidizing the electrode in one of several ways and then chemically dissolving the surface oxides. The solution was then analyzed for dissolved platinum. In addition, a cathodic chronopotentiogram was run in the electrode to determine any oxides remaining undissolved. The sum of the dissolved and remaining oxides was then compared with the value derived from the electrochemical measurements [application of Eq. (7-1)]. Within experimental error, the sum of the Pt(II) and Pt(IV) plus oxide remaining on the surface agreed with the total equivalents of oxide calculated from the chronopotentiometry. The oxides dissolved from the surface corresponded to what would be expected from PtO and PtO₂ in a molar ratio of about 6:1. Similar results were obtained whether the electrode was oxidized electrolytically or by chemical action [i.e., Ag(II), Ce(IV)]. With more severe oxidizing conditions, the total amount of oxides increased but the relative proportions of PtO and PtO₂ remained the same. This work seemingly constitutes direct, chemical proof of the stepwise formation of surface oxides on platinum surfaces.

All the work mentioned up to this point treats the chemical and anodic oxidation of platinum (or gold) as involving rather discrete oxides. There is another school of thought which describes the surface oxidation as adsorbed oxygen of varying forms [a thorough review of the pertinent literature is contained in the works by Breiter and Weininger (23), Laitinen and Enke (24), and Feldberg et al. (25)]. The precise description of such systems naturally involves many difficulties. Indeed, the present state of affairs seems to indicate that most workers in the area are becoming less explicit in their definitions of chemical entities involved in the surface oxidation. A rather happy state of affairs can be reached by referring to the surface as covered with an "oxide surface layer" or "oxygen surface layer" and describing the chemical and electrochemical properties of this layer as rigorously as possible. Such an approach may beg the question but so do many of the studies aimed at differentiating between oxide and adsorbed oxygen, for the electrochemical properties of the surface often fit both pictures.

Giner has dismissed comparisons between potentials on anodic charging curves (similar to those obtained by chemical oxidation shown in Fig.7-6) and the potentials of bulk oxide systems (26). Indeed, comparisons between thermodynamic potentials and freshly formed thin film surfaces can, at best, only be expected to show similarities. Laitinen and Enke explain the various potential holdups in the anodic charging curves (chronopotentiograms) in terms of overpotentials of the various surface reactions (24). Nevertheless, in the chemical oxidations (for instance, the data of Table 7-2) it is quite clear that the observed potential holdups are completely independent of the chemical oxidant. While Ce(IV) and MnO_4^- may, under some circumstances, liberate oxygen from acidic aqueous solutions, analytical chemists know that dichromate solutions are extremely stable and show no such tendencies. All these oxidants very well may *attack a water molecule in the presence of some suitable form of a platinum surface*. This heterogeneous reaction presumably requires participation of the platinum (as opposed to attack of the oxidant on a water molecule, liberating some oxygen intermediate, which then finds a platinum surface with which to collide). It is difficult to pinpoint the chemical (or anodic) oxidation much beyond the point of saying the platinum plays a real role in the process.

Breiter and Weininger (23) have questioned the identification of Pt(II) and Pt(IV) chlorides as conclusive evidence of PtO_2 and PtO_4 in an oxidized platinum surface [work of Anson and Lingane (22)]. The former workers propose that the dissolution can take place in the sense of the following reactions:

(a) $\qquad Pt{-}O + 2H^+ + 2e \xrightarrow{I_1} Pt + H_2O$
(b) $\qquad Pt + 4Cl^- \xrightarrow{I_2} PtCl_4^{2-} + 2e$
(c) $\qquad Pt + 6Cl^- \xrightarrow{I_3} PtCl_6^{2-} + 4e$

These are electrochemical (as opposed to chemical) dissolution processes with the requirement that the total cathodic current (I_1) be equal to the sum of the two anodic currents. The cathodic reaction takes place on the surface oxygen layer and the anodic steps on free platinum surfaces. Further, the mechanism requires that the dissolution take place in the potential range of the surface oxygen layer, i.e., in the region ca. 0.7–1.0 V. The experimental data supported this electrochemical dissolution mechanism but, again, it was concluded that such a process could occur on a discrete oxide surface; i.e., the electrochemical dissolution does not differentiate between oxides vs. surface oxygen. However, Breiter and Weininger claim that the identification of $PtCl_4^{2-}$ and $PtCl_6^{2-}$ in such experiments is not sufficient proof of pure chemical dissolution and hence discrete PtO and PtO_2. This frustrating ambiguity is found in most all studies. *Experimentally*, the results resemble hydrous oxide formation or chemisorbed oxygen surface states. In fact, the distinction between a *freshly formed* hydrous oxide and a chemisorbed oxygen layer may involve some semantics.

Probably even more important than the exact nature of the oxygen surface layer is the role it plays in electrochemical reactions. Most electrode reactions appear to be retarded by oxidized platinum surfaces, although a few show little or no effect (*27–35*). For an up to date review of the platinum oxide situation and its effects on electrode reactions the reader is referred to the review by Gilman (*58*).

If an oxidized platinum electrode is reduced chemically or electrochemically, the freshly formed platinum surface states may resemble platinized platinum. Anson showed that the properties of such an electrode may tend toward those of platinized platinum (*28*). Thus it is necessary to classify platinum electrodes more carefully than with the simple designations of preoxidized or prereduced. The following descriptions are pertinent.

1. An oxidized platinum electrode (Oxid Pt) is one which has been subjected to anodic potentials for times sufficient to film it with a surface oxygen layer (without distinguishing between oxides or chemisorbed oxygen). A similar Oxid Pt is obtained via chemical oxidation, i.e., treatment with dichromate–sulfuric acid cleaning solution, nitric acid, Ce(IV) solutions, or other strong oxidants. One almost always begins an experiment with an Oxid Pt, owing to cleaning operations. Electrical oxidation to this stage may be accomplished by constant current for a fixed time interval, or to a given anodic potential, or by potentiostatic preelectrolysis.

2. The Oxid Pt above can be converted to a reduced platinum electrode

(Red Pt) by cathodic constant-current or potentiostatic electrolysis. This Red Pt has some deposit of platinized platinum on it from the reduction process (a later discussion deals with an alternative interpretation of the platinization). The Red Pt surface can also be obtained by chemical reduction of the Oxid Pt by any reductant which is of sufficient reducing power and which does not dissolve, in any way, platinum oxide–type compounds. Typical reductants are ferrous ion, ferrocyanide, *p*-phenylenediamine, etc. How rapidly a Red Pt surface is obtained may depend on the kinetics of interaction of the particular reductant and the Oxid Pt. It is easy to see then that an Oxid Pt immersed in a solution of ferrous becomes, at least partially, a Red Pt before any anodic reactions can be studied at its surface (the reduction in this case is quite rapid). It can be remarked that mass transport may play a role in the time it takes to convert a heavily oxidized Oxid Pt to Red Pt—i.e., rotated electrodes or stirred solutions may result in more rapid reduction.

3. A second type of reduced platinum surface can be obtained from Oxid Pt but only by reduction with chemical systems which dissolve platinum oxide–type surfaces. Such an electrode is called a stripped electrode (Stripped Pt) since the surface oxygen layer is dissolved or stripped off. This is accomplished, for instance, by dissolving the Oxid Pt in iodide solutions or hot 12 M hydrochloric acid. The surface oxygen film is dissolved as the corresponding platinum iodides or chlorides. This Stripped Pt does not have any platinization and differs in this respect from Red Pt. (For it usage, following the chemical stripping, it is rinsed with water and then immersed in the test solution.)

In addition to the three types of platinum surfaces distinguished above, there may be additional modifications due to age effects (it is assumed that the surfaces described were freshly prepared). It is clear that aged Red Pt may differ from their freshly formed counterparts and aged Oxid Pt also differ. When aged, it is assumed that the particular electrode surface has been allowed to sit in some particular supporting electrolyte for a specified time. If these solutions are air- or oxygen-saturated, additional questions may arise as to their surface condition (*37,38*).

The platinization concept is a very fruitful one in explaining the activity of various platinum surfaces. It provides a far more satisfying explanation of increased reversibility in certain reductions than does the surface oxide electron bridge theory advanced previously (*35,39,40*). In slightly different terms, other workers have described surface changes which resemble platinization (*41–43*). French and Kuwana investigated the lifetimes of some of the activated states using cyclic voltammetry (*44*).

Anson and co-workers have indicated that the surface oxidation may play a distinct role in prohibiting adsorption of various species. In some cases preadsorption seems to be a step without which the overall electrode reaction cannot occur. Hence oxidized surfaces seriously inhibit the electrode reaction. According to Anson and Schultz, this is the case with the oxidation of oxalic acid in acid media. The adsorption of oxalic acid is inhibited not only by surface oxidation of the platinum but by readily adsorbed extraneous species such as amyl alcohol, chloride, or thiocyanate ions (*30*). Anson has also recently discussed the inhibition of surface oxidation by specifically adsorbed anions such as bromide ion. The adsorbed bromide ion prevents surface oxidation presumably by preferential adsorption. This, in turn, increases the rate of an oxidation such as that of Co(II) EDTA, which is inhibited by surface oxide layers. It would appear that the bromide also exerts a specific catalytic effect on this reaction, since other adsorbable species, such as amyl alcohol, which inhibit surface oxidation, do not, however, increase the rate for the oxidation of Co(II) EDTA. It was somewhat surprising that no bromide was incorporated in the Co(III) complexes, showing that the bromide did not participate in an electron bridge transfer, as might have been expected (*34*). Further examples of inhibition of surface oxidation by a variety of organic species have been given by Breiter (*45*).

The study of Feldberg et al. presents an interpretation of surface oxidation and reduction entirely different from that of the platinization picture (*25*). These workers examined specifically the coulombs required to form surface oxygen layers (Qa) compared to the coulombs used in cathodic dissolution (Qc). Vetter and Berndt had shown a ratio Qa/Qc of ca. 2 using constant-current methods (*46*), whereas the first cyclic voltammetry studies of Will and Knorr (*47*) and Bold and Breiter (*48*) had indicated a ratio of 1. Feldberg et al. found a maximum ratio $Qa/Qc = 2$ with long periods of anodic pretreatment. With repeated cycling techniques (constant-current or triangular wave potentiostatic) the ratio decreased to ca. 1. In summary, Feldberg et al. concluded that the results could be explained best by considering both the film formation and dissolution to consist of two one-electron stages, as shown below:

Surface oxidation:

slow: $\quad\quad\quad Pt + xH_2O \rightarrow Pt(OH)_x + xH^+ + xe$

fast: $\quad\quad\quad Pt(OH)_x \rightarrow Pt(O)_x + xH^+ + xe$

Cathodic reduction:

fast: $Pt(O)_x + xH^+ + xe \rightarrow Pt(OH)_x$
slow: $Pt(OH)_x + xH^+ + xe \rightarrow Pt + xH_2O$

Thus the first stage of the reduction is just the reverse of the second stage of surface oxidation and it is fast. The second stage of reduction is the slow step (reverse of the first step of formation). With constant-current reduction of the oxidized surface, the slow step does not occur and the surface becomes half-reduced; i.e., $Qa/Qc = 2$.

Three distinct surface states can thus be defined. A clean electrode (no surface oxidation) is obtained by holding the potential of an oxidized electrode at sufficiently cathodic potentials for lengthy periods of time. (In the work of Feldberg et al. these "cathodic" potentials were 0.7 V for 10 hr or +0.6 V for 3 hr. Clean electrodes have also been obtained by using more drastic cathodic potentials for shorter times.) Clean electrodes have a Qa/Qc ratio of 2.

An oxidized electrode is obtained by anodization at applied potentials of +1.0 V or greater. A half-reduced surface is obtained by constant-current reduction (to a potential of +0.5 V) or via constant-potential electrolysis at +0.6–0.7 V for short periods. The half-reduced electrode, corresponding to the form $Pt(OH)_x$, has a ratio Qa/Qc of 1.

The most interesting point in this work is that Feldberg et al. contend it is this half-reduced surface which enhances electrode reactions (i.e., it is the counterpart of Anson's platinized surface.) Although the data of Feldberg et al. are very convincing in terms of formation of the half-reduced state, it is difficult to see what specific property of such a surface is instrumental in facilitating electron-transfer processes. This interpretation brings up additional questions regarding the nature of the surface state. On the other hand, the properties of platinized platinum in this regard have been known for years and provide a somewhat more immediately satisfying physical picture of the enhancement effects.

Veselovskii and co-workers have interpreted the role of surface oxides in electrode reactions. Using ^{18}O tracers, Rosental and Veselovskii showed that the surface oxides formed at low potentials (0.8–1.2 V) did not take part in subsequent oxygen evolution on the surface. Surface oxygen layers formed at higher potentials (1.8 and 1.9 V) did, however, participate in the oxygen evolution (49). In a series of papers, Veselovskii and co-workers developed a concept where the oxide layers carried out chemical oxidations in certain instances, with the current in the electrode reaction being used to replenish the surface oxides. For relatively low potential reactions, such as Fe(II) oxidation, direct electron transfer was assumed (*50,51*). Very recently Kuwana has indicated it may be possible to precisely pinpoint potentials at which oxygen evolution occurs on oxidized platinum surfaces by observing the light emission in the electro-oxidation of luminol (*52*).

It is evident from the discussions of the previous pages that the nature and role of surface oxidation on noble metal electrodes is far from settled. These questions are the subject of vigorous research at present. It is likely that some of the work cited herein may well be shown to be incorrect by the time this manuscript is in print. Nonetheless, the material presented should serve as a background and framework for evaluating future developments in the area.

7-4. EFFECT OF ELECTRODE HISTORY

Although the previous two sections might logically be included with past history of the electrode, the effects considered here are other than hydrogen and oxide films which may result from previous electrolyses or handling of the electrode.

Ignition of platinum electrodes is sometimes used as a cleaning procedure. This may produce some surface oxidation. Its effect has not been carefully studied, but Griess et al. reported shifts of nearly 100 mV in the deposition potential of traces of silver following ignition of platinum electrodes (*53*). Other workers had noted similar effects (*54*). Rogers and co-workers found that the original behavior was completely restored again after two anodic treatments of the electrode in cyanide media.

Especially in the oxidation of organic compounds, products of the electrode reaction may be deposited as films on the electrode surface. Such effects have been frequently reported (*6,55–57*). In most cases these films are readily removed by treatment with chromic or nitric acid. Some deposits may necessitate ignition of the electrode followed by an acid wash. In general these films are non-conducting and, unless removed, render the electrode useless for further electrochemistry.

It is possible to incorporate impurities into the electrode matrix from previous depositions. This may be the result of intentional deposition reactions. It also can be the result of stringent cathodic pretreatment. Once deposited, these materials may be very difficult to remove. Griess et al. (*53*) reported that in some cases platinum electrodes tenaciously retained traces of radiosilver. The activity could not be removed completely even by strong anodization in cyanide media. In view of the known trace metal contamination of even the most carefully controlled aqueous preparations, prolonged cathodic pretreatment at highly negative potentials should be used with caution. In this connection it should be remarked that mercury plated on a platinum electrode is very difficult to remove. Strange residual currents and other anomalies are often found with such electrodes even after prolonged chemical and anodic cleaning. Removing mercury from gold surfaces is not feasible.

7-5. OPERATING PROCEDURES FOR PLATINUM ELECTRODES

It should be evident from the discussion of this chapter that it is imperative to adopt some standard set of experimental procedures when using platinum electrodes. The total procedure includes strict attention to details of (1) electrode cleaning, (2) electrical or chemical pretreatment, and (3) compatibility of the pretreatment steps with the actual electroactive test system. At first glance this seems like a great deal of intricate detail. In reality it involves no more than the ordinary points of technique in any chemical procedure designed for quantitative operation. It is true that until recently much of this technique with platinum electrodes was largely "black magic." It is only in the last few years that the nuances of surface conditions have been understood at least well enough for pretreatment procedures to be appreciated.

One of the undisputed points with regard to platinum electrode methodology (most of the following discussion also applies to gold and noble metal electrodes in general) is that *reproducible* results are almost always obtained provided the general pretreatment of the electrode is duplicated each time. (Results, while reproducible, may not always be understood completely under these conditions.) Hence it is most sensible to "begin at the beginning" and clean every platinum electrode prior to using it (for instance, at the beginning of a daily period of usage). Thus the practice of storing electrodes overnight or for longer periods in air or solution and then applying pretreatments directly (without prior cleaning) seems unwise to the writer. With the usual cleaning methods one knows what will happen to the surface (it will become heavily oxidized), and variable history effects from storage are immediately removed. The time required for the cleaning procedure is minor.

Several cleaning procedures are favored with various workers. The writer prefers immersion in strong dichromate–sulfuric acid "cleaning solution" for several minutes followed by rinsing with tap water and finally three portions of distilled water. Objections to "cleaning solution" logically include the comment that "dichromate adsorption will occur." This may be true but it is the writer's opinion that in almost any cleaning procedure something will surely be adsorbed. In the very practical sense adsorption is difficult to eliminate, and one may as well live with a routine situation in which this adsorption can be always suspect. The advantages of "cleaning solution" are that it removes grease and oil as well as many of the organic films prevalent in the oxidation of aromatic compounds. It oxidizes the platinum surface strongly, and this is a good starting point for pretreatments.

Hot nitric acid has been recommended and is presumably satisfactory. Anson and King have made a good point for the use of hot aqua regia to bring electrodes to a standard condition before pretreatment (36). Aqua regia dissolves any finely divided platinum, which "cleaning solution" does not. Cleaning procedures which include drying with acetone or the use of alcohol have nothing to recommend them. The purpose of cleaning is to oxidatively destroy any organic materials on the surface (including grease, dust, etc.) and, at the same time, to strongly oxidize the electrode surface. The latter state is one which can now be readjusted by a variety of pretreatments.

The pretreatments vary with the purpose for which the electrode is to be used. In ordinary voltammetry, if one is mainly interested in reproducible i–E curves, the following procedure is useful. Starting with a dichromate–sulfuric acid cleaning and rinsing, the electrode is immersed in the test solution. Assuming that an anodic oxidation is the experiment to be run, the heavily oxidized surface will normally begin to be reduced by the reductant in solution. To hasten and ensure completion of this process it is wise to add electrical pretreatment. The applied potential is adjusted to a value where the surface oxide is reduced, even to the point where hydrogen would normally be evolved. (This potential obviously varies with the test system, but a typical value of E_{app} would be ca. -0.2 V vs. SCE.) With the polarograph connected one can observe the extent of the cathodic processes. After a sufficient time to reduce surface oxides (if the value of E_{app} is < hydrogen evolution, this is the time when the cathodic current decays to zero), the potential is now moved *slightly* anodic (ca. $+0.05$ V vs. SCE) to reoxidize any sorbed hydrogen now on the surface. The completion of this process is evident when the anodic current for hydrogen dissolution decays to close to zero. The electrode is now free of surface oxide and, in addition, has little or no sorbed hydrogen. The test solution is now agitated to remove any concentration gradients of electroactive species at the electrode if stationary polarograms are to be run. This type of pretreatment obviously leaves a film of platinization on the surface if one follows the picture given by Anson et al.

Oxidized surfaces also may be reduced with constant current to predetermined potentials or for fixed periods of time. Finally, chemical reduction (with platinization) is obtained if the cleaned electrode is immersed in ferrous sulfate solutions, then washed and immersed in the test solution. Removal of all oxides is obtained, as mentioned, by treatment with iodide solutions or better, hot 12 M hydrochloric acid. Individual electrode preparation can vary from one experimenter to another and still result in comparable surfaces. Thus starting with a heavily oxidized surface (cleaning), followed by strong cathodization, and finally anodic stripping of sorbed hydrogen results in essentially identical surfaces. The important point is to be able to specify the nature of the surface as closely as possible just prior to use with the test system. The original literature contains further details on pretreatments.

REFERENCES

1. I. M. Kolthoff and N. Tanaka, *Anal. Chem.*, **26**, 632 (1954).
2. L. B. Rogers, H. H. Miller, R. B. Goodrich, and A. F. Stehney, *Anal. Chem.*, **21**, 777 (1949).
3. The terminology "forward & backward" polarograms is avoided here; for definitions see Refs. *1, 4–8*.
4. I. M. Kolthoff and J. Jordan, *J. Am. Chem. Soc.*, **74**, 382 (1952); **75**, 1571 (1953).
5. T. Tsukamoto, T. Kambara, and I. Tachi, *Proc. 1st Polarog. Congr., Prague, 1951*, p. 540.
6. S. S. Lord and L. B. Rogers, *Anal. Chem.*, **26**, 284 (1954).
7. O. H. Müller, *J. Am. Chem. Soc.*, **69**, 2992 (1947).
8. D. B. Julian and W. R. Ruby, *J. Am. Chem. Soc.*, **72**, 4719 (1950).
9. F. Bauman and I. Shain, *Anal. Chem.*, **29**, 303 (1957).
10. F. P. Bowden, *Proc. Roy. Soc. (London)*, **A125**, 446 (1929).
11. A. Hickling, *Trans. Faraday Soc.*, **41**, 333 (1945).
12. A. Hickling and W. H. Wilson, *J. Electrochem. Soc.*, **98**, 425 (1951).
13. S. E. S. El Waakad and S. H. Emara, *J. Chem. Soc.*, **1952**, 461.
14. S. E. S. El Waakad and Shams El Din, *J. Chem. Soc.*, **1954**, 3098.
15. A. Hickling, *Trans. Faraday Soc.*, **42**, 518 (1946).
16. J. A. V. Butler (ed.), *Electrical Phenomena At Interfaces*, Methuen, London, 1951.
17. W. M. Latimer, *Oxidation States of the Elements and Their Potentials in Aqueous Solution*, Prentice-Hall, Englewood Cliffs, N.J., 2nd ed., 1952, p. 205.
18. J. W. Ross and I. Shain, *Anal. Chem.*, **28**, 548 (1956).
19. J. K. Lee, Ph.D. thesis, Princeton Univ., Princeton, N.J., 1955.
20. J. K. Lee, R. N. Adams, and C. E. Bricker, *Anal. Chim. Acta*, **17**, 321 (1957).
21. S. Altman and R. H. Busch, *Trans. Faraday Soc.*, **45**, 720 (1949).

REFERENCES

22. F. C. Anson and J. J. Lingane, *J. Am. Chem. Soc.*, **79**, 4901 (1957).
23. M. W. Breiter and J. L. Weininger, *J. Electrochem. Soc.*, **109**, 1135 (1962).
24. H. A. Laitinen and C. G. Enke, *J. Electrochem. Soc.*, **107**, 773 (1960).
25. S. W. Feldberg, C. G. Enke, and C. E. Bricker, *J. Electrochem. Soc.*, **110**, 826 (1963).
26. J. Giner, *Z. Elektrochem.*, **63**, 386 (1959).
27. B. Baker and W. MacNevin, *J. Am. Chem. Soc.*, **75**, 1476 (1953).
28. F. C. Anson, *Anal. Chem.*, **33**, 934 (1961).
29. M. Breiter, *J. Electrochem. Soc.*, **109**, 425 (1962).
30. F. C. Anson and F. A. Schultz, *Anal. Chem.*, **35**, 1114 (1963).
31. J. Giner, *Electrochim. Acta*, **4**, 42 (1961).
32. J. J. Lingane, *J. Electroanal. Chem.*, **1**, 379 (1960).
33. F. C. Anson and J. J. Lingane, *J. Am. Chem. Soc.*, **79**, 1015 (1957).
34. F. C. Anson, *J. Electrochem. Soc.*, **110**, 436 (1963).
35. F. C. Anson, *J. Am. Chem. Soc.*, **81**, 1554 (1959).
36. F. C. Anson and D. M. King, *Anal. Chem.*, **34**, 362 (1962).
37. J. J. Lingane, *J. Electroanal. Chem.*, **2**, 296 (1961).
38. D. T. Sawyer and L. V. Interrante, *J. Electroanal. Chem.*, **2**, 310 (1961).
39. D. G. Davis, *Talanta*, **3**, 335 (1960).
40. I. M. Kolthoff and E. R. Nightengale, *Anal. Chim. Acta*, **17**, 329 (1957).
41. M. Bonnemay, *Z. Elektrochem.*, **59**, 798 (1955).
42. S. Shibata, *Nippon Kagaku Zasshi*, **79**, 239 (1958); **80**, 453 (1959).
43. S. Shibata, *Kogyo Kagaku Zasshi*, **36**, 525 (1963).
44. W. G. French and T. Kuwana, *J. Phys. Chem.*, **68**, 1279 (1964).
45. M. Breiter, *J. Electrochem. Soc.*, **109**, 425 (1962).
46. K. J. Vetter and D. Berndt, *Z. Elektrochem.*, **62**, 378 (1958).
47. F. G. Will and C. A. Knorr, *Z. Elektrochem.*, **64**, 258 (1960).
48. W. Bold and M. Breiter, *Electrochim. Acta*, **5**, 145 (1961).
49. K. I. Rozental and V. I. Veselovskii, *Dokl. Akad. Nauk SSSR*, **111**, 637 (1956).
50. T. I. Borisova and V. I. Veselovskii, *Zh. Fiz. Khim.*, **27**, 1195 (1953).
51. K. I. Rozental and V. I. Veselovskii, *Zh. Fiz. Khim.*, **27**, 1163 (1953).
52. T. Kuwana, *J. Electroanal. Chem.*, **6**, 164 (1963).
53. J. C. Griess, J. T. Byrne, and L. B. Rogers, *J. Electrochem. Soc.*, **98**, 447 (1951).
54. M. T. Simnad and V. R. Evans, *Trans. Faraday Soc.*, **46**, 175 (1950).
55. J. F. Hedenberg and H. Greiser, *Anal. Chem.*, **25**, 1355 (1953).
56. R. E. Parker and R. N. Adams, *Anal. Chem.*, **28**, 828 (1956).
57. R. N. Adams, J. H. McClure, and J. H. Morris, *Anal. Chem.*, **30**, 471 (1958).
58. S. Gilman, in Electroanalytical Chemistry, Vol. 2 (A. J. Bard, ed.), Dekker, New York, 1967, p. 111.

PART 2
EXPERIMENTAL AND APPLICATIONS

8 INVESTIGATION OF ELECTRODE PROCESSES

8-1.	Introduction	213
8-2.	Evaluation of Diffusion Coefficients	214
8-3.	Correlation of Electroanalytical Techniques: Electrode Reactions Without Chemical Complications	231
8-4.	Determination of Heterogeneous Rate Constants	240
8-5.	Electrode Processes with Coupled Homogeneous Chemical Reactions	244
8-6.	Physicochemical Methods for Studying Electrode Reactions . .	255
References		262

8-1. INTRODUCTION

The title of this chapter encompasses a wide range of topics and yet is too restrictive. The purpose of this material is to examine techniques and approaches used in investigating electrode processes at solid electrodes. A correlation of electrochemical methods is included to illustrate pitfalls in solid electrode studies. An attempt is made to critically evaluate the information derived from the various approaches. Finally, some measurements which are preliminary to the study of the actual electrode process are properly included here.

First, consider the bare minimum of problems to be examined in the study of a general electrochemical reaction at any electrode surface. There are at least three finite stages to every electrode reaction:

1. Mass transfer of reactants to the electrode surface.
2. Electron- or charge-transfer step.
3. Mass transfer of products away from the electrode.

This picture is very oversimplified. Reactants (as well as products) may be adsorbed on the electrode surface. The extent to which adsorption plays a part, with the attendant inequality of concentrations and electro-activities of adsorbed and solution species, plays an important role in many processes. Prior to step 2, one frequently has homogeneous chemical reactions or precursor reactions in which it is necessary that the bulk species undergo a chemical transformation to an electroactive form. After step 2 are follow-up chemical reactions which may consume the initial product and produce electrochemically inactive materials or regenerate starting materials (catalytic reactions). Finally, the specific effects of the solid electrode on the above processes may involve the actual surface conditions which exist during the electrolysis.

A complete analysis of an electrode mechanism involves a great deal more than the general descriptions above. A thorough picture would include, among other studies, careful evaluation of the rates of each of the steps mentioned, measurements of the various electrochemical parameters which characterize the charge transfer, and the influence of the solution environment on these properties. In practice one must be satisfied with far less than a complete study. In fact, in many cases one is delighted even to be able to evaluate n_T, the total number of electrons transferred. Hence the following pages will discuss partial or general examinations of the overall picture of various electrode reactions. There are very few examples in the literature which can be called complete in all respects.

8-2. EVALUATION OF DIFFUSION COEFFICIENTS

The determination of diffusion coefficients (D values) is not normally considered to be a part of electrode mechanism studies. It is clear, however, that much information about an electrode reaction is derived from limiting current measurements. For evaluation of n_T from a limiting current one should have an accurate D value. Furthermore, the electrode area, A, must also be known. Unless ones uses the geometric A, it must be evaluated from a limiting current for a substance whose D is known. A few orbits around this circular path with practical solid electrode systems convinces one that there are difficulties involved. However, several test systems exist with D values known precisely enough to carry out satisfactory area determinations. A suitable method for determining A was discussed in Chapter 3, Section 3-2. Areas are also effectively determined via chronopotentiometry.

In theory, knowing the electrode area (and n) one could evaluate D from any of the following electroanalytical procedures:

1. $it^{1/2}$ curves at constant potential.
2. i_p measurement from peak voltammetry.
3. Chronopotentiometry.
4. i_L measurement at a RDE.

Method 2, using peak voltammetry, is the least desirable of the four techniques. The *experimental* values of the Randles–Sevcik constant at various electrodes vary from 2.72×10^{-5} to as low as 2.52. Since errors in absolute values of i_p are squared in the calculation of D, this range of 2.52–2.72 can give rise to a 16% error in calculating D. An average uncertainty of about 6% (corresponding to a Randles–Sevcik constant of 2.64) is very likely. Until more data on the experimental deviations of the Randles–Sevcik constant are available, peak voltammetry is obviously a poor approach to correct D-value determinations. In addition, even slight departures from reversibility cause peak currents to be completely unreliable.

The RDE is ideally suited to determining D values. The equation for i_L at the RDE is rigorously derived from hydrodynamics, and it is known that the dependence of i_L on several factors in the equation can be verified experimentally to $\pm 1\%$ or better. When one questions absolute values determined with the RDE the facts are not clear. Presumably one should use the Riddiford correction to the Levich equation. Second, one has the usual problems in measuring the area of the disk. Any serious wobble in a RDE introduces questions as to absolute i_L values. A critical evaluation of "ordinary" RDE's in current use is necessary to evaluate the reliability of D-value measurements by this method.

How difficult it is to decipher the sources of the small errors in these measurements is indicated in Table 8-1. These data can be considered as representative—they were done with extreme care but on unshielded electrodes with the usual electroanalytical techniques. It is surprising to note that RDE values agree best with $D(i_p)$. This must be viewed with caution however. $D(\text{RDE})$ would equal $D(it^{1/2})$ if the area of the RDE were 0.121 cm² instead of the 0.126 cm² used in the calculations. This corresponds to a measured diameter of 0.385 cm rather than 0.400 cm. Such an error is indeed greater than the experimental error in measuring the size, but the disk was not ideal, having some observable eccentricity. A further check of D for the 0.52 M sulfuric acid solution showed a value

TABLE 8-1
Determination of Diffusion Coefficients of o-Dianisidine by Various Methods[a]

[H₂SO₄]	Viscosity	$D(i_p)\eta$	$D(\text{RDE})\eta$	$D(it^{1/2})\eta$
2.18 M	1.334	0.474	0.468	0.478
1.91 M	1.286	0.460	0.445	0.490
1.61 M	1.219	0.437	0.453	0.475
1.09 M	1.091	0.457	0.443	0.478
0.52 M	0.989	0.452	0.452	0.480
Mean		0.452 ± 0.008	0.452 ± 0.006	0.480 ± 0.004

[a] In varying solutions of sulfuric acid; $D(i_p)$, measured from peak voltammetry, 2.00 V/min scan rate; $D(\text{RDE})$, at 40 rps; $D(it^{1/2})$, at $E_{app} = 0.750$ V vs. SCE. All D values multiplied by η, viscosity (in cP) of given solution, all $D\eta \times 10^5$ cm²-cP/sec. All measurements at carbon paste electrodes. Data from Zimmerman, (4).

of 0.485 cm²/sec via chronopotentiometry. This checks well with the $D(it^{1/2})$ from Table 8-1 ($D/\eta = 0.480/0.989 = 0.485$ cm²/sec).

With the limited experience on RDE systems, the most reliable measurement of D seems to be via $it^{1/2}$ curves. What is believed to be the most reliable diffusion-coefficient data at solid electrodes was obtained in this manner by von Stackelberg and co-workers (1). Although area measurements are still critical, problems associated with charging current are absent since the double layer is charged in a time far shorter than the time in which one begins to follow the current–time curve. The following general procedure is used in the writer's laboratory for unshielded electrodes where semiinfinite linear diffusion is, at best, only approached. Using a controlled-potential polarograph (or applying an uncontrolled voltage so that, regardless of the initial large iR drop, the instantaneous current will always be maximal), the current–time curve is followed for about 50 sec. Values of $(it^{1/2})$ at about 5-sec intervals are plotted vs. time. This plot, for perfect linear diffusion, should be a horizontal straight line, but this is rarely the case with the unshielded electrode. The plot of $it^{1/2}$ vs. t is extrapolated to $(it^{1/2})_0$ at zero time and this value used to calculate D via the appropriate form of Eq. (3-16). A more convenient form of Eq. (3-16) in which the current is expressed in microamperes and the concentration in millimoles/liter (i.e., if the solution is 3.2 mM, the substitution in the equation is numerically 3.2) is given by

$$it^{1/2} = 54.5 \times 10^3 nAD^{1/2}C^b \qquad (8\text{-}1)$$

In Eq. (8-1), t is still in seconds, A in cm^2, and D in cm^2/sec. The multiplier (54.5 × 10^3) is equal to the Faraday, 96.5 × 10^3, divided by $\pi^{1/2}$. Rearranged for convenient calculation of D, Eq. (8-1) becomes

$$D^{1/2} = \frac{(it^{1/2})_0}{54.5 \times 10^3 nAC^b} \qquad (8\text{-}1a)$$

with the units mentioned above.

Shain and Martin obtained excellent results for D values from analysis of i–t curves (2). Their work involved the hanging mercury drop electrode, but the equation for instantaneous current varies from Eq. (3-16) only by a term involving the radius of the drop. Shain and Martin plotted i vs. $1/t^{1/2}$ and evaluated D from the slope of the straight line.

In the chronopotentiometric evaluation of D, one measures the τ at several values of current density and checks that the product $i\tau^{1/2}$ is independent of i or i_0. The value of D is then calculated from Eq. (6-6) or the rearranged form,

$$D^{1/2} = \frac{i_0 \tau^{1/2}}{8.55 \times 10^4 nC^b}$$

where i_0 is in microamperes per square centimeter and again concentration is in millimoles per liter. All precautions pertinent to chronopotentiometry should be used; i.e., concentrations should be 5–10 mM and transition times for unshielded electrodes restricted to ca. 5–30 sec.

All the above evaluations assume that D is independent of the concentration of diffusing species. This is, of course, not true, but since concentrations of electroactive species are usually all in the millimolar range, not too much trouble is encountered in this regard. The D values obtained will not agree with those calculated for infinite dilution. The change in D of the diffusing species with type and concentration of supporting electrolyte is an important variation.

The $it^{1/2}$ measurement in nonaqueous solvents, especially acetonitrile, has serious problems. This is well illustrated by some recent data of Papouchado shown in Fig. 8-1. In low-viscosity solvents such as acetonitrile, the $it^{1/2}$ product is by no means constant, but increases very rapidly during the entire course of the normal measurement. Extrapolation to $t = 0$ is a tenuous procedure in the case of acetonitrile or N,N-dimethylformamide. On the other hand, more viscous solvents, such as dimethylsulfoxide and especially propylene carbonate, are seen to obey theory and are excellent for this type of study. A cross check of the D value of

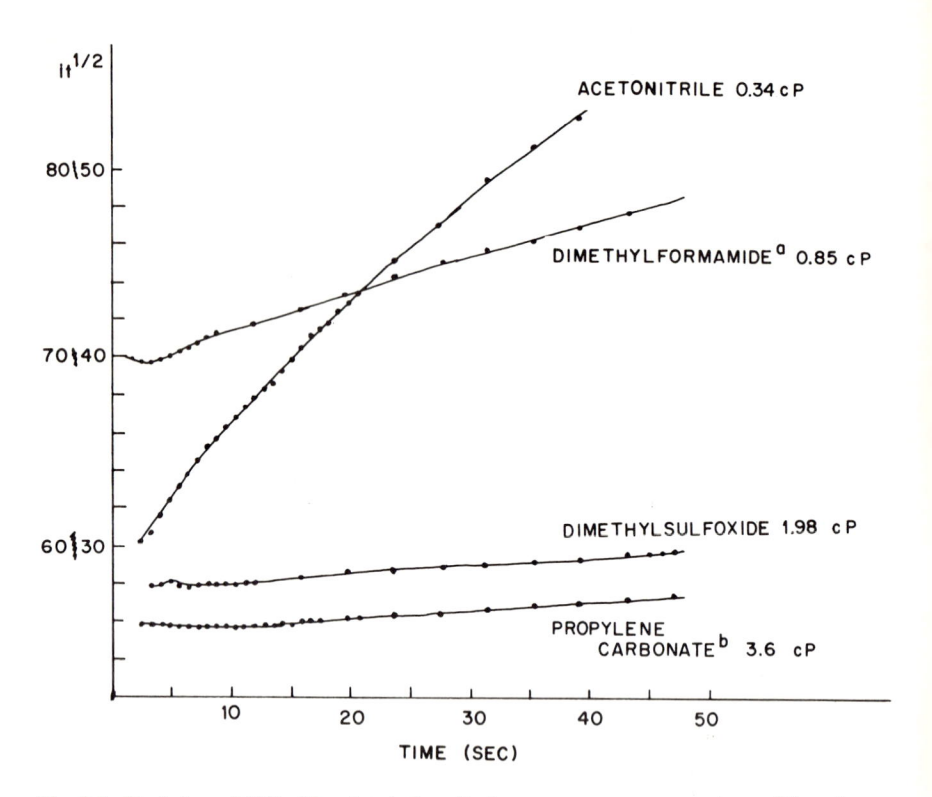

Fig. 8-1. Variation of $it^{1/2}$ with solvent viscosity for some nonaqueous systems. Viscosities from standard tables except as noted: a, viscosity measured with 0.1 M tetraethylammonium perchlorate as supporting electrolyte; b, viscosity measured with 0.5 M tetraethylammonium perchlorate as supporting electrolyte.

trianisylamine in propylene carbonate (0.1 M tetraethylammonium perchlorate) via $it^{1/2}$ gave $0.18_6 \times 10^{-5}$, whereas the RDE value was $0.17_8 \times 10^{-5}$ cm²/sec. These numbers were obtained on two different platinum electrodes (Beckmann) whose areas were calibrated via ferrocyanide in 2 M KCl (99).

The report of von Stackelberg et al. is classical in the area of D values determined from i–t curves. The results for silver ion and ferrocyanide have been cross-checked by several electroanalytical techniques. The value of ferrocyanide in 1–3 M potassium chloride provides a reference standard for area calculations, etc. It is recommended that all electrode area measurements be calculated from an $it^{1/2}$ determination on ferrocyanide in 2 M KCl, using the D value of von Stackelberg et al. listed in

EVALUATION OF DIFFUSION COEFFICIENTS 219

Table 8-2. Additional D values for Pb(II), Cd(II), Zn(II), Tl(I), IO$_3^-$, and quinone are contained in the original literature (1).

In Table 8-3 are collected a considerable number of D values from the literature. The table is selective in that electroactive species normally only of interest to DME polarography are omitted. It is critical in that

TABLE 8-2
Diffusion Coefficients of Several Systems from i–$t^{1/2}$ Curves[a]

Electroactive species	Supporting electrolyte	$D \times 10^5$ cm²/sec
Ferrocyanide ion[b]	0.05 M KCl	0.662 ± 0.004
	0.1 M KCl	0.650 ± 0.002
	0.5 M KCl	0.639 ± 0.003
	1.0 M KCl	0.632 ± 0.003
	2.0 M KCl	0.629 ± 0.003
	3.0 M KCl	0.621 ± 0.003
Ferricyanide ion[c]	0.05 M KCl	0.765 ± 0.001
	0.1 M KCl	0.763 ± 0.005
	0.5 M KCl	0.770 ± 0.002
	1.0 M KCl	0.763 ± 0.002
	2.0 M KCl	0.750 ± 0.002
	3.0 M KCl	0.739 ± 0.003
Silver ion[d]	0.01 M KNO$_3$	1.605 ± 0.008
	0.05 M KNO$_3$	1.566 ± 0.003
	0.1 M KNO$_3$	1.553 ± 0.005
	0.5 M KNO$_3$	1.559 ± 0.004
	1.0 M KNO$_3$	1.571 ± 0.002
	2.0 M KNO$_3$	1.541 ± 0.006

[a] Data selected from von Stackelberg et al. (1).
[b] Ferrocyanide concentration 4.0 mM, upward diffusion.
[c] Ferricyanide concentration 0.4 mM for 0.05 M KCl, others 4 mM ferricyanide; downward diffusion because of density gradient developed with lighter ferrocyanide.
[d] Silver ion concentration variable 0.25–4 mM.

only values which have been checked several ways are generally listed. Where other values are known, but out of line, they are not given. In the chronopotentiometric studies, both shielded and unshielded electrodes have been used. Nor are the electrode material or rotation speed in the case of RDE values identified in the table, and it is necessary to check the original literature for these details.

TABLE 8-3
Selected Diffusion Coefficients of Various Electroactive Species

Electroactive species	Supporting electrolyte	Measurement technique[a]	$D \times 10^5$ cm²/sec[b]	Ref.
Inorganic substances				
Ag(I)	0.1 M KNO₃	A1	1.55₃[c]	1
	0.1 M KNO₃	A1	1.73₀	3
	0.2 M KNO₃	A1	1.53₆	4
Tl(I)	1 M KNO₃	E	1.65	1
	1 M KCl	E	1.57	1
	1 M KCl	C	1.53	5
	1 M NaOH	A1	2.13	6
	1 M NaOH	B	2.19	6
[Fe(CN)₆]⁴⁻	1 M KCl	A1	0.63₂[c]	1
	2 M KCl	A1	0.62₉[c]	1
	3 M KCl	A1	0.62₁[c]	1
	2 M NaOH	F	0.39₀	11
[Fe(CN)₆]³⁻	1 M KCl	A1	0.76₃[c]	1
	2 M KCl	A1	0.75₀[c]	1
	3 M KCl	A1	0.73₉[c]	1
	2 M NaOH	F	0.45₄	11
Pb(II)	1 M KCl	A1	0.91₇	1
	1 M KCl	E	0.92₃	1
	1 M KCl	C	0.89₅	5
	0.1 M KNO₃	B	1.00	7
Cd(II)	1 M KCl	A1	0.80₈	1
	1 M KCl	E	0.78₉	1
	1 M KCl	C	0.77₃	5
	0.1 M KNO₃	B	0.73	7
Fe(III)	1 M H₂SO₄	G	0.38	10
	1 M H₂SO₄	A1	0.30	6
	1 M HClO₄	A1	0.55	6
Fe(II)	1 M H₂SO₄	G	0.53	10
	1 M H₂SO₄	A1	0.50	6
	1 M HClO₄	A1	0.65	6
I₃⁻	0.1 M KI	C	1.02 (25.7°C)	13
	0.08 M KI	G	1.14	30
	0.1 M KI	C	1.13	31
I⁻	0.125 M H₂SO₄	A2	1.99	14
O₂	0.1 M KCl, pH 7, gelatin	E	2.12	23
	0.025 M H₂SO₄	C	1.93 (20°C)	24

EVALUATION OF DIFFUSION COEFFICIENTS 221

Electroactive species	Supporting electrolyte	Measurement technique[a]	$D \times 10^5$ cm^2/sec[b]	Ref.
H$^+$	1.00 M KCl	G	7.73	*25*
	1.00 M KCl	A1	7.32	*32*
	1.00 M KCl	C	7.22	*33*
	1.01 M LiCl	G	6.28	*25*
	0.1 M Na$_2$SO$_4$	C	10.0	*26*
OH$^-$	0.1 M Na$_2$SO$_4$	C	5.6 (20°C)	*26*
	1 M KNO$_3$	C	4.45	*33*
H$_2$	2.5 M H$_2$SO$_4$	C	1.4	*26*
	0.5 M H$_2$SO$_4$	C	3.83 (23°C)	*28*
Organic substances				
Anthracene	MeCN,[d] 0.1 M TEAP,[e]	F	2.55	*51*
	MeCN, 0.1 M LiClO$_4$, (0.37 M pyridine)	B	2.3	*48*
	MeCN	E	2.65	*100*
Naphthalene	MeCN, 0.1 M TEAP	F	2.74	*51*
	DMF,[f] 0.1 M TEAP	F	1.27	*51*
	75/25% dioxane–H$_2$O, 0.1 M TEAP	F	0.56	*51*
	dioxane H$_2$O[g]	E	0.62	*101, 102*
	DMF, 0.1 M TEAP	F	1.27	*51*
Biphenyl	MeCN, 0.1 M TEAP	F	2.48	*51*
	DMF, 0.1 M TEAP	F	1.20	*51*
Ferrocene	MeCN, 0.2 M LiClO$_4$	B	2.4	*53*
Ruthenocene	MeCN, 0.2 M LiClO$_4$	B	2.2	*53*
Osmocene	MeCN, 0.2 M LiClO$_4$	B	2.2	*53*
Hydroquinone	0.1 M KCl, 0.0005 M H$_2$SO$_4$	C	0.85 (15°C)	*54*
	0.1 M KNO$_3$	E	0.74	*59*
Quinone	2 M KCl	C	1.27 (26°C)	*54*
	0.1 M KNO$_3$	E	0.86	*59*
p-Methoxyphenol	2.0 M H$_2$SO$_4$	A2	0.70	*61*
		B	0.69	*61*
p-Ethoxyphenol	2.0 M H$_2$SO$_4$	A2	0.62	*61*
		B	0.62	*61*
p-Phenylene-diamine (PPD)	Buffer, pH 2.65	B	0.72	*48*
	Buffer, pH 2.5	B	0.73	*55*
	MeCN, 0.1 M LiClO$_4$	B	2.0	*48*
N,N-Dimethyl-p-phenylenediamine (DPP)	Buffer, pH 2.5	A1	0.84	*6*
	Buffer, pH 2.65	B	0.76	*48*
	MeCN, 0.1 M LiClO$_4$	B	2.3	*48*

TABLE 8-3 (contd)

Electroactive species	Supporting electrolyte	Measurement technique[a]	$D \times 10^5$ cm^2/sec[b]	Ref.
p-Aminophenol	0.51 M H$_2$SO$_4$	B	0.79	61
2,6-Dichloro-p-aminophenol	2.0 M H$_2$SO$_4$	A2	0.48	61
		B	0.49	61
2,6-Dibromo-p-aminophenol	2.0 M H$_2$SO$_4$	A2	0.50	61
		B	0.48	61
3,3'-Dimethoxybenzidine (o-dianisidine, o-DIA)	1.09 M H$_2$SO$_4$	D	0.41	56
	1.09 M H$_2$SO$_4$	A2	0.44	56
	1.09 M H$_2$SO$_4$	C	0.41	56
	1 M H$_2$SO$_4$	F	0.47	57
	2 M H$_2$SO$_4$	C	0.37	58
	2 M H$_2$SO$_4$	A1	0.37	58
	2 M H$_2$SO$_4$	D	0.37	58
	pH 2.5	A2	0.42	61
	pH 2.5	B	0.42	61

[a] A1, linear diffusion, $it^{1/2}$ curve analysis, shielded electrode; A2, linear diffusion, $it^{1/2}$ curve analysis, unshielded electrode; B, chronopotentiometry; C, rotated disk electrode; D, peak voltammetry; E, DME polarography; F, tracer diffusion from capillaries; G, other method—see original literature.
[b] All D values are for 25°C unless noted otherwise.
[c] Used in the writer's laboratory as "standards."
[d] Acetonitrile.
[e] Tetraethylammonium perchlorate.
[f] Dimethylformamide.
[g] See original literature.

A. Inorganic Substances

1. Ag(I), Tl(I)

The 0.1 M KNO$_3$ data by von Stackelberg et al. can be considered to be a standard reference value. The value in 0.2 M KNO$_3$ is also very reliable, since it was determined from exceptionally stable i–t curves. A value of about 2.1 × 10^{-5} for 0.2 M KNO$_3$–0.01 M HNO$_3$ which can be calculated from the corrected chronopotentiometric data of Bard (8) seems almost surely too high. The effect of gelatin on the value in 0.2 M KNO$_3$ is presumably reflected in the value $1.41_6 \times 10^{-5}$ obtained at the RDE by Kraichman and Hogge (12).

The Tl(I) data in 1 M NaOH refers to the oxidation reaction leading to Tl_2O_3 and were run at a shielded boron carbide electrode. Although the NaOH values are considerably higher than the nonalkaline reaction schemes, they are believed reliable since there is remarkable agreement between the $it^{1/2}$ and chronopotentiometric data. The tracer D for Tl(I) in 0.1 M KCl given by Wang and Polestra (18) (1.8×10^{-5}) compares with that obtained by von Stackelberg et al (1) (1.74×10^{-5}). Polarographic values for Tl(I) in 1 M NaClO$_4$ and Ag(I) in 0.5 M KNO$_3$ were reported by Pavlopoulos and Strickland (21).

2. $[Fe(CN)_6]^{4-}$, $[Fe(CN)_6]^{3-}$

These values (together with those of Table 8-2) are all from the data of von Stackelberg et al. and are believed to be the most reliable D values for checking solid electrode performance. Since the ferri-ferrocyanide reaction is not completely reversible at less than 1–2 M KCl background, the lower KCl values are not of such great interest (the reversibility plays no part in determining the D value, of course).

Eisenberg et al. carried out tracer studies of ferri- and ferrocyanide in 2 M NaOH (11). The ratio of tracer values $D_{ferrocy}/D_{ferricy}$ was 1.16, which agreed well with the ratio in 1 M KCl, where it is 1.21, indicating the relative behavior of the two ions in alkaline media is unchanged. Very recently Bazan and Arvia presented a very thorough RDE study of the ferri-ferrocyanide system in NaOH background (92). At ca. 2 M NaOH they found that $D_{ferrocy} = 0.51$ and $D_{ferricy} = 0.42$ cm^2/sec, although it is probably not justifiable to pick these as individual values from their extensive data. It is more meaningful to indicate the ratio $D\eta/T$ in the NaOH as supporting electrolyte. This value was found to be $(2.54 \pm 0.13) \times 10^{-10}$ for ferricyanide and $(2.09 \pm 0.11) \times 10^{-10}$ for ferrocyanide in the range of 1–2 M NaOH solutions. The units of $D\eta/T$ in this case are cm^2-P/sec-°K. A ratio $D_{ferricy}/D_{ferrocy} = 1.20$ agrees very well with the previous tracer data in NaOH. For further details on individual concentration and ionic strength dependences of D values of ferri- and ferrocyanide the paper by Bazan and Arvia should be consulted.

3. Pb(II), Cd(II)

The results of the RDE seem to compare quite well with $it^{1/2}$ and polarographic measurements on lead and cadmium in 1 M KCl. There is some doubt about the RDE values since they were reported differently in a previous publication (9). Wang has reported a value of 1.00×10^{-5} for Pb(II) in 1 M KCl by the tracer technique with open-end capillaries. Wang plotted the variation of tracer D for Pb(II) as the concentration of KCl varied and correlated the tracer D with polarographic measurements (17). Other correlations of D values by various workers and their interpretation in terms of DME limiting current equations have been examined in detail by Rulfs and Macero (19,20).

4. Fe(III), Fe(II)

D values obtained by Wijnen and Smit (10) involved a square-wave technique under linear diffusion conditions at platinum electrodes. The $it^{1/2}$ values listed were at boron carbide electrodes. The smaller D values in H_2SO_4 may reflect the complexing with sulfate ion. If RDE values for 1 M HClO$_4$ given by Vielstich and Jahn are reversed [i.e., Fe(III) = 0.57 and Fe(II) = 0.65] they are in excellent agreement with the $it^{1/2}$ data. It seems almost certain these values are misprinted in reverse in the literature (35). Lingane has calculated $D = 0.47 \times 10^{-5}$ for Fe(III) in 1 M HCl, which gave the best fit to some experimental chronopotentiometry data (37).

5. Halide Systems

The situation with regard to the iodine–iodide–triiodide species is not very clear. The value of 1.02 found by Hogge and Kraichman (13) using the standard Levich equation at a RDE is claimed by Gregory and Riddiford to be too low. Their value in almost the same iodide background electrolyte concentration is 1.14 and was obtained by diffusion through sintered disks (30) (this technique has been widely used and checked vs. the open-end capillary method in tracer-D-value studies). The value 1.14 when used in conjunction with the Riddiford correction to Levich's equation gives experimental results in good agreement with theory. Further, Newson and Riddiford found a value of 1.13 via RDE in later studies which include much information on the D value of I_3^- at various temperatures and in sucrose-containing solutions (31). The modified Levich equation was used in these studies. It seems most likely that the Hogge and Kraichman value for I_3^- is too low and 1.13–1.14 is the most reliable.

Beilby and Crittenden's data for I^- was obtained under essentially linear diffusion conditions (14). Kolthoff and Jordan give 2.05×10^{-5} for iodide ion in 0.1 M HClO$_4$ (15), but it is not clear if this is calculated via conductance at infinite dilution. They also found the addition of 0.1 M KI to a dilute I_2 solution in 0.1 M NaClO$_4$ or HClO$_4$ did not alter the cathodic i_L, from which it can be concluded that $D_{I_2} = D_{I_3^-}$.

Morgan et al. reported Br$^-$ = 2.12 × 10^{-5} and Br$_2$ = 1.47 × 10^{-5}, both in 0.05 M H$_2$SO$_4$. These values were determined via their charge-integration technique, carried out under essentially linear diffusion conditions (16). Davis and Everhart have reported Br$^-$ = 2.3 × 10^{-5} via chronopotentiometry at carbon paste electrodes (38).

6. Oxygen

In spite of the wealth of papers on the polarographic reduction of oxygen, there is a lack of reliable information on the D value. One difficulty in comparing measurements is that gelatin is sometimes used in the polarographic solutions. The papers by Jordan and co-workers give a critical account of polarographic values in viscous media (23). The value of 2.6 × 10^{-5} in 0.1 M KNO$_3$ given by Kolthoff and Miller (22) was calculated via the usual Ilkovic equation. According to Lingane this becomes 2.0 × 10^{-5} when the Lingane–Loveridge modification of the Ilkovic equation is used (36). Lingane also determined D_{O_2} in 1 M H$_2$SO$_4$ (2.02) and 1 M NaOH (1.79) using the extended polarographic equation (36). The latter value is to be compared with 2.7$_2$ reported by Kolthoff and Jordan in 0.1 M NaOH [calculated by comparison with i_L for Tl(I) at a RPE (29)]. One of these values is clearly in error. The only RDE value is for 0.025 M H$_2$SO$_4$ at 20°C but compares well with Lingane's polarographic data in 1 M H$_2$SO$_4$.

7. H$^+$, OH$^-$, H$_2$

As is well known, the D value of H$^+$ far exceeds that of any other ions. Reasonable agreement exists among the various investigations, although slight differences in technique presumably result in a larger difference in D. The tracer work of Woolf actually employed chemical titrations and extrapolation to zero concentration—hence the designation tracer diffusion coefficient. The value for 0.48 M KCl of 7.93 × 10^{-5} reported by Woolf (25) agrees well with a polarographic determination in 0.33 M KCl given as 7.90 × 10^{-5} (20°C) by Bagotsky (27). An even closer comparison is afforded by the first three entries in Table 8-3 for H$^+$, all in 1 M KCl. The agreement between the linear diffusion data of von Stackelberg and Pilgram (7.32) and the RDE work of Landsberg et al. is excellent, and both are quite close to Woolf's open-end capillary tracer data (7.73). The work of von Stackelberg and Pilgram contains data for D_{H^+} in KCl varying between 0.1 and 3 M. It is interesting to observe that D is lowered to 5.9 × 10⁻⁵ in 3 M KCl (32). The studies of Landsberg et al. cover the range 0.001–1.5 M KCl. This work was done at the azobenzene–graphite RDE (33).

The value of D_{H_2} in H_2SO_4 is apparently very sensitive to H_2SO_4 concentration, as shown in the studies of Breiter and Hoffman given in Table 8-3. The latter authors investigated D_{H_2} over the temperature range −10 to 60°C (the value for 2.5 M H_2SO_4 in Table 8-3 is extrapolated from their temperature data). Lewis and Ruetschi, using a RDE, report D_{H_2} to be 18.2, 13.8, 9.9, and 4.0 × 10⁻⁵ cm²/sec in 0.1, 1.0, 3.0, and 5.0 M KOH, respectively (30°C). They also found the D value for deuterium to be ca. 3.5 times less than that of hydrogen (since the solubility of deuterium in KOH is unknown the values could only be expressed in this relative fashion). Lewis and Ruetschi observed that their values were higher than previous data (although done in different solutions with different techniques) and suggested that the platinized surface of their RDE with its inherent roughness might account for increased mass transport (34).

8. Miscellaneous Inorganic Systems

Sanborn and Orlemann compared tracer D of Ni(II) in $NaClO_4$–$HClO_4$ and found it to agree well with the polarographic values. In 0.1 M $NaClO_4$–0.001 M $HClO_4$, D is 0.60 × 10⁻⁵ by the tracer technique (39). Wang and co-workers examined the self-diffusion of water using deuterium, tritium, and ¹⁸O tracers (40).

Sawyer and Day suggest $D = 2.4 \times 10^{-5}$ for cyanide ion in 0.25 M NaCl and 0.01 M NaOH, but this is obtained from a combination of DME and chronopotentiometry (41). Bodnar and Himmelblau, using ¹⁴CO_2, determined tracer D_{CO_2} to be 2.15 × 10⁻⁵ in water at 30°C (42). Munson reports D of carbon monoxide (in ca. 0.1 N $HClO_4$) as 2.5 ± 0.1 × 10⁻⁵ cm²/sec from chronopotentiometric measurements (94).

Landsberg and co-workers (93) showed that the D value of manganate ion (MnO_4^{2-}) decreased from 0.83 to 0.64 × 10⁻⁵ cm²/sec in 1 and 3 N KOH, respectively. These measurements were made at a platinum RDE employing the Gregory and Riddiford corrections and are for 20°C.

Davis found hydroxylamine to be 1.41 × 10⁻⁵ in 1 M H_2SO_4 via chronopotentiometry. This value agrees well with that determined polarographically (43). In dilute aqueous solution Karp and Meites calculated $D_{N_2H_4}$ to be 1.39 (44). Using this value, Bard estimated $D_{N_2H_4} = 1.47$ in 1.44 M H_2SO_4 (45).

Voorhies and Schurdak reported that the D value of Fe(III) in nitromethane with tetraethylammonium chloride (0.05 M) as supporting electrolyte was 1.6×10^{-5} (27°C). This was determined chronopotentiometrically and, after assessing various errors which contribute to too large a D value, the authors concluded that the experimental D was still far greater than in aqueous media (46).

B. Organic Substances

There are very few organic materials of interest to solid electrode voltammetry for which reliable D values exist. The individual compounds are discussed below.

1. Hydrocarbons

An early chronopotentiometric study of anthracene in acetonitrile (0.1 M NaClO$_4$) by Voorhies and Furman reported a D value of 6.6×10^{-5}—almost 10 times faster than most species in aqueous media (47). Kuwana obtained 5.9×10^{-5} in acetonitrile with 0.1 M LiClO$_4$ as supporting electrolyte, using an extrapolated value of $i\tau^{1/2}$ to zero transition time. However, with the same medium, in the presence of 0.37 M pyridine, 2.3×10^{-5} was found (48). Kuwana used pyridine since Lund and Friend and Ohnesorge had previously shown that the oxidation of anthracene under these conditions yielded 9,10-dihydroanthranyl dipyridinium diperchlorate—a clear two-electron oxidation (49,50). Kuwana's interpretation of the large D values in the absence of pyridine was that further oxidative reactions were occurring and $n_T = 2$ was not a correct substitution in the Sand equation. This is in complete accord with recent studies of hydrocarbon oxidations in inert solvents. With certain substituted hydrocarbons, clean one-electron oxidations can be observed (103,104, 122). In the case of anthracene, stabilization of a two-electron product is obtained in the presence of pyridine. In the absence of stabilizing substituents on the hydrocarbon (or added stabilizing reagents), the hydrocarbon oxidations progress to higher apparent n values, and evaluations of D are totally unreliable.

The electrochemical value for anthracene in the presence of pyridine is completely substantiated by the tracer data in Table 8-3. Using anthracene with ring-labeled tritium, the D_{Tr} in acetonitrile–0.1 M TEAP is 2.55×10^{-5} cm^2/sec. The tracer D values of anthracene and several other hydrocarbons included in Table 8-3 show a good constancy of $(D_{Tr}\eta)$ over a wide range of solvent viscosities (51). Ward's chronopotentiometric studies of antioxidants suggests that the D values or organic molecules such as hydroquinone and phenylenediamines are nearly the same in water or acetonitrile. Actually the values of $i\tau^{1/2}/C$ given show that D could vary by a factor of 2 in the two solvents (*52*). It appears clear from the tracer and electrochemical data available that most organic materials have D values about a factor of 2–3 greater in acetonitrile than in water. The metallocenes studies by Kuwana et al. are in agreement with this picture (*53*).

2. Hydroquinone–Quinone

It is unfortunate that the RDE values for hydroquinone and quinone are at temperatures other than 25°C (*54*). Both values appear high with respect to the only check value, that of Kolthoff and Orlemann in 1 M KNO_3 (*59*). The latter values were used by Peters and Lingane to calculate theoretical values of $i\tau^{1/2}/C$ for their chronopotentiometry at cylindrical electrodes (*60*). Since their experimental data agreed within a few per cent with the theoretical $i\tau^{1/2}/C$ plots, it seems that the DME data are reliable. The Peters and Lingane work was done in 1 M H_2SO_4, which indicates little change in D between KNO_3 and H_2SO_4 media.

3. Aromatic Amines

The D value for *p*-phenylenediamine (PPD) shows excellent agreement at about pH 2.5, measured in one case on platinum (*48*) and in the other on carbon paste (*55*). The value in acetonitrile is again a factor of 2–3 larger. The questionable part of the aqueous data is that, at the concentrations used, electrode filming during the oxidation is certainly present. The N,N-dimethyl derivative (DPP) is not much different from PPD, and its behavior is parallel in acetonitrile.

The D value of *o*-dianisidine (*o*-DIA) is probably the best checked of the oxidizable organic molecules with special relevance to solid electrode work. There is surprisingly good agreement in 1 M H_2SO_4 between the tracer values and those obtained via $it^{1/2}$ curves, peak voltammetry, and the RDE. Excellent agreement via electrical methods is also observed in 2 M H_2SO_4 and the results cross-check reasonably well with regard to constancy of the product ($D\eta$). Very careful measurements at pH 2.5 show good agreement between $it^{1/2}$ and chronopotentiometry. At this pH, varying the *o*-DIA concentration from 1.16 to 5.33 mM results in a decrease of D as indicated in Table 8-4 (*61*). Results such as these can be obtained for *o*-DIA because it is such a well-behaved system upon oxidation at platinum or carbon paste electrodes. It is oxidized in an apparently reversible two-electron process with no evidence of chemical follow-up reactions and there is little tendency toward electrode filming. It is unique among oxidizable aromatic amines in these respects.

In contrast with o-DIA, the determination of correct D values for N,N-dimethylaniline (DMA) is very difficult. The electrode reaction is known to be complex, and filming is clearly present during oxidation. For systems such as DMA it is apparent that the determination of D values under conditions used in the electrochemistry, but independent of the electrochemical measuring technique, would be valuable. A program of measuring tracer diffusion coefficients of selected molecules has been in progress

TABLE 8-4

Concn. o-DIA, mM	$it^{1/2}$	$D \times 10^5$ cm²/sec chronopotentiometry
1.16	0.424	0.418
2.17	0.417	0.420
3.35	0.410	0.412
4.90	0.387	0.388
5.33	0.386	0.394

in the writer's laboratory for some time. Tritium-labeled aromatic molecules have been used with the open-end capillary techniques of Wang and Polestra (62–64). The procedures for DMA, o-DIA, and hydrocarbons have been described (51,57).

The D values for tritiated DMA were measured over the pH range 2.5–5.6, a region in which data from various electroanalytical methods were available. In particular, the results were compared with RDE measurements. These are shown in Fig. 8-2, and it is clearly seen that tracer and RDE values not only fail to agree, but that RDE values have a far smaller pH dependence. The tracer values have a standard deviation of ca. ±0.07 over most of the pH range, while the reliability of RDE values can probably be given as ±0.01. Allowing for the poorer precision in the tracer data, there still is no question about the lack of agreement between the two methods.

There are several possible explanations of the "high" tracer values. First, the DMA was generally labeled, which gives some tritium activity in the N-methyl groups as well as on the ring. It could be supposed that some chemical decomposition such as deamination occurred during the 3- to 5-day diffusion period. This might result in, say, a dimethylamine fragment with a higher D value. This would contribute a greater loss in

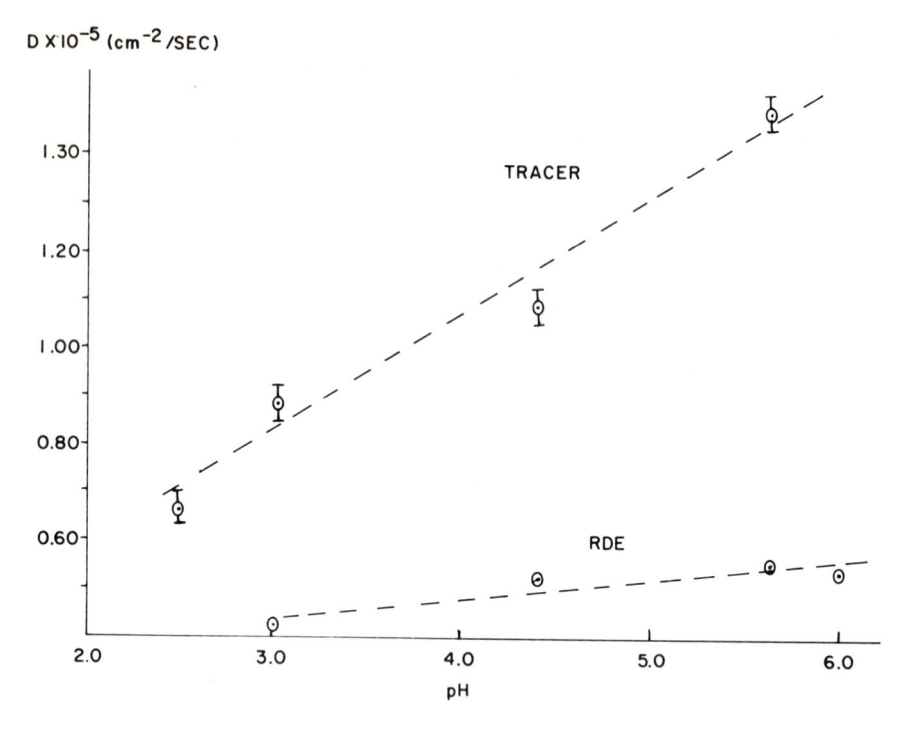

Fig. 8-2. Comparison of tracer and RDE diffusion coefficients for N,N-dimethylaniline.

activity and hence a measured D greater than the true value for DMA. However, a monitoring of the UV spectrum of a DMA solution exposed in the same fashion as the diffusing medium showed a constancy within the limits of spectrophotometric reproducibility. No indications of such decomposition were found. Further, the tracer values are in good agreement with electrochemical methods (including the RDE) for a well-behaved system such as o-DIA (see Table 8-3).

It is believed the tracer values are correct and that the discrepancy lies in the nature of the anodic oxidation of DMA. First, the DMA oxidation involves a fast follow-up chemical reaction which renders i_L questionable at all but high rotation rates. It can be definitely established that some degree of filming of the electrode surface occurs during the DMA oxidation. Thus a repeat polarogram on the same electrode surface always gives a lesser limiting current in a DMA oxidation. Further, this filming increases with increasing pH, and hence the magnitude of i_L error for DMA oxidation becomes larger as pH increases. While i_L increases as pH is raised (due to increased D), apparently the falloff of i_L due to filming brings about a compensation so that the plot of apparent D vs. pH shows only a very small acidity dependence, as seen in Fig. 8-2. The nature of the organic film deposited during oxidation is not known, but such effects are widespread in anodic oxidations of aromatic molecules. A systematic and thorough comparison of tracer and electrochemical D values for organic systems will do much to clarify some of the discrepancies now existing in this area.

4. Miscellaneous Organic Systems

Excellent data for the self-diffusion coefficients of a number of hydrocarbons, alcohols, etc., have been determined via tracer studies by Fishman and Vassiliades (*65,66*) and NMR spin-echo techniques by McCall et al. (*67*). (The data are naturally not directly applicable when these substances are used as electroactive species in electrolyte solutions.) Druschel and Miller estimated D values of a series of aliphatic sulfides via peak voltammetry at wire electrodes. While the absolute values are in doubt, the utility of a plot of molecular weight vs. apparent D for a series of related compounds is well illustrated by their study (*68*). Bard estimated a D value for riboflavin to be 0.52×10^{-5} in pH 2 buffer (*69*). Solon and Bard have reported diphenylpicrylhydrazyl (DPPH) to have $D = 1.2 \times 10^{-5}$ in 1 M NaClO$_4$–acetonitrile solution (*70*). From rather involved chronopotentiometric measurements, Testa and Reinmuth reported $D = 0.82 \times 10^{-5}$ for *o*-nitrophenol and 1.0×10^{-5} for nitrobenzene, respectively, in 50% ethanol–water at an apparent pH of 6.2 (*71*).

The D values of a large number of substituted triphenylamines have been measured in acetonitrile by $it^{1/2}$, chronopotentiometry, and the RDE. The values from the various techniques do not agree well enough to be given in detail but an approximate value of 1.3×10^{-5} cm^2/sec can be quoted [among 20 substituted triphenylamines, D does not vary by more than $\pm 20\%$ (*143*)].

8-3. CORRELATION OF ELECTROANALYTICAL TECHNIQUES: ELECTRODE REACTIONS WITHOUT CHEMICAL COMPLICATIONS

All enthusiasm one has for solid electrodes must be tempered with the knowledge that there are definite limitations and discrepancies which are peculiar to their use. These show up most clearly when attempts are made to correlate various electroanalytical techniques. The remarks of this section pertain only to simple electron-transfer processes with no chemical complications.

There are several ways of relating the major electroanalytical methods. One which has not been used widely illustrates the response of the various methods to variations in intrinsic electron-transfer rates. The electron- or charge-transfer rate can be measured in terms of the heterogeneous rate constant. This rate constant is given the symbol k_{el} (electrochemical) here but has acquired a variety of subscripts and superscripts in its development. In any event, it has the units of cm/sec, since it describes a heterogeneous process and is a quantitative measure of "reversibility." For a recent and authoritative treatment of electron-transfer rates the reader is referred to the monograph *Double Layer and Electrode Kinetics* by Delahay (95).

For the present discussion one only needs the widely accepted premise that the rate of an electron-transfer reaction is an exponential function of the applied potential. It follows that at sufficiently large cathodic or anodic potentials (for reduction and oxidation, respectively) the electron-transfer rate is maximal. This maximal rate in chronopotentiometry corresponds to potentials at or beyond the transition time. For peak voltammetry it is slightly beyond the peak potential. Similarly, for a potentiostatic i–t measurement, a potential past the peak would give a maximal rate of electron transfer. Finally, for a convective system such as the RDE, potentials corresponding to the limiting current plateau would be used.

It is not simple to describe reversible–irreversible behavior in quantitative terms of k_{el} since the characteristics obviously depend on the relative magnitudes of both electron- and mass-transfer rates. The observable response of the system can be dominated by whichever process is the slower and there will be mixed control when electron- and mass-transfer rates are competitive. One has limited experimental control of the electron-transfer rate by operating at the maximum k_{el} (large applied potentials) or lesser values (i.e., at potentials on ascending portions of chronopotentiograms, current–voltage curves, etc.). Obviously the mass-transfer rate can be controlled over a wide range from quiet solution to, for instance, RDE conditions. In certain instances, where the mass-transfer rate is rather fixed by the electrode system, as in DME polarography, one can define "reversibility limits" in terms of k_{el}. One commonly finds quoted that, for the DME, systems with $k_{el} >$ ca. 10^{-2} cm/sec are, for all practical purposes, reversible.

CORRELATION OF ELECTROANALYTICAL TECHNIQUES 233

Figure 8-3 compares the response characteristics when both the electron-transfer and mass-transfer rates vary experimentally, as they do in the various electroanalytical techniques. Using chronopotentiometry (Fig. 8-3A) for a reversible process, a plot of $i_0\tau^{1/2}$ vs. increasing i_0 is a horizontal straight line. This is still true for a reversible reaction if one plots $it_E^{1/2}$, where t_E is the time at some potential less than that at which τ is measured

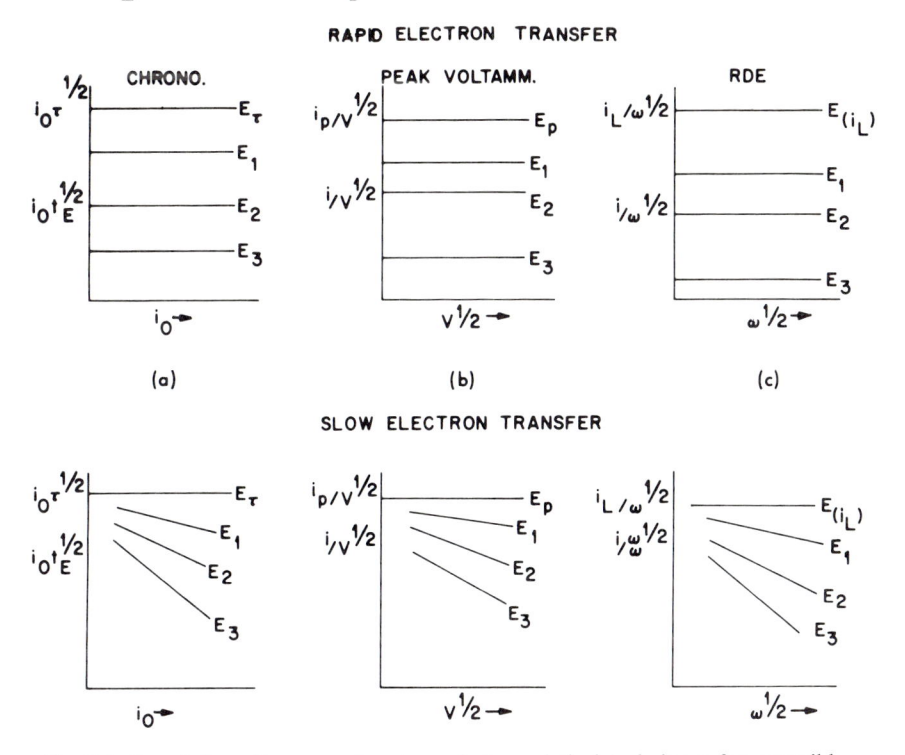

Fig. 8-3. Correlation diagrams of various electroanalytical techniques for reversible and irreversible systems.

(i.e., at potentials on the ascending portion of the chronopotentiogram). Hence $it_{E1,2,3}^{1/2}$ are also horizontal straight lines, where $E_3 < E_2 < E_1 < E_\tau$. Now, if the charge transfer is slower, then $i_0\tau^{1/2}$ will still be a horizontal straight line, but all $i_0 t_E^{1/2}$ will show negative slopes. The line $i_0\tau^{1/2}$ slopes downward with increasing i_0 only when a chemical reaction preceding the electron transfer is rate-determining (it can have a positive slope if adsorption occurs). This type of analysis was shown first by Gierst and Hurwitz *(72,76)*. Alternatively, one can plot $i_0\tau^{1/2}$ vs. $1/\tau^{1/2}$ *(77)*.

Exactly the same type of graphical analysis can be used with peak voltammetry if one plots $i_p/V^{1/2}$ vs. $V^{1/2}$ (Fig. 8-3B). Again $i_p/V^{1/2}$ vs. $V^{1/2}$ will be a horizontal line except for preceding chemical reactions. The plots of $i_p/V^{1/2}$ vs. $V^{1/2}$, where i's are measured at E's less than E_p, will be horizontal if electron transfer is fast but slope downward when it becomes rate-determining. (Actually, a potential greater than E_p is necessary if the heterogeneous rate constant is small.)

Similar analyses exist for $it^{1/2}$ curves at constant potential and for $i_L/\omega^{1/2}$ in RDE work (Fig. 8-3C). The $it^{1/2}$ vs. $1/t^{1/2}$ plots are not particularly advantageous since, if the electron-transfer reaction is only moderately slow ("slightly irreversible"), the value of $it^{1/2}$ approaches that of pure diffusion control for times about 15–20 sec or greater. Hence the

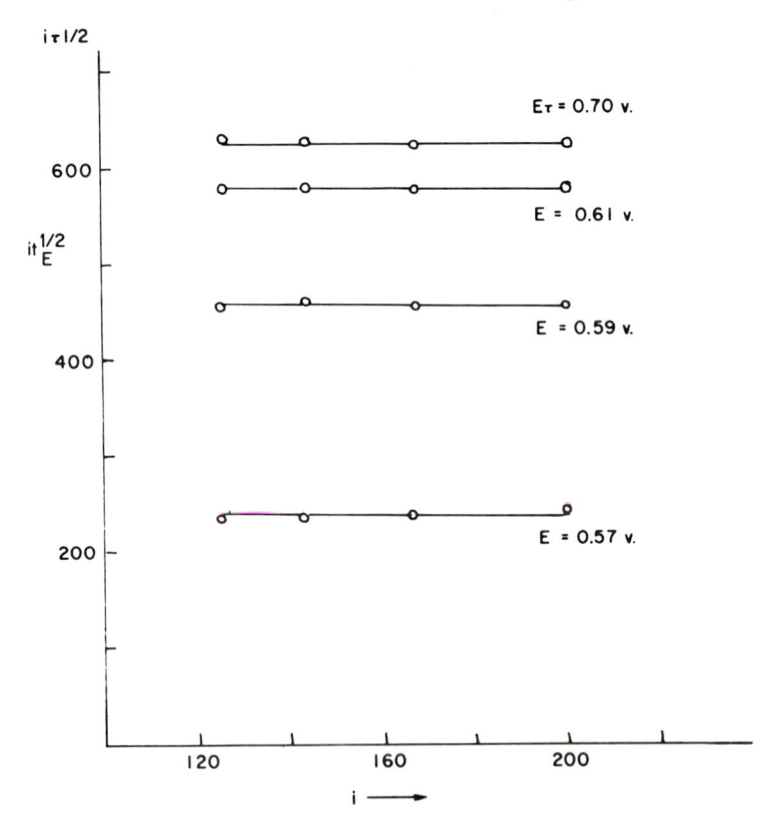

Fig. 8-4. Chronopotentiometric behavior of *o*-dianisidine in 1 M sulfuric acid. Carbon paste electrode, usual chronopotentiometric conditions.

CORRELATION OF ELECTROANALYTICAL TECHNIQUES 235

plot of $i t^{1/2}$ vs $1/t^{1/2}$ practically coincides with the horizontal line of the reversible case for small $i/t^{1/2}$ values (78) It is probably more revealing to plot increasing $t^{1/2}$ as the x axis in this case.

Some *practical examples* of these graphical analyses at solid electrodes are seen in Fig. 8-4 through 8-8. In Fig. 8-4 one sees that o-DIA in H_2SO_4 shows all the characteristics of a pure diffusion-controlled process

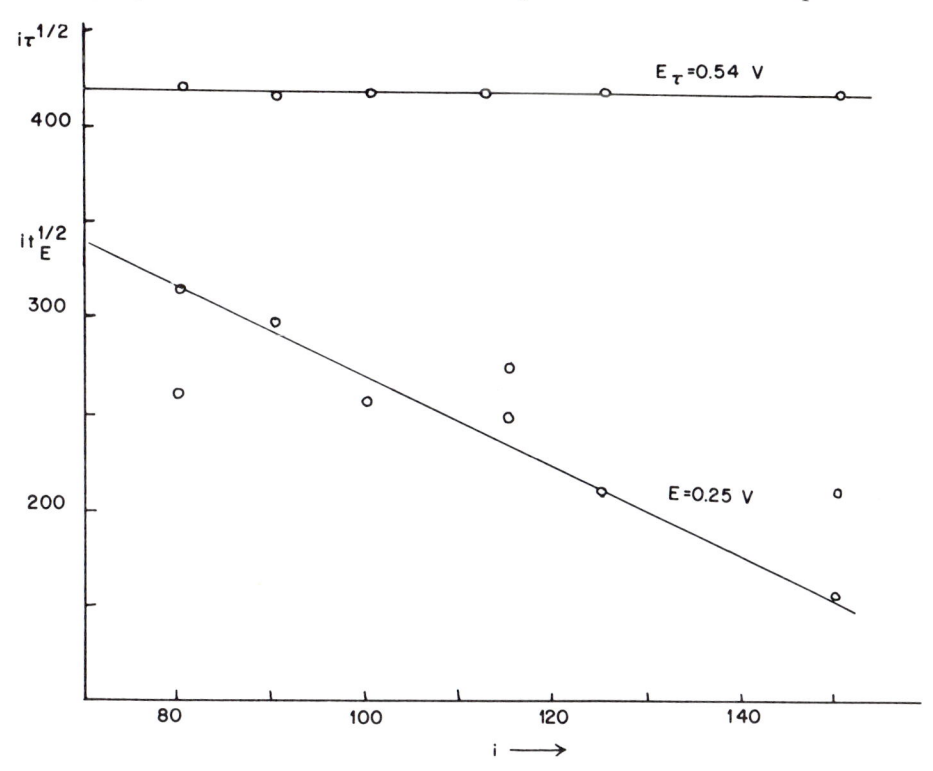

Fig. 8-5. Chronopotentiometric behavior of ferrocyanide oxidation in 0.1 M KCl. Carbon paste electrode, usual chronopotentiometric conditions.

when examined by chronopotentiometry. On the other hand, the oxidation of ferrocyanide in 0.1 M KCl is clearly charge-transfer-controlled, as seen in Fig. 8-5. As predicted, the plot of $i\tau^{1/2}$ (where τ is measured at 0.54 V) is horizontal, but the line corresponding to $it_E^{1/2}$ where $E = 0.25$ V deviates markedly. These data were taken at a carbon paste electrode, and it is interesting to note that the reproducibility is excellent for the $i\tau^{1/2}$ line but the deviations are considerable when charge-transfer control predominates. This is an interesting and perplexing property of the carbon paste electrode.

The plot of $i_p/V^{1/2}$ vs. $V^{1/2}$ was apparently first suggested by Galus and Adams in connection with preceding chemical reactions (73). Figure 8-6 shows that o-DIA again measures up to reversible characteristics. The various horizontal lines are for the indicated potentials up to and including

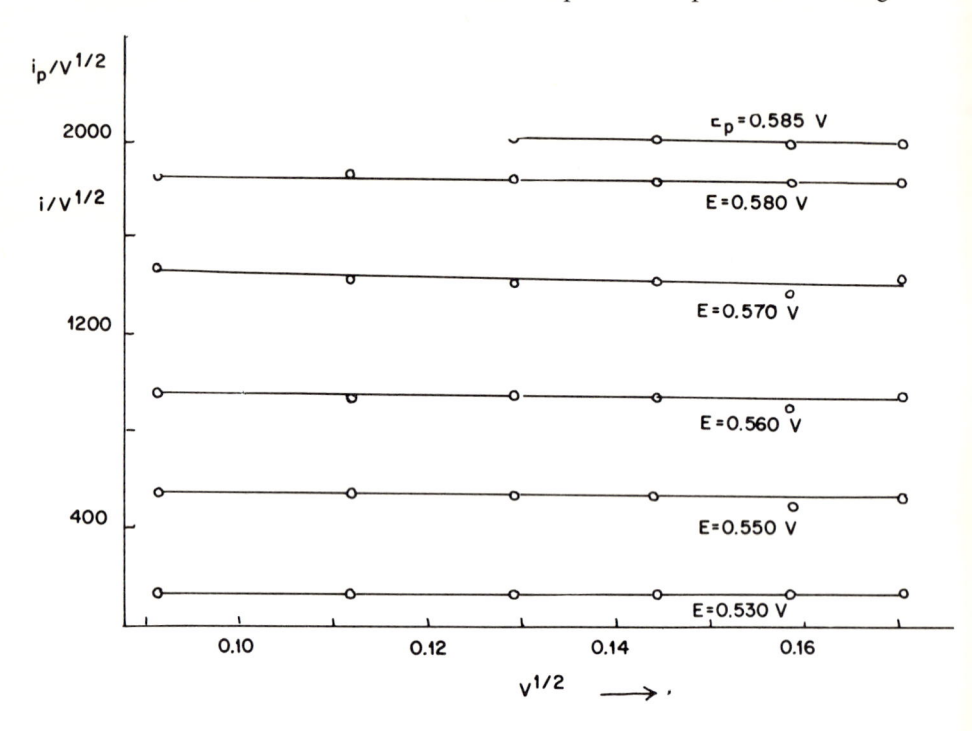

Fig. 8-6. Peak voltammetry behavior of o-dianisidine oxidation in sulfuric acid solution. Carbon paste electrode, 5.3×10^{-3} M o-dianisidine in 0.65 M H_2SO_4.

$E_p = \pm 0.585$ V. Figure 8-7 for ferrocyanide oxidation in 1 M KCl at carbon paste electrodes is revealing. The plot is actually $i_p/V^{1/2}$ vs. $V^{1/2}$. The ferrocyanide oxidation is in the "quasireversible" area and presumably i_p is controlled both by diffusion and electron transfer, hence the change in $i_p/V^{1/2}$ with scan rate.

With the RDE, a plot of $i_L/\omega^{1/2}$ vs. $\omega^{1/2}$ for a system like the oxidation of Fe(II) in HCl (with $k = $ ca. 10^{-4} cm/sec) shows the expected results,

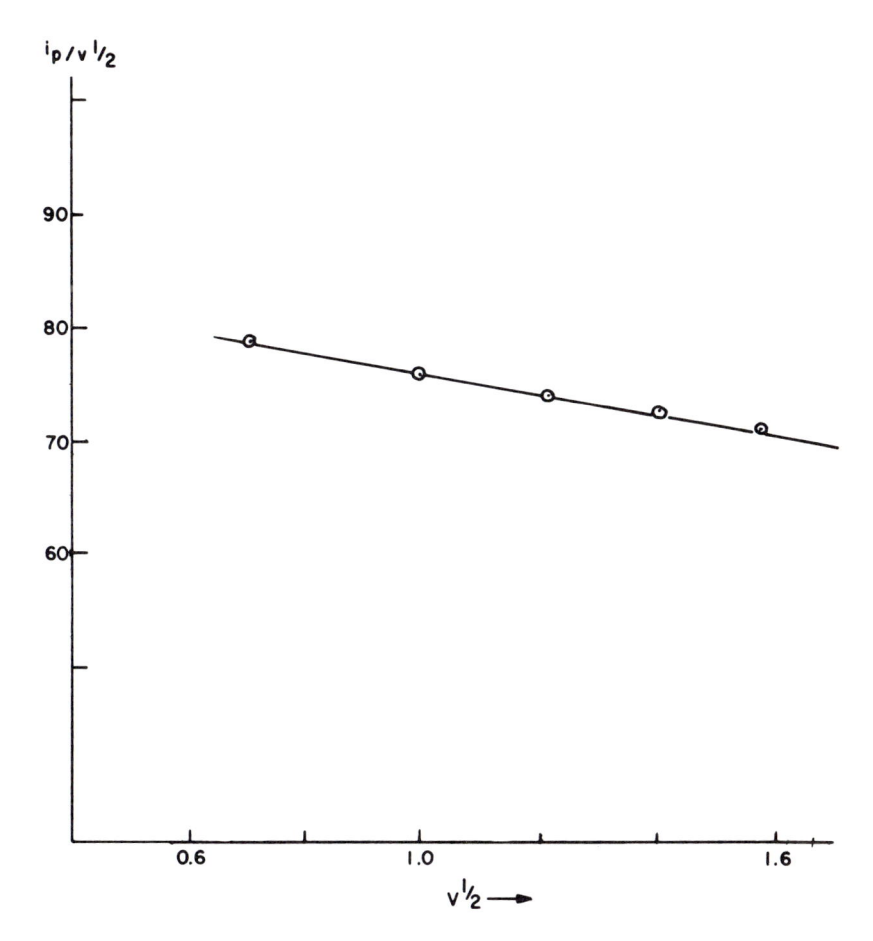

CORRELATION OF ELECTROANALYTICAL TECHNIQUES

Fig. 8-7. Peak voltammetry behavior of ferrocyanide oxidation in 1 M KCl. Carbon paste electrode.

as seen in Fig. 8-8. These data were taken from the work of Galus and Adams on the measurement of heterogeneous rate constants of moderately rapid electrode reactions (74).

All the above analyses presuppose no chemical complications which can be rate-determining. It should be noted especially that the ECE process gives a plot of $i_L/\omega^{1/2}$ vs. $\omega^{1/2}$ similar to that of a slow electron transfer.

Clearly, from the above discussion, one cannot a priori expect an electroactive system to behave identically when examined by a variety of techniques. If the heterogeneous rate constant is "borderline," then variations may appear due to the range of electron- and mass-transfer rates accessible to experimental control. However, if one has reason to believe the electron transfer is very rapid, then one should expect similar behavior from different techniques if the experiments are carried out under the maximal rate conditions. It appears, in practice, that only a very few organic systems (such as o-DIA) behave this ideally at solid electrodes in aqueous solution.

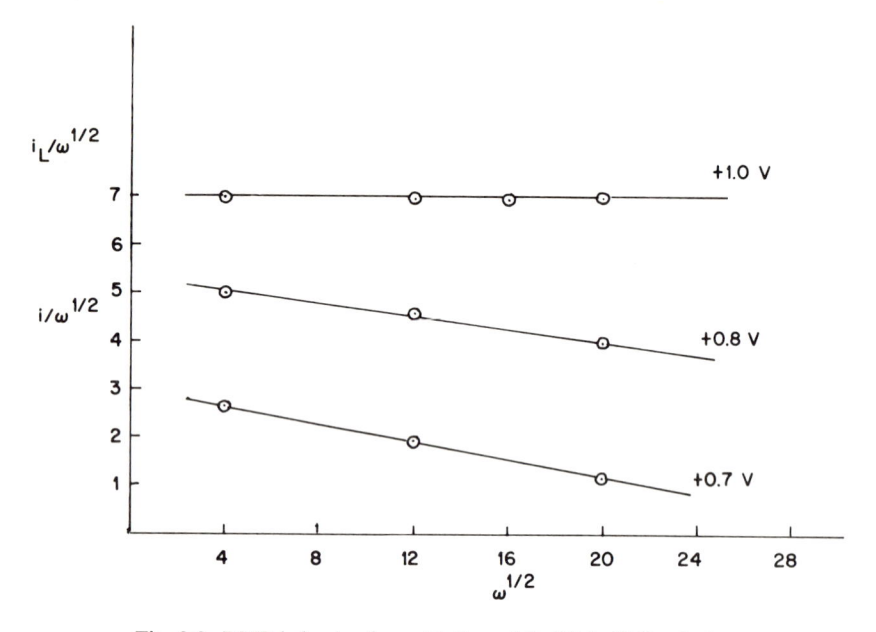

Fig. 8-8. RDE behavior for oxidation of Fe(II) in HCl solution.

In many cases it is not experimentally possible to cross-check techniques under conditions where one would expect similar behavior. Some of these problems have been mentioned in Chapters 5 and 6.

Another means of correlating those electroanalytical techniques which employ linear diffusion was first pointed out by Mueller and Adams (75). Equations (3-16), (5-31), and (6-5) represent respectively, the equations for the current–time curves (chronoamperometry), peak polarograms, and chronopotentiograms of a mass-transfer-controlled process under conditions of semiinfinite linear diffusion. The initial and boundary conditions for the solution of the differential equations leading to Eqs.

CORRELATION OF ELECTROANALYTICAL TECHNIQUES 239

(3-16), (5-31), and (6-5) are very similar. The similarities in these equations should permit correlations of the three electroanalytical techniques.

The quantities i_p and $it^{1/2}$ may be related (for reversible systems) if the potential of the electrode for the determination of $it^{1/2}$ is set at a value more anodic (or more cathodic for a reduction) than the potential at which the limiting current plateau is obtained in a normal polarogram. Under these conditions, simultaneous solution of Eqs. (3-16) and (5-31) leads to

$$\frac{i_p}{it^{1/2}} = \frac{n^{1/2}\pi^{1/2}V^{1/2}}{F} \cdot K \qquad (8\text{-}2)$$

where K is the Randles–Sevcik constant. Thus, having run an $it^{1/2}$ curve at a particular electrode, one can predict the peak current which will be obtained at any scan rate. Since C^b and A are eliminated in the simultaneous solution, this comparison assumes that the same solution and electrode are used. More important, as was pointed out in Chapter 5, the Randles–Sevcik constant can be evaluated without knowledge of C^b, A, or even the diffusion coefficient. The frailities of this evaluation were also indicated in Chapter 5.

The chronopotentiometric product ($i\tau^{1/2}$) can be related to the corresponding chronoamperometric quantity ($it^{1/2}$) via Eqs. (3-16) and (6-5) to give the remarkably simple result

$$i\tau^{1/2} = it^{1/2}\frac{\pi}{2} \qquad (8\text{-}3)$$

While this experimental determination of π will hardly give enough significant figures for astronomical calculations, the comparison of the two techniques is useful. In some instances very close correlations are obtained. Thus the oxidation of a 3.582 mM solution of ferrocyanide in 2 M KCl at carbon paste gave an average $it^{1/2} = 113.6 \pm 0.3$ μA-sec$^{1/2}$. The mean value of three chronopotentiograms run on this solution gave a calculated value [via Eq. (8-3)] of 112.9 μA-sec$^{1/2}$. Again, on the basis of an $it^{1/2}$ value for a given ferrocyanide solution in 2 M KCl at a platinum linear diffusion electrode, an $i\tau^{1/2}$ of 261.0 was predicted. The observed values was 260.5. However, for the same system, calculation of the Randles–Sevcik constant from the correlation between Eqs. (5-31) and (6-5), which is

$$i_p = \frac{(i\tau^{1/2})2\, V^{1/2}n^{1/2}K}{\pi^{1/2}F} \qquad (8\text{-}4)$$

gave $K = 2.64$, which is lower than expected. In the reduction of ferricyanide in 2 M KCl at the platinum electrode the $it^{1/2}$ product predicted an $i\tau^{1/2}$ of 281.6. The observed value was 281.5 ± 1.3. The Randles–Sevcik constant for the system was 2.59 (4).

A variety of other correlations are possible and some have been indicated previously. All these correlations may be initial steps in investigating electrode reaction mechanisms, and the results, in general, are reliable. Where two or more electrochemical techniques do not agree, the differences are often serious enough to cast considerable doubt on the results obtained by either method. Difficulties may be encountered also when one compares the same reaction at different electrode surfaces.

One point is quite clear from the extended discussion above. Even with relatively simple electrode reactions one can place little reliability on solid electrode data taken by a single electrochemical technique. On the other hand, for analytical applications, where conditions will necessarily be standardized, reliable results can be expected. The literature is well documented with practical analytical results on complex systems using a single electrochemical technique.

8-4. DETERMINATION OF HETEROGENEOUS RATE CONSTANTS (k_{el})

A detailed description of heterogeneous electrode kinetics is beyond the scope of this monograph. However, since correlations of electroanalytical techniques in which the principal variable was the rate of electron transfer were discussed in the previous section, a summary of methods for determining k_{el} is given now. In the following discussion all homogeneous chemical reactions are assumed absent.

The general idea of an electrode reaction with mixed control by mass transfer and electron transfer was introduced in Chapter 3, Section 3-1. If we are interested in measuring the electron-transfer rate, it is clear we must make the mass transfer very rapid in order that the former may be the slower step and kinetically observable. Actually, all techniques for measuring k_{el} depend on this idea. There are two general approaches: (1) relaxation techniques and (2) steady-state or dc measurements. In the relaxation methods, the system is perturbed from equilibrium and the potential or current is followed for a very short period of time. The perturbing input may be a single pulse (chronopotentiometric, chronoamperometric methods, etc.) or may be a sinusoidal (or other) time function, as in faradaic impedance and rectification techniques. The high mass-transport rate is achieved by virtue of the very steep concentration gradients attendant with the short time during which the observation is made. This short time interval is an inherent feature and also a disadvantage of the relaxation methods. It is evident that a reaction will be, practically speaking:

Mass-transfer-controlled if $V_{mt} \ll k_{el}$

Electron (charge)-transfer-controlled if $V_{mt} \gg k_{el}$

where V_{mt} is the average velocity of mass transport (or mass-transport coefficient) in cm/sec and k_{el} has identical units.

In a typical potentiostatic measurement, with linear diffusion operative, the average V_{mt} can be expressed as $V_{mt} = 2(D/\pi t)^{1/2}$. If one is attempting to determine a fast electron-transfer rate (i.e., $k_{el} = 1$ cm/sec), assuming the usual value of $D = 10^{-5}$ cm²/sec, it is easily seen that a time of ca. 10^{-5} sec is required to achieve $V_{mt} = 1$. These short times place stringent requirements on the potentiostats and assorted circuitry. In addition, cell resistances must be very low to keep the RC time constant short and not a limiting factor. These problems have been solved in an elegant fashion by many workers. Relaxation methods have been discussed thoroughly by Delahay (95,124). Excellent reviews of recent developments have been presented by Reinmuth (125) and these references should be consulted for further details.

In the steady-state methods, the high mass transport is obtained by some form of stirring, and a dc current–potential curve is analyzed. The classical form of this technique is, of course, the Tafel plot. Here strong stirring is used and one evaluates the log i–E curve. This approach, and especially the variations of Vetter (126), in which Tafel plots are examined as a function of concentration of reacting species to establish reaction orders, should undoubtedly be more widely appreciated by electroanalytical chemists.

In summary form, the steady-state methods can be discussed best from the hydrodynamic voltammetry viewpoint of Jordan. It was pointed out in Chapter 4 that voltammetric currents can always be expressed as

$$i = nFAM(C^b - C^e)$$

where m is a mass-transport coefficient with the units of cm/sec, and C^b and C^e are concentrations in the bulk solution and at the electrode surface respectively. For the particular case where $C^e = 0$, the current is the limiting current, i_L. Jordan (127) used relations similar to those of Eqs.

(5-1)–(5-4) and introduced expressions for the heterogeneous rate constants as

$$\text{Ox} + ne \underset{k_b}{\overset{k_f}{\rightleftarrows}} \text{Red}$$

$$k_f = k^{\circ\prime} \exp\left[\frac{-\alpha nF(E - E^{\circ\prime})}{RT}\right]$$

and

$$k_b = k^{\circ\prime} \exp\left[(1 - \alpha)\frac{nF}{RT}(E - E^{\circ\prime})\right]$$

where $k^{\circ\prime}$ is the apparent formal rate constant at the formal potential $E^{\circ\prime}$ and the other symbols have their usual significance. The value $k^{\circ\prime}$ is equivalent to k_{el} mentioned previously. The Jordan derivation yields an equation for the current–voltage curve with mixed control by convective mass transport and electron transfer.

Malachesky has shown that this equation, in slightly different form, can be reduced to all the various "name" methods for treating heterogeneous rate-constant data (118). The expression for a cathodic wave (i.e., $C_R^b = 0$) is (118,127)

$$\frac{i}{i_{L,c}} = \frac{1}{\dfrac{m_O}{k^{\circ\prime} \exp[(-\alpha nF/RT)(E - E^{\circ\prime})]} + \dfrac{m_O}{m_R} \exp\left[\dfrac{nF}{RT}(E - E^{\circ\prime})\right] + 1} \quad (8\text{-}5)$$

Here m_O and m_R are mass-transport coefficients for the oxidized and reduced species, respectively. Experimentally, they often are approximated as equal.

Various limiting forms of Eq. (8-5) result, depending on the relative magnitudes of m_O and $k^{\circ\prime}$. If $m_O \ll k^{\circ\prime}$, it is clear that

$$\frac{m_O}{k^{\circ\prime} \exp[(-\alpha nF/RT)(E - E^{\circ\prime})]} \ll \frac{m_O}{m_R} \exp\left[\frac{nF}{RT}(E - E^{\circ\prime})\right] + 1$$

and Eq. (8-5) reduces to

$$\frac{i}{i_{L,c}} = \frac{1}{(m_O/m_R) \exp[(nF/RT)(E - E^{\circ\prime})]}$$

Rearranging, taking logs, and solving for E gives

$$E = E^{\circ\prime} - \frac{RT}{nF} \ln \frac{m_O}{m_R} + \frac{RT}{nF} \ln \frac{i_{L,c} - i}{i} \quad (8\text{-}6)$$

DETERMINATION OF HETEROGENEOUS RATE CONSTANTS (k_{el}) 243

Equation (8-6) precisely describes a "reversible" cathodic polarogram. It is analogous to that for, say, the DME, only the ratio of $(D_O/D_R)^{1/2}$ is replaced by a ratio of mass-transport coefficients for convective diffusion. Of course, a reversible polarogram is just what one expects when the mass transport is the slower of the competing processes.

If, however, $m_O \gg k^{\circ\prime}$, Eq. (8-5), after substitution of $nFAm_OC_O$ for $i_{L,c}$, becomes

$$i = nFAk^{\circ\prime}C_O \exp\left[-\frac{\alpha nF}{RT}(E - E^{\circ\prime})\right]$$

If this is expressed in logarithmic form, it is seen to be the standard Tafel relationship:

$$\ln i = \ln(nFAk^{\circ\prime}C_O) - \frac{RT}{\alpha nF}(E - E^{\circ\prime}) \tag{8-7}$$

Finally, when m_O and $k^{\circ\prime}$ are of the same order of magnitude, Eq. (8-5) must be used as is. It can be rearranged conveniently, to calculate $k^{\circ\prime}$, as

$$k_f = \frac{1}{\dfrac{1}{m_O}\dfrac{i_{L,c} - i}{i} - \dfrac{1}{m_R}\exp\dfrac{nF}{RT}(E - E^{\circ\prime})} \tag{8-8}$$

Experimentally, m_O and m_R can be determined from the limiting currents. One then measures i at various E's along the wave and calculates k_f as a function of $E - E^{\circ\prime}$. Then a plot of $\log k_f$ vs. $E - E^{\circ\prime}$, i.e., the overpotential, yields $k^{\circ\prime}$ as intercept and a slope proportional to the transfer coefficient α.

The application of the RDE to determinations of $k^{\circ\prime}$ was discussed in Chapter 4. According to Levich, the applicable equation [Eqs. (4-24) and (4-25)] for a reduction process is

$$i = \frac{nFAD_OC_O^b}{\delta + (D_O/k_f)} \tag{4-25}$$

where δ is the thickness of the diffusion layer at the RDE. If this is rearranged to the form

$$i + i\frac{D_O}{\delta k_f} = \frac{nFAD_OC_O^b}{\delta} \tag{8-9}$$

the right side is seen to be $i_{L,c}$, since D_O/δ is the mass-transport coefficient for a RDE. Hence Eq. (8-9) can be written

$$\frac{D_O}{\delta k_f} = \frac{i_{L,c} - i}{i}$$

or

$$k_f = \cfrac{1}{\cfrac{1}{m_O}\left(\cfrac{i_{L,c} - i}{i}\right)} \qquad (8\text{-}10)$$

Actually, this is a form of the general equation (8-8) when $E - E^{\circ\prime} >$ ca. 100 mV. For a cathodic process, $E - E^{\circ\prime}$ is a negative quantity and the exponential-containing second term in the denominator of Eq. (8-8) becomes negligible. Thus the Levich-type equation is a form of the generalized hydrodynamic voltammetry treatment valid for large overpotentials.

Finally, it can also be shown that the Jordon treatment yields the Randles equation (128) for treating mixed anodic–cathodic polarograms (118). The Jordan treatment can be applied to any type of convective mass-transport system. Turbulent flow conditions can be used provided the turbulence is reproducible. It should be noted, however, that to evaluate fast reactions ($k^{\circ\prime} \sim 1$ cm/sec), the hydrodynamic treatment requires totally unreasonable mass-transport rates. Even with the flow tubes used by Jordon and co-workers, the maximum mass-transport rate was $V =$ ca. 0.13 cm/sec. Hence, the steady-state methods are mainly applicable to measurement of moderately rapid electron-transfer rates in the region 10^{-3}–10^{-1} cm/sec. A very thorough review of heterogeneous rate constants has been compiled by Tanaka and Tamamushi (129). As illustrated by this review, there is a considerable body of information on $k^{\circ\prime}$ for inorganic redox systems but almost nothing on organic compounds. A few measurements on aromatic hydrocarbons have been made by Aten and Hoijtink using a-c polarography (130), and aromatic nitro compounds and hydrocarbons were studied by Malachesky et al. using low-temperature hydrodynamic voltammetry (118,131). As yet, not enough information is available for meaningful discussion.

8-5. ELECTRODE PROCESSES WITH COUPLED HOMOGENEOUS CHEMICAL REACTIONS

Most of the electrode processes discussed in this section concern aromatic organic compounds and, in addition, involve oxidations at relatively low current densities, i.e., under conditions ordinarily encountered in electroanalytical studies. This does not reflect any intention to slight inorganic processes, but stems from the writer's background and interests.

There are differences in complexity between inorganic and organic processes, especially with regard to chemical follow-up reactions. In the charge-transfer process

$$Fe^{3+} + e \leftrightarrow Fe^{2+}$$

one realizes that the process is not just the simple injection of an electron into a ferric ion. The solvation atmospheres of ferric and ferrous are markedly different, and a considerable rearrangement of bonds accompanies this apparently elementary act. The resulting ferric ion is, nevertheless, easily recognized as an iron species—only now in a different oxidation state and with a new solvation sheath. But the addition or removal of one or more electrons in an organic molecule usually has a more profound influence on the molecular properties. A species may result which has entirely different stability and chemical characteristics from those of its parent.

The oxidation of *p*-phenylenediamine (PPD) in aqueous media is an excellent illustration of the spectrum of chemical complications which can develop in an organic electrode process.

In moderately strong acid (0.3 M HClO$_4$, etc.) the oxidation of PPD involves deprotonation of the di- and monoprotonated species, H$_2$PPD^{2+} and HPPD$^+$. These preceding or precursor reactions were examined by Mark and Anson (*96*) and more recently by Hawley (*61*). In this acidic medium one can show that $i_0\tau^{1/2}$ decreases markedly with increasing i_0, from which information on the rate of the precursor reactions can be obtained.

Far more involved chemical reactions enter the picture in the range pH 1–3. The overall charge transfer process is a two-electron oxidation of the diamine to *p*-phenylenediimine (PDI)—although the one-electron product, the cation radical, is stable and exists in this pH range. The generalized reaction can be written

PPD $\rightleftharpoons$ PDI $+ 2H^+ + 2e$

The diimine is indeed the conjugate oxidant of the parent PPD, but there the similarities in the chemical properties of the two compounds end. The diimine is a very reactive species. The relatively simple electron transfer written above can be followed by a complex series of fast chemical reactions (EC processes). The extent and rate of these are highly pH-dependent. In this moderate-pH range, the PDI undergoes rapid hydrolysis to the quinoneimine (QI). The QI, in turn, further hydrolyzes to produce *p*-benzoquinone (Q) as

$$\text{PDI} \xrightarrow[H_2O]{k_1} \text{QI} \xrightarrow[H_2O]{k_2} \text{Q}$$

Thus an oxidized solution of PPD rapidly acquires *three oxidants, only one of which is conjugate to the starting reductant.* The rates of these EC processes can be determined, for instance, by reverse-current chronopotentiometry.

The complexity, unfortunately, does not end at this point. The quinoidal products of the hydrolysis reactions are well known to undergo 1,4 addition (Michael) reactions with various nucleophiles. In the present case the nucleophile can be the excess starting diamine, PPD. Using the quinone as an example, the typical 1,4 addition reaction can be formulated:

$$Q + :NH_2{-}C_6H_4{-}NH_2 \longrightarrow \text{hydroquinone-NH-}C_6H_4{-}NH_2$$

The product of the 1,4 addition is now a substituted hydroquinone which is oxidized more easily than the PPD. Hence at the applied potential,

$$\text{HO-}C_6H_3(\text{OH}){-}NH{-}C_6H_4{-}NH_2 \longrightarrow \text{O=}C_6H_3(\text{=O}){-}NH{-}C_6H_4{-}NH_2 + 2H^+ + 2e$$

with further 1,4 additions now possible. All the quinoidal intermediates, including PDI and QI, are capable of the addition reactions (including double additions). The latter stages of the PPD reaction are seen to be ECE processes. The result is a cascade of coupled electrooxidations and chemical reactions. It is small wonder that the electrochemical oxidation of PPD appears to be an excellent means of preparing red, blue, or purple dyestuff solutions!

Depending on the type of mass transfer, the variety of products may or may not be seen as part of the overall electrode reaction. Thus time enters as a factor to be considered in stating the course of the reactions. Chemical reactions following charge transfer are certainly very predominate in organic systems and may lead quickly to extreme complications. The remainder of this section is concerned with follow-up reactions, especially the ECE variety.

The PPD example shows how quinoidal systems readily undergo follow-up chemical reactions. In general, one can expect similar behavior from the following electrochemically generated species:

1. Quinoidal compounds and similar unsaturated systems (quinone-methides).
2. Cation radicals ⎫ highly reactive toward solvent and other species.
3. Anion radicals ⎭
4. Various charged intermediates.

Clearly, in electrooxidations we can expect widespread occurrence of follow-up reactions. Furthermore, wherever these involve nucleophilic reactions with electron-donating entities, one can reasonably expect the products to be at least as oxidizable as the initial materials, and "cascade"-style reactions will prevail. In the electrooxidation of PPD and similar compounds, it is often the case that the product of the chemical reaction is more easily oxidized than the starting material, hence a "cascade"-type process. It should be emphasized that the ECE process simply designates a chemical reaction interposed between two electron transfers. The product of the intervening chemical reaction may be easier or more difficult to oxidize (reduce). This point will be examined more fully in the discussion of potentiostatic measurements of the chemical rates.

There are a variety of techniques which can be used to detect and measure the rates of ECE processes. All of them have a common approach, which is:

1. The initial electron transfer is carried out.
2. One provides a suitable "time gate" allowing the chemical reaction to occur.
3. One then examines:
   a. Electrochemically, the effect of the chemical reaction on the initial and subsequent electron transfers.
   b. By physical or other means, the products of the chemical and electrochemical reactions.

The time gate is the focal point of any of the methods. If the time gate is short, then *in the limit* one sees no effect of the chemical reaction—only the initial electron transfer. If the time gate is long with respect to the rate of the follow-up reactions, one sees their maximum effects. The time interval is easily provided in the fast sweep electrochemical techniques—or it may be adjusted via mass-transport conditions.

A general summary of the available techniques is outlined below. Many of the methods and some of their applications to these problems have been discussed individually in Chapters 4–6.

Electrochemical Techniques for Studying ECE Processes
1. Voltammetry at stationary electrodes
   a. Single sweep, i_p–$V^{1/2}$ variation.
   b. Cyclic sweep, direct observation of intermediates possible (cyclic voltammetry is excellent for qualitative studies; rate measurements are possible but difficult in practice).
2. Potentiostatic (chronoamperometric) method
   a. Measure $it^{1/2}$ (observed) and compare to $it^{1/2}$ (no chemical reaction). Working curves of $it^{1/2}$ ratio vs. kt.
   b. Coulometry at controlled potential—study i–t curves and examine products.
3. Chronopotentiometry
   a. Time gate adjustable via i_0. Measure τ ratios and $i_0\tau^{1/2}$ vs. i_0 variations.
   b. Current programs (current reversal).
   c. Thin-layer chronopotentiometry.
4. Rotated disk electrodes
   a. Examine i_L vs. $\omega^{1/2}$. Time gate is adjusted via ω. Recent treatments available for quantitative work.
   b. Ring-disk electrode—quantitative and direct observation of intermediates possible.

Cyclic voltammetry is, without question, the best technique for qualitative investigations and initial surveys. Direct evidence of intermediates is often possible. Nicholson and Shain have summarized many of the possible combinations of cyclic ECE patterns (97). It should be observed that in the Nicholson and Shain treatments irreversible charge transfers are those which exhibit no reverse current (i.e., an irreversible oxidation shows *no* cathodic current on the reverse sweep). This situation (like

ELECTRODE PROCESSES

the totally reversible counterpart) is seldom seen in practice. Quasi-reversible charge transfers are more frequent. This factor is to be considered in comparing experimental cyclic polarograms to the Nicholson and Shain patterns. Experimentally, irreversible charge transfers may well have some reverse currents.

The potentiostatic or chronoamperometric method is excellent for determining the intervening chemical reaction rates. Following the general treatment given by Alberts and Shain (98) it is intuitively easy to see how the chemical rate, k, can be measured. The general ECE process is denoted as

$$A \underset{}{\overset{\pm n_1 e}{\rightleftharpoons}} B$$

$$B + Z \xrightarrow{k} C$$

$$C \underset{}{\overset{\pm n_2 e}{\rightleftharpoons}} D$$

where Z (solvent, other nonelectroactive reactant, etc.) is most often considered to be present in high enough concentration that the chemical reaction is pseudo-first-order.

If either the time gate of k is very small, then it is clear little chemical reaction can occur. In the limit for $k = 0$, the current will simply obey the ordinary $it^{1/2}$ dependence as

$$i_{k=0} = \frac{n_1 F A D^{1/2} C_A^b}{\pi^{1/2} t^{1/2}}$$

If, on the other hand, the time interval is very long, or k infinitely large,

$$i_{k=\infty} = \frac{(n_1 + n_2) F A D^{1/2} C_A^b}{\pi^{1/2} t^{1/2}}$$

For the intermediate range, Alberts and Shain expressed the ratio of the observed current-time product $it^{1/2}$ to that for $k = \infty$ as

$$\frac{(it^{1/2})_{obs}}{(it^{1/2})_\infty} = 1 - \frac{n_2}{n_1 + n_2} e^{-kt}$$

Since, for many systems, $n_1 = n_2$, this reduces to the simple expression

$$\frac{(it^{1/2})_{obs}}{(it^{1/2})_\infty} = 1 - 0.5 e^{-kt}$$

Assigning numerical values to the parameter kt, a dimensionless working curve can be constructed of $(it^{1/2})_{obs}/(it^{1/2})_\infty$ vs. kt (usually log kt).

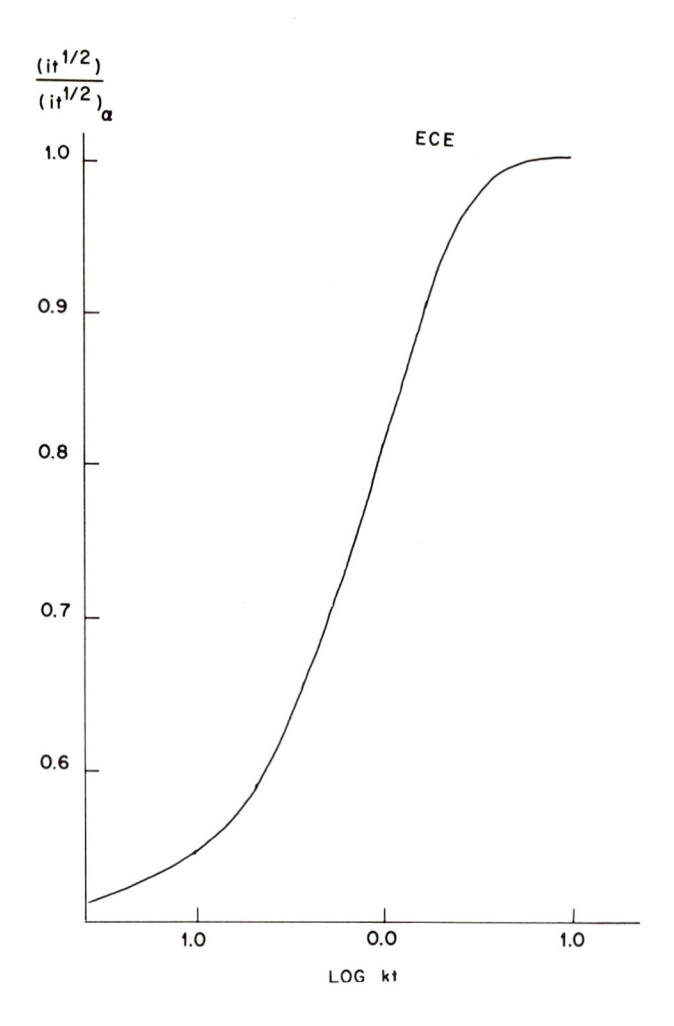

Fig. 8-9. Working curve for potentiostatic treatment of ECE processes.

A typical plot is seen in Fig. 8-9. Experimentally, k is determined as follows. The potentiostatic measurement is made at a value of E_{app} such that C_A and C_C at the electrode surface are zero. The resulting it curve is recorded and supplies the value $(it^{1/2})_{obs}$. The corresponding quantity $(it^{1/2})_\infty$ (for small k) obviously would require evaluation at large values of t. Departures from linear diffusion can be expected under these circumstances. A preferable procedure is to adjust solution conditions

ELECTRODE PROCESSES

so $k = 0$ (via pH control, etc.) and measure $(it^{1/2})_0$. The value of $(it^{1/2})_\infty$ is obviously a simple multiple (usually 2) of $(it^{1/2})_0$. (This assumes no gross change in the diffusion coefficient of the species over the wide pH change, but this problem does not appear to be significant.) From the ratio $(it^{1/2})_{obs}/(it^{1/2})_\infty$ one obtains directly the value of kt from the working curve. Evaluating several such points, one can now plot kt vs. t, which should be a straight line whose slope gives k (*61*).

Feldberg first pointed out a somewhat subtle but important discrepancy in the formulation presented above. The general scheme for an ECE reaction is rewritten here specifically for an oxidation:

$$A \rightleftharpoons B + n_1 e \qquad E$$
$$B + Z \xrightarrow{k} C \qquad C$$
$$C \rightleftharpoons D + n_2 e \qquad E$$

When the product of the chemical reaction, C, is more easily oxidized than the starting compound, A, it is clear that electrochemical oxidation of C is possible at the applied potential. However, it is also implicit from the formal potentials of the redox couples B,A and D,C that the oxidant B can *chemically* oxidize C. (Oxidation examples are used here but entirely analogous arguments hold for reductions.) In fact, even if $E^{\circ\prime}$ of the system D,C is about the same or even slightly more anodic than that of B,A, there still exists the equilibrium expression for the chemical interaction:

$$B + C \rightleftharpoons D + A$$

The usual calculations of the ECE process really only hold for the case where the forward and reverse rates of this equilibrium are zero. This is, in reality, a very unlikely situation. If both the redox couples B,A and D,C are reversible systems, one would predict their cross-reaction rate to also be rapid.

Electrochemically, the *net effect* is exactly the same as if C were oxidized via $n_2 e$'s at the electrode surface, but the process should be written as ECC (two subsequent chemical steps):

$$A \rightleftharpoons B + n_1 e \qquad E$$
$$B + Z \xrightarrow{k} C \qquad C$$
$$n_2 B + n_1 C \rightleftharpoons n_1 D + n_2 A \qquad C$$

When this additional chemical equilibrium is included in the kinetic problem, Hawley and Feldberg showed that the working curves, while agreeing with the ECE calculations at low values of kt, were significantly different at higher values *(141)*. Feldberg and co-workers have discussed these nuances of the ECE process in detail with experimental verifications *(141,142)*.

To clear up any confusion developed by the ECE nuances one can write the overall process as

$$A \rightleftharpoons B + n_1 e \quad \text{at potential } E_1$$

$$B + Z \xrightarrow{k} C$$

$$C \rightleftharpoons D + n_2 e \quad \text{at potential } E_2$$

$$(B + C \rightleftharpoons D + A)$$

Now, one extreme case is where E_2 is considerably more anodic (ca. 0.2 V) than E_1. In this case, to study the overall reaction as an ECE requires an applied potential of E_2. If only E_1 is applied, the reaction will simply be an EC process consuming $n_1 e$'s (the chemical conversion of B → C still occurs but will not be seen chronoamperometrically if E_1 is applied).

If, however, E_2 is only slightly more anodic, equal to, or cathodic of E_1, then in all cases the "second" chemical reaction $B + C \rightleftharpoons D + A$ is at least competitive with the second electrochemical step and should be included in the calculations. The extent of its participation in the calculational results will depend on K_{eq} (and therefore on the difference in formal potentials of the redox systems B,A and D,C). It should be observed in passing that Feldberg's notation states the forward direction of the chemical equilibrium in reverse to that used here. Hence small values of K_{eq} correspond to an overriding ECC reaction.

In spite of the reality and calculational importance of the "second" chemical step, it seems preferable to retain the shorthand ECE for all electrochemical reactions with a chemical step interposed between two charge transfers. The ECE designation seems preferable since it indicates that additional electrons are observed as current flow experimentally, regardless of whether the detailed reaction is ECE or ECC. In actual fact, it would appear that all but a few such reactions proceed via the ECC mode and should be calculated accordingly. Feldberg has calculated potentiostatic working curves for a variety of processes, including second-order (dimerization) reactions *(107)*.

Some promising new developments in chronoamperometric techniques for follow-up reactions include the double-potential-step method of

Schwarz and Shain (*144*) and the double-potential-step chronocoulometry technique of Christie and Lingane (*145,146*).

Chronopotentiometry is also quite useful in studying ECE processes and was applied initially by Testa and Reinmuth in 1961 (*71*). Alberts and Shain compared their potentiostatic measurements of k with chronopotentiometric data (*98*). Herman and Bard discussed the utility of programmed current chronopotentiometry (reverse, current decrease, and cyclic) to the ECC problem (*105,106*). The general idea of current programmed chronopotentiometry applied to electrode processes with chemical complications were discussed in Chapter 6 and further details can be found there and in the original literature.

The newest technique, and one which has exciting possibilities especially for ECE processes, employs the rotated disk electrode (RDE). A relatively thorough treatment of the RDE was given in Chapter 4 and only details pertinent to electron transfers with coupled chemical reactions are presented here. Actually, fairly early in RDE history, Levich and Koutecky described the behavior of the RDE for a variety of coupled chemical processes (*114*). Galus and Adams treated similar processes using reaction-layer approximations (*115*). In addition, a variety of workers, including especially Vielstich and his group, examined precursor reactions. This work was described in detail in Chapter 4. Haberland and Landsberg have recently studied catalytic processes using the system $Fe(III)/Fe(II)-H_2O_2$ as a test (*116*). Tong et al. looked at reactions of interest to the photograplic process by oxidizing N,N-dialkyl-*p*-phenylenediamines at the RDE in the presence of various coupling agents, etc. (*117*).

The application of the RDE to ECE processes is intuitively straightforward and it is surprising it has not been used until now. Imagine an ECE mechanism with a moderate rate for the intermediate chemical step. One applies a potential on the limiting current plateau and increases the rotation rate of the RDE. It may, in the accesible range of ω, be possible to move the intermediate product (C, in the previous nomenclature) away from the electrode surface too rapidly for it to undergo the second electron-transfer step. Hence i_L will correspond to an n_1 electron process. Conversely, at low rotation rates, most or all of C is around the electrode surface long enough to show i_L corresponding to $n_1 + n_2$ electrons.

A plot of i_L vs. $\omega^{1/2}$ is identical in its form to that for the potentiostatic case given by Alberts and Shain (Ref. *98*, Fig. 1). Fundamentally, a potentiostatic experiment is being carried out at the RDE, but it is operated under convective diffusion conditions rather than semiinfinite linear diffusion. In the RDE method, the rotation rate (mass-transfer time gate) is the analogue of the time dependence of current in the potentiostatic treatment.

This analogy was used by Malachesky et al. (*118,119*) to develop an approximate quantitative treatment for the RDE. The time for a molecule to diffuse from the electrode surface ($x = 0$) to the limit of the diffusion-layer thickness ($x = \delta$) at the RDE was assumed to be given by $t = \delta^2/\pi D$. This, then, allows substitution of RDE variables into the chronoamperometric (potentiostatic) equation for the flux of intermediate C. A final solution in terms of n_{app}/n_1 or the ratio of $i_{L(obs)}/i_{L(k=0)}$ allows calculation of k (*119*). The equations were verified experimentally via the reduction of *p*-nitrosophenol (a well-known ECE reaction).

The success of this approximate treatment led Feldberg to apply the digital simulation techniques used for potentiostatic ECE and ECC processes to the RDE. The general approach of the digital simulation technique has been given by Feldberg and Auerbach (*120*). When applied

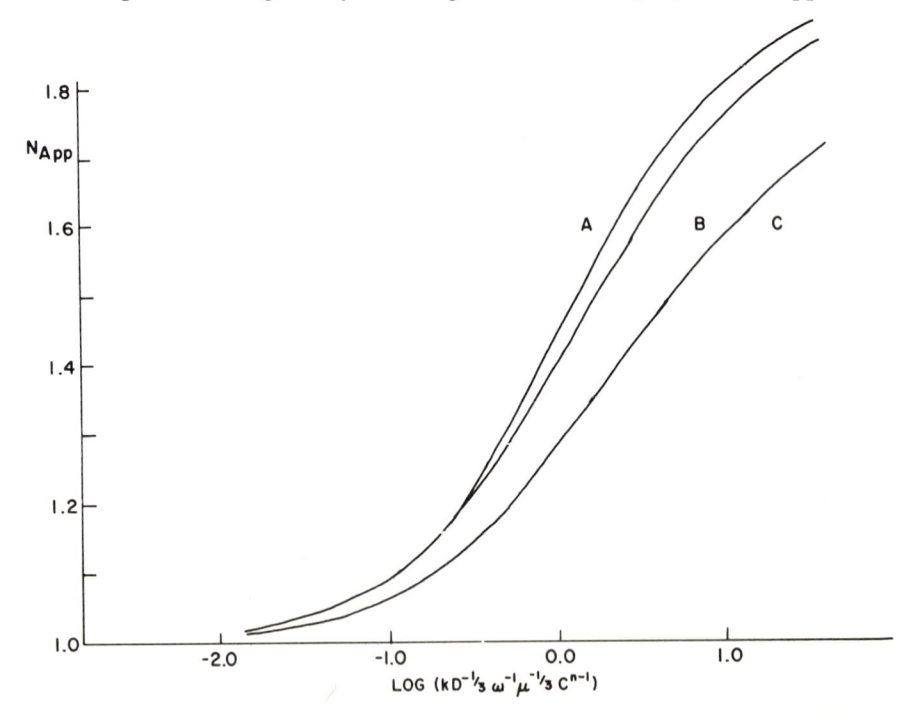

Fig. 8-10. Working curves for RDE treatment of ECE processes. A, First-order ECE; B, first-order ECC, $K = 0$; C, second-order ECE.

to the simple case of electron transfer with no chemical complications at the RDE, it gives results in excellent agreement with Levich's equation and the numerical calculations of Hale (*121*). The inclusion of the ECE and ECC complications leads to working curves of n_{app}/n_1 vs. a parameter related to the RDE, $\log k D^{-1/3} \omega^{-1} v^{1/3} C^{n-1}$, where the variable C^{n-1} allows plotting first- and second-order chemical reactions on the same graph. The other RDE symbols have their usual significance.

Figure 8-10 shows typical working curves for several processes. Case A is the usual first-order ECE (without the ECC nuance). Case B is the ECC calculated for an equilibrium constant $K = 0$. As in the potentiostatic case, there are significant differences at $n_{app}/n_1 >$ ca. 1.2. Case C is an important one, that of a second-order intermediate chemical step:

Case C:
$$A \xrightarrow{\pm n_1 e} B$$
$$2B \xrightarrow{k} B_2$$
$$B_2 \xrightarrow{\pm 2 n_1 e} D$$

The working curve in Fig. 8-10 for case C is a straight ECE calculation. Experimental verification of the case C behavior has been obtained by Marcoux for the oxidation of substituted triphenylamines (*122*). Here the initial electron transfer gives the cation radical. Dimerization of two cations yields the tetraphenylbenzidene, which undergoes further electrooxidation. Typically, these dimerization rates are of the order of 10^3–10^4 M^{-1}/sec and were measured with ease and precision at the RDE (*123*). The potential applications of the RDE to chemical follow-up reactions are very promising.

8-6. PHYSICOCHEMICAL METHODS FOR STUDYING ELECTRODE REACTIONS

The approach to the study of electrode mechanisms outlined so far in this chapter essentially employs one or more electrochemical methods to evaluate various kinetic parameters of the electrochemical reaction. For want of a better name it might be called the electrode kinetics method. It relies heavily on deductions made only from electrochemical measurements. *With the uncertainties which exist in solid electrode practice, it is the writer's opinion that this method alone cannot be trusted.* Over a period of several years' experience with organic oxidation reactions we have decided that the only safe approach is to use first as many of the different electrochemical techniques as possible. All these results should not be trusted until they can be supported, preferably by physicochemical methods essentially independent of the electrochemistry. An organic electrode reaction is, after all, only an organic reaction which is made to proceed via the influence of an electrical potential. As such, it has the added complexities of the electrical process, but it loses none of the subtleties of an organic molecule interacting with its environment. A proper interpretation of the electrode reaction must satisfy the basic tenets of sophomore organic chemistry. It is rare today that the organic chemist relies strictly on melting points and elemental analysis to verify products of his reactions. Every physicochemical technique available is in widespread use in organic reaction mechanism studies. It seems even more pertinent that they be used in the study of organic electrode processes. This attitude is not unique in the writer's laboratory—it has been used very successfully by a variety of workers in the area.

The independent physicochemical techniques which are very useful include controlled-potential coulometry, visible, UV, and IR spectroscopy (applied directly to the electrode reaction or to products following electrolysis), electron paramagnetic resonance (EPR) methods, trace identification of products, etc. While controlled-potential coulometry is clearly an electrochemical technique, it differs from those methods considered under the electrode kinetics approach in that, directly or indirectly, it attempts to identify intermediate and final products of the electrode reaction. Some of these methods will be discussed briefly in the next sections.

Ordinary optical spectroscopy methods require little discussion. Monitoring of an electrolysis by periodic examination of the solution via visible, UV, or IR spectroscopy is straightforward. Rapid-scan spectrophotometers are available and one can set up the entire electrolysis cell in the spectrophotometer and monitor moderately short-lived intermediates or other products. In many cases, even slow scan equipment used concurrently with the electrolysis, can give valuable information. As an example one can cite the work of Kemula and Sioda on the nitrobenzene anion radical. Here the visible spectrum of the radical ion was observed and at the same time voltammetry measured the concentration of nitrobenzene and its negative ion in the optical-electrochemical cell (*80*). Fluorescene spectrophotometry is often a valuable aid in identifying products with aromatic and heterocyclic systems. Secondary chemical reactions can be identified via spectrophotometry (79). Infrared identification, although it has not been used widely, is often quite valuable (81).

A spectroscopic technique of great potential was introduced by Kuwana et al. (108). They showed that tin oxide–coated glass surfaces could be used as electrodes and bulk concentrations of electrolysis products and intermediates could be monitered via absorption spectrophotometry. The most exciting development has been the use of such electrodes in conjunction with internal reflection spectrometry to measure concentrations very near the electrode surface (109,110). Similar studies at germanium surfaces in the infrared region and on thin films of platinum and palladium deposited on glass and optical crystals have been made by Mark and co-workers (111,112). These techniques offer great possibilities toward studying absorbing intermediates at or near the actual electrode surface.

Controlled-potential coulometry was pointed out as a very effective tool in studying electrode reactions by Geske and Bard (82). The effect of secondary chemical reactions on the concentration–time curves and the apparent number of electrons involved in the reaction was evaluated. Their mathematical treatment of single and parallel secondary chemical reactions as well as catalytic regeneration schemes allows evaluation of n and, in many cases, the rates of the chemical follow-up reactions. Geske and Bard consider secondary reactions in terms of interaction of the initial charge-transfer product, R, with the starting substance, C, or a nonelectroactive species, Z. Two of the simplest cases which one encounters are:

Case I: $C \pm ne \rightarrow R$

Case II: (a) $R + Z \rightarrow X$
          (b) $R + C \rightarrow Y$ (X, Y electroinactive)
          (c) $R + Z \rightarrow C$

Case I is, of course, merely a charge-transfer process. Controlled-potential coulometry gives current–time curves which decrease exponentially with time and integral n values for the reaction are obtained. The utility of controlled-potential coulometry for this type of process is well documented (83).

Case IIa is like the simple charge-transfer reaction except that the final product, instead of being R, is transformed into X. Case IIb, however, consumes intial starting material C by reaction with product R. Here the apparent n value will usually be nonintegral and will vary with the concentration of starting material C. If the reaction of case IIb has a rate which is very large or very small, then integral final n values will be observed. The notation n_{app}^0 was used by Geske and Bard to signify the coulometric n values at the end of electrolysis when C approaches zero concentration. If the rate of IIb is extremely small, the system reverts to case I and $n_{app}^0 = n$. If the rate of IIb is very rapid, $n_{app}^0 = n/2$. For intermediate values of the chemical rate constant, n_{app}^0 is a function of the initial concentration of C. Details of the treatment to evaluate the rate constant are given in the original literature along with more complicated reaction schemes (*82*).

In continuing work Bard and co-workers have extended this type of treatment to include follow-up chemical reactions which generate electroactive products (*84*) and chemical reactions which precede or run simultaneously with the charge-transfer process (*85*). In the preceding chemical reaction case, where C is formed reversibly from a precursor Y, in the evaluation of n_{app}^0 all of C and Y are electrolyzed. Hence $n_{app}^0 = n$ and the presence of the reaction can be discerned only from the current-time curves.

Bard has pointed out the advantages of high-speed controlled-potential coulometry. By using total electrolysis times of ca. 100 sec or less, competing side reactions can sometimes be eliminated or their effect better studied (*86*). A suitable apparatus with high rates of convective mass transport was constructed for this purpose by Bard. For very short electrolysis times the output capacity of the potentiostat must be quite high. For integration of the i–t curve, the current is fed to a standard dropping resistor which is the input to a voltage to frequency converter. A counter (scaler) records the output of the converter and can be arranged to read out in coulombs (*87*).

Determination of n values using short electrolysis times has also been accomplished by Christensen and Anson in their study of chronopotentiometry in thin layers of solution (*88*). In this technique, the chronopotentiometry is carried out at an electrode with the solution bounded a short distance (l) away (ca. 10^{-3} cm) by an inert wall. If the transition time is chosen so that the quantity $l^2/3D$ is negligible compared to τ, then τ is given by (*88*)

$$\tau = \frac{nFAlC^b}{i}$$

To determine n for an unknown system, one calibrates by determining τ for a known system [i.e., reduction of Fe(III)]. A determination of τ for an equal concentration of the unknown system gives a direct measure of n. Alternatively, n can be calculated via the ratio of unequal concentrations.

Thin-layer chronopotentiometry is, in a sense, a coulometric method. McClure and Maricle recently demonstrated the utility of a new dip-type thin-layer cell for n-value determinations (*113*).

Meites and co-workers have contributed heavily to controlled-potential coulometry in the study of electrode reactions (*89,90*). An excellent review of the application of controlled-potential electrolysis has been given by Meites (*91*).

Within the span of 7–8 years electron paramagnetic resonance (EPR) techniques have become an important adjunct of organic electrochemistry. EPR is concerned with the detection and identification of free radicals and radical ions (although there are many other applications which will not be considered here). No attempt will be made to present the principles of EPR. Very thorough and authoritative treatments are available (*132–134*). Two excellent reviews by Carrington (*135*) and Horsfield (*136*) cover the applications to chemical problems.

The elegant work of Maki and Geske mated the diverse areas of electrochemistry and EPR. They carried out the electrolysis directly in the microwave cavity by internal generation (IG) and showed that a variety of radical ions could be produced in nonaqueous media. The technique was extended to aqueous solutions by the writer and co-workers. External generation (EG) at large electrode surfaces and pumping of the radical species to the EPR cavity by various means was largely developed by Fraekel's group. These techniques, their applications, and relative merits have been reviewed (*137*).

There are two main areas in which EPR has contributed to organic electrochemistry. The most obvious is in the detection and identification of radical intermediates (or stable products) in electrode reactions. Second, EPR provides information about solution interactions of radical ions which is closely correlated with the electrochemistry. While the former area is well documented with examples of practical applications, the latter has not been exploited fully yet.

When a well-defined EPR spectrum is obtained, the EPR method is, without question, the best method of identifying a radical intermediate in a complex organic electrode reaction! The electrochemical methods reflect only secondary properties of the intermediate (i.e., redox potential, diffusion coefficients, etc.). Other physicochemical techniques (UV, visible spectra, etc.) usually give a generalized structural picture. EPR provides positive identification of the radical species. This is so because the detailed nature of the hyperfine interaction between the unpaired electron and the various magnetic nuclei in the radical species produce an intimate picture of the molecular structure. Although EPR spectra are not absolutely unique, it is rare that a well-resolved spectrum cannot provide positive identification in electrochemical studies.

In this connection, the IG technique would seem preferable, since short-lived intermediates are generated directly in the EPR cavity. This is certainly the preferred initial experimental approach, but the IG method is fraught with considerable difficulties. The peculiarities of EPR cavity design dictate that most of the useful EPR signal originates from a very small portion of the thin electrolysis cell used. Hence accurate positioning of the cell is required for reproducible results. Further, while quantitative calculations of radical ion concentration profiles can be made from diffusion theory, in practice linear diffusion theory simply does not hold. Gross material convection (sometimes advantageous) develops in the thin electrolysis cells. (The volume and shape of the electrochemical cell which can be inserted in an EPR cavity is restricted by the requirements of the microwave absorption.) The IG technique develops a radical ion concentration which generally is in equilibrium with a large excess of the parent compound and broadening of the EPR hyperfine via homogeneous electron transfer can be a detriment to obtaining well-resolved spectra. While the IG technique was eminently successful (and continues to be so) in the identification of radical intermediates, it is often especially difficult to work with in the quantitative sense.

The EG method might seem to be of little value in detecting intermediates but, of course, many organic radical ions have relatively great stability. Where follow-up chemical reactions ensue, the EG technique is often of value in identifying secondary or final products. In fact, combined with controlled-potential coulometry, external generation is a very powerful tool for checking overall reaction schemes. For instance, frequently substituted quinones are the end result of aromatic oxidations. These quinones are not those which result from degradative oxidation at high anodic potentials. Rather, they develop as a result of a series of follow-up chemical reactions (hydrolyses, etc.) after initial electron transfer. Now, semiquinone anions are extremely well documented in EPR. Furthermore, wherever one has an aqueous solution of an aromatic quinone or hydroquinone, it is frequently possible to obtain a relatively stable solution of the paramagnetic semiquinone by carefully adjusting the pH to ca. 8–10. (The exact nature of this reaction is not well established. However, it generates a surprisingly high concentration of semiquinone radical ion and is almost always successful. Care must be taken to not raise the pH too high or polymeric products result.) Thus an excellent means of establishing overall products in several aromatic oxidations involves EG at controlled potential (with coulometric integrations to yield n_T) followed by examination in the EPR spectrometer for the final quinone product. Any identification one can make is superior and less time consuming than chemical separation and identification.

While EPR has not fulfilled the hopes of some investigators in detecting very short-lived intermediates in electrode reactions, it has provided a wealth of data to elucidate especially complex organic processes. Chemical follow-up reactions are often immediately confirmed by the presence of a radical ion which is not the conjugate one-electron intermediate of the starting material (137–139).

There is a great deal of valuable information yet to be obtained on the chemical and physical interaction of radical ions which will play an important role in understanding the solution electrochemistry. The utility of EPR in this respect is already documented. In particular, comparisons of homogeneous and heterogeneous electron-transfer rates of organic systems is an interesting new area (118,131). Interactions of ion radicals with solvent systems and other solution species should provide much data of value to electrochemistry.

The more conventional chemical and physicochemical separation techniques should be emphasized in electrode mechanism studies. One must be careful in this regard that the electrochemistry is not run either on such a large-scale or long-time basis that the final products bear little relation to the detailed course of the reaction. However, with high-speed coulometry, quantities of products sufficient for identification can be made easily in relatively short times. Column, paper, and gas chromatography as well as thin-layer chromatography and electrophoretic techniques can be applied easily to such simple sizes. At lower product levels, tracer techniques are sometimes advantageous, although they have not been widely used (140).

Although this section on nonelectrochemical methods is relatively short, it is hoped that it appears in proper perspective. These supporting techniques are indispensable in verifying the nature of a complex organic electrode reaction. As pointed out in the preceding sections, the purely electrochemical information is often tenuous (especially with solid electrode surfaces). In complex organic reactions there may well be sets of electrode kinetics parameters which will satisfy several interpretations. In the final analysis, however, electrons are transferred, intermediates and products are formed, and positive identification of these, independent of the electrochemical numbers game is most satisfying.

REFERENCES

1. M. von Stackelberg, M. Pilgram, and V. Toome, *Z. Elektrochem.*, **57**, 342 (1953).
2. I. Shain and K. J. Martin, *J. Phys. Chem.*, **65**, 254 (1961).
3. H. A. Laitinen and I. M. Kolthoff, *J. Am. Chem. Soc.*, **61**, 3344 (1939).
4. J. F. Zimmerman, Ph.D. thesis, Univ. Kansas, Lawrence, 1964.
5. V. G. Levich, *Physicochemical Hydrodynamics*, Prentice-Hall, Englewood Cliffs, N.J., 1962, p. 326.
6. T. R. Mueller, Ph.D. thesis, Univ. Kansas, Lawrence, 1961.
7. T. O. Rouse, Ph.D. thesis, Univ. Minnesota, Minneapolis, 1961.
8. A. J. Bard, *Anal. Chem.*, **35**, 340 (1963).
9. A. I. Fedorova, G. L. Vidovich, L. I. Boguslavskii, and V. D. Yukhtanova, *Soviet Electrochemistry, Proc. Congr. Electrochem. (English Transl.) 4th, Moscow, 1956*, **1**, 153 (1961).
10. M. D. Wijnen and W. M. Smit, *Rec. Trav. Chim.*, **79**, 289 (1960).
11. M. Eisenberg, C. W. Tobias, and C. R. Wilke, *J. Electrochem. Soc.*, **101**, 306 (1954).
12. M. B. Kraichman and E. A. Hogge, *J. Phys. Chem.*, **59**, 986 (1955).
13. E. A. Hogge and M. B. Kraichman, *J. Am. Chem. Soc.*, **76**, 1431 (1954).
14. A. L. Beilby and A. L. Crittenden, *J. Phys. Chem.*, **64**, 177 (1960).
15. I. M. Kolthoff and J. Jordan, *J. Am. Chem. Soc.*, **75**, 1571 (1953).
16. E. Morgan, J. E. Harras, and A. L. Crittenden, *Anal. Chem.*, **32**, 756 (1960).
17. J. H. Wang, *J. Am. Chem. Soc.*, **76**, 1528 (1954).
18. J. H. Wang and F. M. Polestra, *J. Am. Chem. Soc.*, **76**, 1584 (1954).
19. C. L. Rulfs, *J. Am. Chem. Soc.*, **76**, 2071 (1954).
20. D. J. Macero and C. L. Rulfs, *J. Am. Chem. Soc.*, **81**, 2942, 2944 (1959).
21. T. Pavlopoulos and J. D. H. Strickland, *J. Electrochem. Soc.*, **104**, 116 (1957).
22. I. M. Kolthoff and C. S. Miller, *J. Am. Chem. Soc.*, **63**, 1013 (1941).
23. J. Jordan, E. Ackerman, and R. L. Berger, *J. Am. Chem. Soc.*, **78**, 2979 (1956).
24. Ref. 5, p. 325.
25. L. A. Woolf, *J. Phys. Chem.*, **64**, 481 (1960).
26. M. Breiter and K. Hoffman, *Z. Elektrochem.*, **64**, 462 (1960).
27. V. Bagotzky, *Zh. Fiz. Khim.*, **22**, 1466 (1948).
28. E. A. Aykazyan and A. I. Fedorova, *Dokl. Akad. Nauk SSSR*, **86**, 1137 (1952); see also Ref. 5, pp. 311, 325.
29. I. M. Kolthoff and J. Jordan, *J. Am. Chem. Soc.*, **74**, 4801 (1952).
30. D. P. Gregory and A. C. Riddiford, *J. Chem. Soc.*, **1956**, 3756.
31. J. D. Newson and A. C. Riddiford, *J. Electrochem. Soc.*, **108**, 695 (1961).
32. M. von Stackelberg and M. Pilgram, *Collection Czech. Chem. Commun.*, **25**, 2974 (1960).
33. R. Landsberg, W. Geissler, and S. Müller, *Z. Chem.*, **1961**, 169.
34. G. P. Lewis and P. Ruetschi, *J. Phys. Chem.*, **67**, 65 (1963).
35. D. Jahn and W. Vielstich, *J. Electrochem. Soc.*, **109**, 849 (1962).
36. J. J. Lingane, *J. Electroanal. Chem.*, **2**, 296 (1961).
37. J. J. Lingane, *J. Electroanal. Chem.*, **2**, 46 (1961).
38. D. G. Davis and M. E. Everhart, *Anal. Chem.*, **36**, 38 (1964).
39. R. H. Sanborn and E. F. Orlemann, *J. Am. Chem. Soc.*, **77**, 3726 (1955).
40. J. H. Wang, C. V. Robinson, and I. S. Edelman, *J. Am. Chem. Soc.*, **75**, 466 (1953).

REFERENCES

41. D. T. Sawyer and R. J. Day, *J. Electroanal. Chem.*, **5,** 195 (1963).
42. L. H. Bodnar and D. M. Himmelblau, *J. Appl. Radiation Isotopes*, **13,** 1 (1962).
43. D. G. Davis, *Anal. Chem.*, **35,** 764 (1963).
44. S. Karp and L. Meites, *J. Am. Chem. Soc.*, **84,** 906 (1962).
45. A. J. Bard, *Anal. Chem.*, **35,** 1602 (1963).
46. J. D. Voorhies and E. J. Schurdak, *Anal. Chem.*, **34,** 939 (1962).
47. J. D. Voorhies and N. H. Furman, *Anal. Chem.*, **31,** 381 (1959).
48. T. Kuwana, Ph.D. thesis, Univ. Kansas, Lawrence, 1959.
49. H. Lund, *Acta Chim. Scand.*, **11,** 1323 (1957).
50. K. E. Friend and W. E. Ohnesorge, *J. Org. Chem.*, **28,** 2435 (1963).
51. T. A. Miller, B. Prater, J. K. Lee, and R. N. Adams, *J. Am. Chem. Soc.*, **87,** 121 (1965).
52. G. A. Ward, *Talanta*, **10,** 261 (1963).
53. T. Kuwana, D. E. Bublitz, and G. Hoh, *J. Am. Chem. Soc.*, **82,** 5811 (1960).
54. E. A. Aykazyan and Yu. V. Pleskov, *Zh. Fiz. Khim.*, **31,** 205 (1957); also see Ref. 5, p. 325.
55. H. Y. Lee, M.S. thesis, Univ. Kansas, Lawrence, 1962.
56. J. Zimmerman, D. Hawley, and R. N. Adams, unpublished data, 1964.
57. T. A. Miller, B. Lamb, K. Prater, J. K. Lee, and R. N. Adams, *Anal. Chem.*, **36,** 418 (1964).
58. Z. Galus, C. Olson, H. Y. Lee, and R. N. Adams, *Anal. Chem.*, **34,** 164 (1962).
59. I. M. Kolthoff and E. Orlemann, *J. Am. Chem. Soc.*, **63,** 664 (1941).
60. D. G. Peters and J. J. Lingane, *J. Electroanal. Chem.*, **2,** 1 (1961).
61. D. Hawley, Ph.D. thesis, Univ. Kansas, Lawrence, 1965.
62. J. H. Wang, *J. Am. Chem. Soc.*, **73,** 510, 4181 (1951).
63. J. H. Wang, *J. Am. Chem. Soc.*, **76,** 1528 (1954).
64. J. H. Wang and F. M. Polestra, *J. Am. Chem. Soc.*, **76,** 1584, (1954).
65. E. Fishman, *J. Phys. Chem.*, **59,** 469 (1955).
66. E. Fishman and T. Vassiliades, *J. Phys. Chem.*, **63,** 1217 (1959).
67. D. W. McCall, D. C. Douglass, and E. W. Anderson, *J. Chem. Phys.*, **31,** 1555 (1959).
68. H. V. Druschel and J. F. Miller, *Anal. Chem.*, **29,** 1456 (1957).
69. S. V. Tatwawdi and A. J. Bard, *Anal. Chem.* **36,** 2 (1964).
70. E. Solon and A. J. Bard, *J. Am. Chem. Soc.*, **86,** 1926 (1964).
71. A. C. Testa and W. H. Reinmuth, *J. Am. Chem. Soc.*, **83,** 784 (1961); *Anal. Chem.*, **33,** 1320 (1960).
72. L. Gierst, *Z. Elektrochem.*, **59,** 784 (1955).
73. Z. Galus and R. N. Adams, *J. Am. Chem. Soc.*, **84,** 2061 (1962).
74. Z. Galus and R. N. Adams, *J. Phys. Chem.*, **67,** 866 (1963).
75. T. R. Mueller and R. N. Adams, *Anal. Chim. Acta*, **25,** 482 (1961).
76. L. Gierst and H. Hurwitz, *Z. Elektrochem.*, **64,** 36 (1960).
77. M. Breiter, M. Kleinerman, and P. Delahay, *J. Am. Chem. Soc.*, **80,** 5111 (1958).
78. P. Delahay, in *Treatise on Analytical Chemistry* Pt. 1, Vol. 4, (I. M. Kolthoff and P. J. Elving, eds.), Wiley (Interscience), New York, 1963, Chap. 44.
79. D. H. Geske, *J. Phys. Chem.*, **63,** 1062 (1959).
80. W. Kemula and R. Sioda, *Bull. Acad. Polon. Sci.*, **10,** 507 (1963); *Nature*, **197,** 588.

81. D. M. Mohilner, R. N. Adams, and W. J. Argersinger, *J. Am. Chem. Soc.*, **84**, 3618 (1962).
82. D. H. Geske and A. J. Bard, *J. Phys. Chem.*, **63**, 1057 (1959).
83. J. J. Lingane, *Electroanalytical Chemistry*, Wiley (Interscience), New York, 2nd ed., 1958.
84. A. J. Bard and J. S. Mayell, *J. Phys. Chem.*, **66**, 2173 (1962).
85. A. J. Bard and E. Solon, *J. Phys. Chem.*, **67**, 2326 (1963).
86. A. J. Bard, *Anal. Chem.*, **35**, 1125 (1963).
87. A. J. Bard and E. Solon, *Anal. Chem.*, **34**, 1181 (1962).
88. C. R. Christensen and F. C. Anson, *Anal. Chem.*, **35**, 205 (1963).
89. L. Meites and S. A. Moros, *Anal. Chem.*, **31**, 23 (1959).
90. S. Karp and L. Meites, *J. Am. Chem. Soc.*, **84**, 906 (1962).
91. L. Meites, in *Technique of Organic Chemistry*, Vol. 1, Pt. 4 (A. Weissberger, ed.), Wiley (Interscience), New York, 3rd ed., 1960.
92. J. C. Bazan and A. J. Arvia, *Electrochim. Acta*, **10**, 1025 (1965).
93. R. Landsberg, W. Müller, and J. Hendel, *J. Electroanal. Chem.*, **2**, 400 (1961).
94. R. A. Munson, *J. Electroanal. Chem.*, **5**, 292 (1963).
95. P. Delahay, *Double Layer and Electrode Kinetics*, Wiley, New York, 1965.
96. H. B. Mark and F. C. Anson, *Anal. Chem.*, **35**, 722 (1963).
97. R. S. Nicholson and I. Shain, *Anal. Chem.*, **37**, 178 (1965).
98. G. S. Alberts and I. Shain, *Anal. Chem.*, **35**, 1859 (1963).
99. L. Papouchado and L. S. Marcoux, unpublished data, 1967.
100. S. Wawzonek, E. W. Blaha, R. Berkey, and M. E. Runner, *J. Electrochem. Soc.*, **102**, 235 (1955).
101. H. A. Laitinen and S. Wawzonek, *J. Am. Chem. Soc.*, **64**, 1765, 2365 (1942).
102. G. J. Hoijtink and J. Van Schooten, *Rec. Trav. Chim.*, **71**, 1089 (1952).
103. M. E. Peover and B. S. White, *J. Electroanal. Chem.*, **13**, 93 (1967).
104. J. Phelps, K. S. V. Santhanam, and A. J. Bard, *J. Am. Chem. Soc.*, **89**, 1752 (1967).
105. H. B. Herman and A. J. Bard, *J. Phys. Chem.*, **70**, 396 (1966).
106. A. J. Bard and H. B. Herman, in *Polarography 1964* (G. J. Hills, ed.), Macmillan, New York, 1966, p. 373.
107. S. W. Feldberg, in *Electroanalytical Chemistry*, Vol. 3 (A. J. Bard, ed.), Dekker, New York, 1969.
108. T. Kuwana, R. K. Darlington, and D. W. Leedy, *Anal. Chem.*, **36**, 2023 (1964).
109. W. N. Hansen, R. A. Osteryoung, and T. Kuwana, *J. Am. Chem. Soc.*, **88**, 1062 (1966).
110. W. N. Hansen, T. Kuwana, and R. A. Osteryoung, *Anal. Chem.*, **38**, 1810 (1966).
111. H. B. Mark and B. S. Pons, *Anal. Chem.*, **38**, 119 (1966).
112. B. S. Pons, J. S. Mattson, L. O. Winstram, and H. B. Mark, *Anal. Chem.*, **39**, 685 (1967).
113. J. E. McClure and D. L. Maricle, *Anal. Chem.*, **39**, 236 (1967).
114. V. G. Levich and J. Koutecky, *Zh. Fiz. Khim.*, **32**, 1569 (1952).
115. Z. Galus and R. N. Adams, *J. Electroanal. Chem.*, **4**, 248 (1962).
116. D. Haberland and R. Landsberg, *Ber. Bunsenges. Physik. Chem.*, **70**, 720 (1966).
117. L. K. J. Tong, K. Liang, and W. R. Ruby, *J. Electroanal. Chem.*, **13**, 245 (1967).
118. P. A. Malachesky, Ph.D. thesis, University of Kansas, Lawrence, 1966.
119. P. A. Malachesky, L. S. Marcoux, and R. N. Adams, *J. Phys. Chem.*, **70**, 4068 (1966).

REFERENCES

120. S. W. Feldberg and C. Auerbach, *Anal. Chem.*, **36**, 505 (1964).
121. J. M. Hale, *J. Electroanal. Chem.*, **8**, 332 (1964).
122. L. S. Marcoux, Ph.D. thesis, Univ. Kansas, Lawrence, 1967.
123. L. S. Marcoux, R. N. Adams, and S. W. Feldberg, to be published.
124. P. Delahay, *Advan. Electrochem.*, *Electrochem. Eng.*, **1**, 263 (1961).
125. W. H. Reinmuth, *Anal. Chem.*, **36**, 211R (1964); **38**, 270R (1966).
126. K. J. Vetter, in *Transactions of the Symposium on Electrode Processes* (E. Yeager, ed.), Wiley, New York, 1961, p. 47.
127. J. Jordan, *Anal. Chem.*, **27**, 1708 (1955).
128. J. E. B. Randles, *Can. J. Chem.*, **37**, 238 (1959).
129. N. Tanaka and R. Tamamushi, *Electrochim. Acta*, **9**, 963 (1964).
130. A. C. Aten and G. J. Hoijtink, in *Advan. Polarog. Proc. Intern. Congr. 2nd, Cambridge, Engl., 1959*, **2**, 777 (1960).
131. P. A. Malachesky, T. A. Miller, J. Layoff, and R. N. Adams, in *Exchange Reactions, Intern. At. Energy Agency Symp. Brookhaven, New York, 1959*, p. 157.
132. D. J. E. Ingram, *Free Radicals as Studied by Electron Spin Resonance*, Academic Press, New York, 1958.
133. R. Bersohn, in *Determination of Organic Structures by Physical Methods*, Vol. 2 (F. S. Nached and W. D. Phillips, eds.), Academic Press, New York, 1962, Chap. 9.
134. A. Carrington and A. D. McLachlan, *Introduction to Magnetic Resonance*, Harper & Row, New York, 1967.
135. A. Carrington, *Quart. Rev.* (*London*), **17**, 67 (1963).
136. A. Horsfield, *Chimia* (*Aarau*), **17**, 42 (1963).
137. R. N. Adams, *J. Electroanal. Chem.*, **8**, 151 (1964).
138. E. T. Seo, R. F. Nelson, J. M. Fritsch, L. S. Marcoux, D. W. Leedy, and R. N. Adams, *J. Am. Chem. Soc.*, **88**, 3498 (1966).
139. R. F. Nelson, D. W. Leedy, E. T. Seo, and R. N. Adams, *Z. Anal. Chem.*, **224**, 184 (1967).
140. Z. Galus, R. M. White, F. S. Rowland, and R. N. Adams, *J. Am. Chem. Soc.*, **84**, 2065 (1962).
141. M. D. Hawley and S. W. Feldberg, *J. Phys. Chem.*, **70**, 3459 (1966).
142. R. N. Adams, M. D. Hawley, and S. W. Feldberg, *J. Phys. Chem.*, **71**, 851 (1967).
143. R. F. Nelson, Ph.D. thesis, Univ. Kansas, Lawrence, 1967.
144. W. M. Schwarz and I. Shain, *J. Phys. Chem.*, **69**, 30 (1965).
145. J. H. Christie, *J. Electroanal. Chem.*, **13**, 79 (1967).
146. P. J. Lingane and J. S. Christie, *J. Electroanal. Chem.*, **13**, 227 (1967).

9 FABRICATION OF ELECTRODE SYSTEMS

9-1. Electrolytic Cells 267
9-2. Working Electrodes 270
9-3. Reference Half-Cells 288
9-4. Instrumentation 291
References 300

9-1. ELECTROLYTIC CELLS

There has been no lack of ingenuity in the design of electrolytic cells. Comprehensive surveys of the types most useful in electroanalytical work are given in the monographs devoted to this area (*1–6*). Most of the cells which are described are intended primarily for use with the DME. Delahay has discussed cells for solid electrode applications in somewhat greater detail (*7*). Each individual ordinarily constructs or adapts an existing cell to meet personal tastes or experimental requirements. Ruby and Tremmel recently have described cells particularly advantageous for chronopotentiometry and cyclic voltammetry (*78*). No detailed treatment is necessary since the literature contains descriptions to meet most any need.

In terms of practicality, there is one factor which dictates the design of a cell for solid electrode work. If platinum or gold are used, the working electrode must be cleaned with chromic acid (or other cleaning agent) usually *after each determination.* It is extremely unwise to use a cell in which the complexity prevents a rapid cleaning operation.

A particularly convenient assembly is shown in Fig. 9-1. The cell is very simply constructed, yet is rugged and adaptable to rapid, routine, operations. The cell itself is a standard "sample bottle" of 50–100 ml capacity with plastic screw cap. As many as five holes can be drilled readily in such a cap. A hole is drilled in the center for the solid electrode stem. On opposite sides of this, near the perimeter of the cap, two holes admit the auxiliary electrode and the reference half-cell. If the size of the side arm on the reference electrode is proper, this may rest on the lid top. A small amount of sealing wax around the drilled hole makes a very stable arrangement. Further openings may be used for a nitrogen bubbler, a second solid electrode, or other devices.

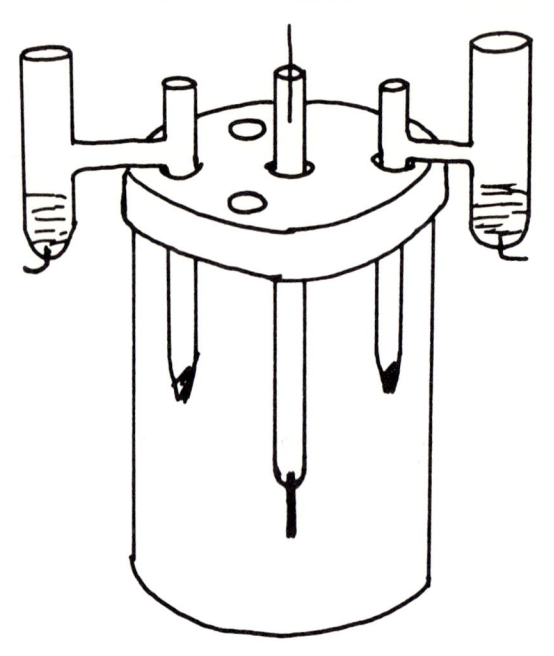

Fig. 9-1. Simple polarographic cell for solid electrodes.

The screw cap is held rigid with an extension clamp about 6–8 in. from the desk top or a constant-temperature bath. This permits the sample bottle to be rapidly removed and replaced from below. Several samples may be kept waiting in identical bottles.

Cleaning is accomplished by raising a small vial of cleaning solution into contact with the working electrode. The initial spacing of the drilled holes is arranged to permit this operation. Also, the tips of calomels and other devices entering the solution are adjusted to be slightly higher than the solid electrode surface to avoid interference with the vial. After

ELECTROLYTIC CELLS 269

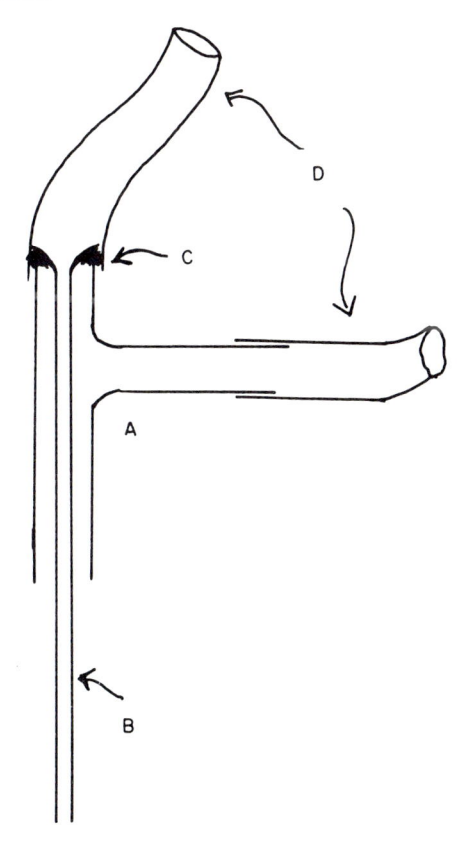

Fig. 9-2. Simple device for nitrogen atmosphere. A, Glass T tube (ca. 6 mm); B, inner tube with flared end (2–3 mm); C, Duco cement; D, tubing to two-way stopcock.

each cleaning, the entire unit can be conveniently rinsed with spray from a polyethylene wash bottle.

A very convenient nitrogen bubbler can be constructed without extensive glassblowing from a T tube as seen in Fig. 9-2. The inner tube is a piece of 2- to 3-mm soft glass tubing, flared at the top. A drop of Duco cement fastens the inner tube to the T tube at the flared joint. The vertical portion of the T tube is cut to extend to within 1/4 in. above the level of solution. A two-way stopcock, which can be located away from the cell, usually controls the flow of nitrogen over or through the solution. Although dissolved oxygen gives no anodic waves in solid electrode polarography, many of the electroactive systems are readily air-oxidized. In any investigation of a new polarographic system, dissolved air should be removed, or it should be established that its presence is not an important variable in the determination. Where the latter is true, of course, nitrogen sweeping should be eliminated for simplicity.

9-2. WORKING ELECTRODES

Although the operational definition of a working electrode given in Chapter 1, Section 1-3, is still retained, it is more convenient in this section to classify working electrodes in terms of physical characteristics. Accordingly, wires, both stationary and rotated, and foils are each discussed individually even though they all may be platinum electrodes. The discussion is restricted to pertinent information on the construction and proper usage of various systems.

A. Stationary Wires

By far the most usual form of stationary electrode is constructed by sealing fairly heavy-gage platinum wire in soft glass. Pyrex is less suitable for sealing platinum. Usually 10- to 16-gage wire is used. Electrical contact is easily made either through mercury or by soldering a copper wire to the platinum as shown in Fig. 9-3. Small battery clips permanently attached to the copper wire make convenient external connections.

Considerable care must be exercised in sealing glass–platinum seals. Improper seals may develop minute cracks, completely invisible to the naked eye, but disastrous to the functioning of the electrode. Repairing cracked electrodes is a waste of time. They should be broken, the glass carefully removed from the wire, and a new seal prepared. The presence of invisible cracks is indicated when the electrode gives anomalously high residual currents, spurious signals in recorded polarograms, or sometimes extremely sluggish response in acquiring expected solution potentials. In extreme cases cracks give rise to visible corrosion products near the seal. One advantage of the soldered copper connection is that such electrodes, if cracked, will develop a green copper film at the solder junction, especially if used in chloride media.

Although it does not ensure against cracks developing, it is worthwhile to reinforce the glass seal on the inside with wax. A small amount of powdered sealing wax (Dennison's, a red sealing wax in bar form, works best, but deKotchinsky cement can be used) is tapped down the glass

WORKING ELECTRODES 271

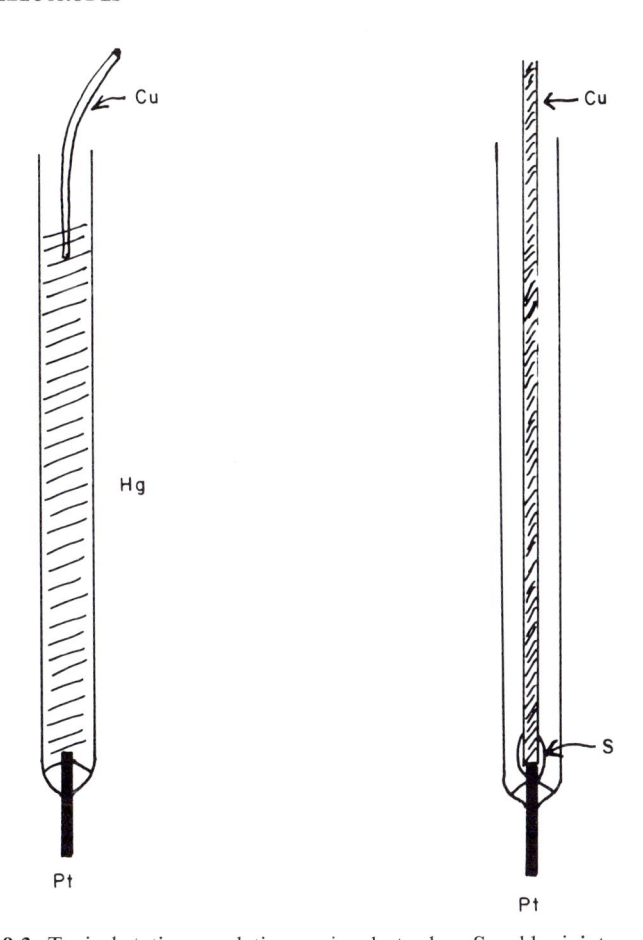

Fig. 9-3. Typical stationary platinum-wire electrodes. S, solder joint.

tube, to a depth of perhaps $\frac{1}{2}$ in. The wax is then melted by passing the tube quickly through a cool flame several times. The wax may be powdered with a flat file. Electrodes prepared in this way have been found to remain free of cracks for very long periods of time.

Materials other than glass tubing have been used to support platinum wires. Lord and Rogers inserted a small piece of platinum wire in a brass rod (8). The wire was held in place with a small set screw. The entire rod and wire was coated with Apiezon wax. The platinum was then scraped clean of wax for the electrode surface. Such electrodes are durable and provide freedom from cracking but seem to be seldom used. Krylon clear lacquer spray is also a convenient source of electrical insulation for this type of electrode.

The vast majority of stationary wire electrodes are built as in Fig. 9-3; i.e., the wire extends vertically downward in the solution. Some workers have preferred a horizontal orientation with the wire sealed near the end of the tube perpendicular to the length (9–11). Julian and Ruby believed horizontal wires more reliable at high current levels (9). No rigorous examination of mass transport to stationary horizontal wires seems to have been made (see Chapter 3). It would appear they should operate generally the same as their vertical counterparts. They are, however, much more prone to accidental damage or bending which may change their limiting current behavior. Shielding arrangements have been devised but there is little to recommend this added complexity when vertical wires work perfectly well.

The preparation or gold wire electrodes presents a real problem. Gold seals poorly to glass, and while satisfactory joints can be made with graded-seal techniques they are not for the timid glassblower. In addition, if mercury is used inside the tube to provide means for external contact, the gold amalgamates and mercury leakage occurs. Kolthoff and Jordan attempted to overcome this difficulty by soldering the gold to copper. The solder joint was then sealed to a glass tube in the usual fashion, with the copper end extending into the tube. Mercury was used as the contacting medium (12,13). Baumann and Shain reported high residual currents due to mercury leakage with this type of electrode (14). The Kolthoff and Jordon electrode was actually built for rotating work and an extended copper lead can replace the mercury for stationary applications.

Gold wire soldered into a drilled brass rod (see Ref. 8) will circumvent the difficulties associated with mercury connections. Baumann and Shain devised a gold wire electrode where the wire was pressure-fitted to a Teflon rod. Although this was a horizontally projecting wire for rotated purposes, a similar device could be made with a vertical wire (14). Some of the new resin kits now available require only simple molding equipment and may prove very useful in fabricating electrode holders.

Where special diffusion studies are involved, the end effects on wire electrodes should be eliminated. Nicholson sealed a small glass bead on the end of the wire in her exacting studies of cylindrical diffusion (15–17). Pavlopoulos and Strickland used a similar arrangement in their study of the kinetics of metal depositions on platinum wires (18). Berzins and Delahay sealed both ends of a platinum wire in capillary tubes for an oscillographic study involving cylindrical diffusion (19).

A commercial platinum electrode has become available which is highly satisfactory for many forms of solid electrode voltammetry. This is the Beckman platinum inlay or "button" electrode (39273). It consists of a circle or disk of platinum sealed flat in the end of a glass tube with a rugged rim of glass flush with the metal surface. This electrode has been widely used for single-sweep and cyclic voltammetry and chronopotentiometry in many laboratories. It gives results in close accord with linear diffusion theory in spite of having no restricting mantle. The approximate electrode area, ca. 0.2 cm^2, is convenient for ordinary concentration ranges. Another advantage is that the surface areas of different electrodes are quite close. Several such electrodes have been in use in the writer's laboratory for several years, which attests to their ruggedness. No evidence has yet been found for leakage at the platinum–glass seal. These electrodes are moderately priced and are far superior to "homemade" devices for general applications.

B. Rotated Wires

The construction of rotated wire electrodes differs from that of stationary only in the external connection, which must, of course, not interfere with the rotation. The original design of Laitinen and Kolthoff consisted of a short platinum wire sealed at right angles near the end of an iron shaft. Near the top of the shaft, an integral wiper disk dropped into a mercury well, providing the electrical contact. The remainder of the rod except the platinum was coated with ceresin wax (20).

Using the mercury-filled glass-tube electrodes described in the previous section, the simplest connection is made by dipping a stiff copper wire into the mercury. The wire is usually fixed to an external binding post and bent so that it dips at least ½ in. into the mercury. The wire preferably should not touch the rotating tube. Many workers have preferred this type of equipment (21–26).

In all the references just mentioned, the platinum wire extended vertically downward. Ferrett and Phillips made a critical study of horizontally rotating platinum wires (27). Using the Laitinen and Kolthoff electrode design (20), they selectively exposed portions of the revolving wire by scraping the wax. Much more reproducible polarograms were obtained using only the leading edge exposed. On reversing the rotation so that the trailing edge was exposed, poorly defined waves were obtained. These discrepancies increased with increasing rates of rotation. The results were explained in terms of turbulence at the trailing edge as the electrode sweeps through the solution. These effects are certainly in accord with the hydrodynamics of such systems. Excellent results have been obtained with the vertical wire (*28*).

There is one bent-platinum-wire configuration which is particularly advantageous. This type of electrode was described by Bruckenstein and Nagai in the form of a mercury-plated surface (*29*). A short length of platinum wire is bent into a U shape and both ends sealed into the usual

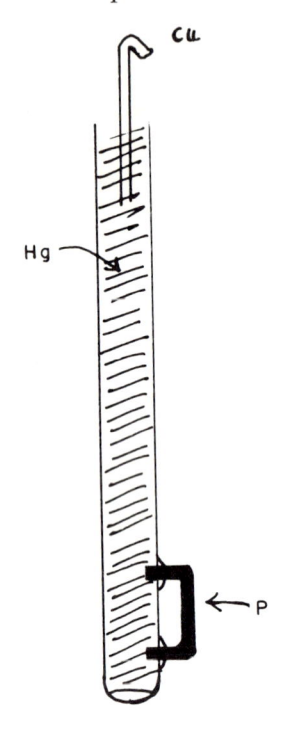

Fig. 9-4. Rotated platinum electrode design of Bruckenstein and Nagai.

soft glass tube, at right angles to the rotation axis of the tube. Electrical contact is made via the usual mercury filling. This design is illustrated in Fig. 9-4. The U shape, sealed at both ends provides an electrode surface which is extremely rigid and difficult to distort by accidental handling. This electrode is simple and ingenious and has much to recommend it for routine applications.

Until recently rotating devices were "do it yourself" propositions. The Sargent synchronous rotator (*30*), designed specifically for rotated electrodes, is a convenient and compact unit. The rotator consists of an 1800-rpm synchronous motor with internal 3:1 gearing to the shaft. With only a moderate amount of adjustments to the top and bottom knurled collars, 6-mm tubing can be positioned without appreciable wobble. A binding post on top of the housing is used for connection of a stiff copper wire to the mercury. The device operates at a fixed speed of 600 rpm, which is a recommended speed for ordinary work. It is somewhat difficult to position auxiliary electrodes, nitrogen lines, etc., between the top of the cell and the bottom of this apparatus because of the wide dimensions which "overshadow" the cell. If the separation is increased by raising the Rotator, the increased electrode length beneath the bottom collar results in some wobble during rotation. Short protrusions on the reference electrode and other lines can be employed to overcome this problem. Although expensive, the Sargent rotator is an excellent device recommended for routine voltammetry.

A simple device which provides considerable free space above the electrolysis cell can be constructed. The electrode is of the glass-tube type and is attached by a set screw to an aluminum pulley wheel which rests on a top bearing block. A lower bearing block prevents wobble but its small size does not interfere with the cell. Any type of bearing block is satisfactory but the ball-bearing shaft hangers manufactured by Servo Corporation of America (31), are particularly convenient and give many hours of trouble-free operation. With their associated clamps they can be easily fastened and accurately positioned to a vertical piece of Pegboard (32) which serves as a backing for the apparatus. The motor can be located on a separate stand and coupling to the aluminum pulley is by rubber band or similar flexible connection. The ball-bearing shaft blocks provide quite stable rotation speeds. The reader is referred to the individual literature for descriptions of other rotating devices (20, 27,33,34).

As a general rule wires about 2 cm in length (vertical height) are satisfactory. It is interesting to see how closely two similar wires can be matched in their limiting currents. Table 9-1 compares two 16-gage platinum wires which were sealed in the usual fashion. The electrodes were then clipped with sharp cutter pliers to the same exposed length. The estimation was made by eye using a small steel rule. When used in a stationary configuration, it is seen that limiting currents probably can be matched to ± 20–30% by this simple technique. This, of course, is satisfactory for the purpose at hand. When these two electrodes were rotated at 600 rpm, however, the differences are magnified manyfold, one

TABLE 9-1
Comparison of Limiting Currents at "Identical"
Platinum-Wire Electrodes

Electrode	i_L stationary, μa	i_L rotated, μa
A	5.2	52.5
B	6.8	21.5

[a] Electrode reaction: oxidation of 2×10^{-4} M iodide in 1 M sulfuric acid background.

electrode giving over twice the limiting current of the other. This divergence naturally arises from the slight surface irregularities, invisible kinks, etc., between the two electrodes. In the test described, the electrodes were rolled between polished steel "flats" in an attempt to minimize bends, etc. It can be seen, then, that no firm rules can be stated with regard to apparent size vs. i_L. This is not to say that i_L is not proportional to electrode area. Rather it emphasizes the fact that a measured area (geometric) and area effective for electrode reaction are quantities not necessarily identical.

The situation is not as serious as implied above. Electrodes can be compared on the basis of sensitivities for a specific electrode reaction:

$$S = \frac{i_L}{C} \tag{9-1}$$

The sensitivity, S, is usually given the units of microamperes per millimole per liter. This sensitivity evaluation, introduced by the schools of Kolthoff and Jordan, is particularly useful.

C. Foil Electrodes

The term foil usually implies a planar, vertical electrode, although curved and variously oriented designs are sometimes used. The discussion here will be limited to the former type. Their construction is simple. Heavy-gage platinum or gold wire is pressure- or spot-welded to the desired size of foil. In the former case the foil is placed on a flat metal block with the end of the wire lying on top near the upper end of the foil. A gas–air flame is directed on the juncture and when both pieces are at red heat, several sharp blows from a small hammer will produce the weld. With but a little practice this operation can be accomplished without mutilating the plane surface. The electrical spot welding usually produces a neater job. In either case, the wire is then sealed in soft glass with any of the connections discussed previously.

The same effect can be achieved by initially cutting the metal foil with a thin supporting projection. The projection can be sealed to glass, eliminating the wire. This method has two disadvantages: It is wasteful of noble metal and it adds considerable edge effects to the electrode unless the support is subsequently waxed. On the other hand, this one-piece foil electrode is easier to make and never suffers from separation of wire and foil.

Commonly used foils range in size from 1–2 cm on edge. As pointed out in Chapter 3, they are singularly unsuited for voltammetry using long observation times, due to interference from natural convection. They have recently become important in chronopotentiometry. Here it is quite useful to define apparent current densities in terms of the geometric electrode area. For this purpose it is well to mask a larger foil electrode to a smaller, predetermined size. This can be conveniently done in the following way. A piece of glazed paper is cut to known dimensions and placed on the surface of the plane electrode. The entire surface is then sprayed with Krylon clear spray (Crystal Clear No. 1301, obtained from Krylon, Inc., Norristown, Pa.). When dry, the paper mask can be removed leaving a known exposed area. The opposite side of the foil is, of course, covered completely with the spray. Krylon appears to stand up well in a large variety of aqueous electrolyte solutions.

Pure gold metal foils are especially soft and present some difficulty in quantitative work, where accidental bending causes calibration data to vary. Actually, pure platinum metal is quite soft, and platinum–iridium alloy very often is substituted.

The commercial platinum inlay electrode described previously is recommended as a replacement for any moderate-sized foil.

D. Vibrating Wires

Two styles of vibrating electrodes have been developed, those which oscillate in an up and down sense and those which vibrate from side to side.

The Harris and Lindsay apparatus is an example of a vertical vibrating wire which has been mentioned in Chapter 4, Section 4-8 (*35,36*). Connection of the electrode to a commercial massage unit was recommended.

Roberts and Meek (*37*) utilized the voice coil of a loudspeaker to power their vertical vibrator.

It is significant that Roberts and Meek matched electrodes by cutting to identical length using a measuring microscope. Variations in limiting currents were almost undetectable in their study. This is in marked contrast to the naked-eye matching data given in Table 9-1.

A very simple vibrating assembly was used by Lund in his study of oxidations in acetonitrile (38). The vertical wire was vibrated by a commercial unit manufacture by Fisher Scientific of Switzerland.

A horizontal vibrating electrode for polarographic use seems to have been described first by Dirscherl and Otto (39). Oehme and Noack improved considerably on the versatility of the above apparatus (40). The vibrating rod was modified with a clamp arrangement so that conventional glass-tube electrodes could be attached. The reader is referred to the original literature for details of the construction of vibrating electrode systems.

E. Carbon Rod Electrodes

Although carbon or graphite anodes were used long before in preparative electrooxidations, analytical interest in carbon rod electrodes was apparently initiated by the work of Lord and Rogers, who suggested that a new and reproducible surface could be obtained after each polarogram by simply breaking off the electrode tip. The outer surface of the electrode was rendered nonconducting with an insulating wax. High-purity spectroscopic rods were used, although medium-hard pencil leads were found satisfactory except for their fragility (8).

Gaylor et al. examined the utility of such electrodes for organic oxidations (41). They used ¼ in. spectroscopic rods brushed with melted ceresin wax. This electrode was found to be satisfactory for the oxidation of a variety of phenolic compounds provided certain experimental conditions were maintained. Further study indicated that sensitivity limitations were imposed by high residual currents. This led Gaylor et al. to investigate carbon electrodes which were fully impregnated or saturated with wax (42). This technique was found to materially decrease residual currents and resulted in a much more satisfactory electrode for analytical purposes. All further work with graphite rods have employed the total impregnation technique, and only this type of electrode will be discussed in detail below.

For practical purposes pencil leads are seldom used, but it would appear they could be selected where size is a determining factor. Usually ¼ in. spectroscopic rods about 6–8 in. in length are employed. All but about

½ in. of the rod is immersed in melted wax and allowed to stand for 2 hr in the wax at 100°C. Gaylor and co-workers used opal wax (E.I. du Pont de Nemours & Company) or castor wax (Baker Castor Oil Company). After this impregnation period the electrode was removed and cooled in a vertical position. The outer surface was then painted with Seal-All (an insulating compound obtainable from Allen Products Corporation) and allowed to dry. Connection to the rod can be made either by a small battery clip at the unwaxed end or with a short piece of polyethylene tubing filled with mercury. With the mercury well and the usual piece of copper wire the electrode can be used in a rotating configuration if desired.

To prepare the electrode for use, Gaylor et al. broke off about ¼ in. from the working end of the rod and sanded the exposed surface lightly with fine sandpaper. The electrode was then inserted in the solution. This procedure was repeated for each polarogram to provide a fresh surface. Using this type of electrode, commercial antioxidants could be determined with a standard deviation of 2% in unstirred solution. Using stirred solutions, as low as about 2 ppm of antioxidant in gasoline samples could be determined.

A comprehensive survey of impregnating agents was made by Gaylor et al. (*41*). Opal, ceresin, and castor were most valuable in terms of decreasing residual currents. Table 9-2, abridged from the data of Ref. *41*,

TABLE 9-2
Comparison of Impregnating Agents for Graphite Electrodes[a]

Impregnating agent	i_L/i_r
None	0.1
Saran resin	1.3
Silicone resin	2.0
Lemon wax	2.5
Ceramid wax	3.7
Silicone 200	6.5–8.7[b]
Opal wax	11.3
Ceresin wax	12.2
Castor wax	20.0

[a] All measurements based on the oxidation of 10^{-4} M dibutyl phenylenediamine in 1:1 alcohol–water–acetate buffer, pH 5.2. Polarograms run in quiet solution at a fixed scan rate of 1.24 mV/sec (*41*).

[b] Varying viscosity.

compares the efficiency of various waxes as measured by the ratio of limiting current to residual current. A saturation period of about 2 hr in the hot impregnating agent is sufficient for all practical purposes. Elving and Krivis have used similar electrodes for both voltammetry and chronopotentiometry of diamines and phenols (*43*).

Morris and Schempf improved materially on the performance of wax-impregnated carbon electrodes by evacuating the porous carbon rod during the wax impregnation. In addition, they designed a small, manual lathe to cut off a fresh surface prior to each run. The results with this type of electrode pretreatment were excellent (*44,45*). In addition to the vacuum impregnation and lathe cutting of the surface, Elving and Smith recently added a significant improvement. They found that one of the major problems with wax-impregnated carbon rods was proper wetting of the electroactive surface. Therefore, prior to each run, the fresh electrode surface was immersed for 1 min in a 0.003% solution of the wetting agent Triton X-100. The surface was then rinsed with a portion of the polarographic test solution before using. Excellent precision was obtained using these electrodes for both chronopotentiometry and voltammetry (*46*). Ward recently used wax-impregnated graphite rods to determine antioxidants in nonaqueous media via chronopotentiometry. A Teflon sleeve was used to cover all but the end of the rod. Ward suggests that the surface wetting is not a problem in acetonitrile and 95% ethanol (*47*).

F. Carbon Paste Electrodes

Carbon paste electrodes developed from an attempt to prepare a fluid of suspended carbon particles to be used in the sense of a dropping electrode for anodic oxidations. Such a mixture can be prepared and dropped from a capillary of internal bore slightly greater than that of the DME. However, the conditions for successful operation of the dropping carbon paste electrode are far from ideal, and it soon became apparent that a thick paste, packed in a pool configuration and used either stationary or rotated, would be far more advantageous. This type of carbon paste electrode is believed to be one of the most simple and practical of solid electrodes and yet is capable of precision in limiting-current measurements rarely matched by noble metal electrodes. Carbon paste electrodes have practically zero residual current over the entire anodic potential range. There are few anomalies such as oxide-dissolution or hydrogen-desorption currents. For routine applications the carbon paste is far superior to platinum or gold.

The carbon paste itself is prepared by simple hand mixing of carbon (usually graphite) with any liquid which is sufficiently nonmiscible with water to keep the electrode matrix from dissolving when immersed in the test solution. A wide choice of liquids is possible, but practical considerations of low volatility, purity with respect to electroactive impurities, and expense narrow the choice to several practical liquids. Those most frequently used to date are Nujol and bromoform. For convenience such electrodes are designated in terms of the pasting liquid as carbon electrode–Nujol paste (CE-NjP), etc.

A practical Nujol paste is prepared by mulling together with a spatula 15 g of carbon (Acheson graphite, Grade 38 or equivalent) and 9 ml of Nujol until the entire mixture appears uniformly wetted. The paste has a consistency about like that of peanut butter. The paste is stored in a wide-mouthed sample bottle. The amount given above will prepare enough electrodes for about 1 year of use. Pastes which employ other liquids are prepared in the same fashion. The choice of carbon is not critical except that gritty-type carbons (Nuchar, etc.) do not work well since they give a rough surface to the pool. While the limiting current at a carbon paste electrode varies with the paste composition, this is marked only when the paste is quite thin (i.e., a high ratio of liquid to carbon). With the proportions given above a variation of ca. ±0.5 ml of Nujol makes only a little difference. In any event the values of i_L are self-consistent for a given batch of paste which, if prepared in the quantity indicated, is sufficient for several thousand polarograms.

To prepare the electrode proper, the paste is hand-packed with a spatula into a well-like depression in a Teflon cup as shown in Fig. 9-5. The Teflon cup is held in a glass tube. Electrical connection to the carbon paste is made via a short platinum contact bent flat against the bottom of the pool. Copper is soldered to the platinum and leads out through the top of the glass tube for external electrical connection. Connections made entirely of copper have been used successfully. If the electrode is to be used in a rotated sense, the copper wire may be brought out for a wiper contact, or the usual mercury connection mentioned previously can be used.

The electrode design given is representative of those most frequently used in the writer's laboratory. Obviously many other satisfactory forms can be devised. For instance, a single piece of glass tubing terminating in a female ball joint has been used. The paste is packed in the ball joint.

Electrodes of this type were used very successfully by Jacobs for anodic stripping following deposition of traces of gold and silver (48).

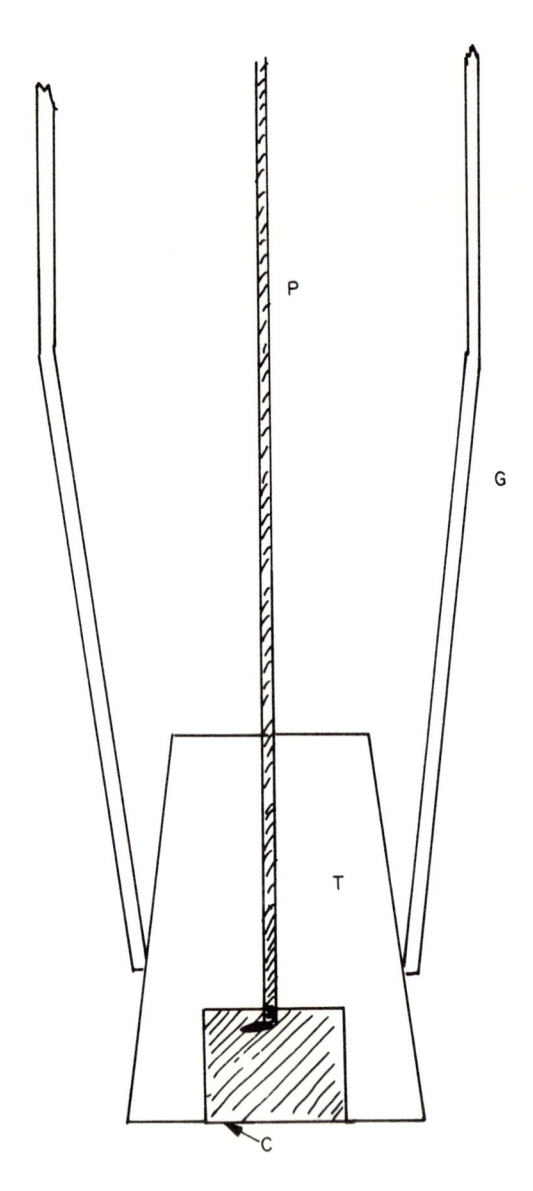

Fig. 9-5. Typical stationary carbon paste electrode. G, Standard-taper glass joint or similar tube; T, Teflon plug with well depression; P, platinum or copper lead to bottom of well; C, carbon paste.

The pool is formed by tamping the paste in the well with a spatula. Care is taken to fill the well, leaving no holes or channels, but the paste is not packed too firmly. Finally, the surface is best smoothed off by rubbing the electrode several times slowly across a glazed paper surface (for instance, an index card on the bench). Any excess paste is carefully wiped free with Kleenex. Like any mechanical procedure, one soon develops a methodical and reproducible way of smoothing the carbon paste surfaces. With a little experience, the pool can be packed in 2–3 min in a fashion which will show limiting-current reproducibility of $\pm 2\%$ (peak polarograms). In practice, using special care, results reproducible to $\pm 5\%$ have been obtained routinely in the writer's laboratory.

Normally, working with organic compounds where films may be deposited during electrolysis, a new surface is prepared for each run. It is necessary to only remove the top $\frac{1}{8}$ in. or so of the paste and refill. Obviously, many refinements could be made in the method of preparing the paste and the packing of the electrodes. However, throughout the development of the electrodes an attempt was made to keep all operations as simple as possible.

A nonaqueous carbon paste, satisfactory for use in acetonitrile, nitromethane, and propylene carbonate, has been described (77). It is prepared, packed, and used in the same fashion as described above for the Nujol pastes.

G. Rotated Disk Electrodes

Fundamentally the construction of a RDE involves nothing beyond that required for any rotated electrode. Since a precise limiting current equation is available for the RDE, it is desirable to construct such electrodes so that quantitative measurements can be made. For this reason a great deal of care is required in building them. It appears that the practical difficulty of preparing RDE's has limited their usage to date.

The simple glass tube–heavy wire disk is made by sealing heavy-gage platinum in thick-wall soft glass tubing. The working end is then cut and ground flat and polished smooth with a Carborundum or similar grinding surface. This type of seal is obviously subject to cracking, both on annealing and grinding, and is perhaps best attempted by an experienced glassblower.

Heavy-gage wire can also be mounted in various resins. The main problem is the centering of the wire on the axis of rotation. Azim and Riddiford accomplished this by soldering or brazing the platinum wire into a center recess in a precision-ground stainless-steel shaft. The end of the shaft and wire were then coated with Araldite resin and shaped down on a lathe (*49*). With proper care in centering and machining, this procedure gives excellent results. [A clear plastic called Quickmount (marketed by E. H. Sargent Co., Chicago) has been found to be very useful for encapsulating electrodes.]

Simple Teflon holders into which a platinum wire is pressure-fitted can be used. These are attached directly to the motor shaft via a threaded connection (the shaft is precut and threaded). An advantage is that one can have several different electrodes which are easily interchanged on the shaft (*50*).

The best platinum RDE's used in the writer's laboratory have been developed recently by Marcoux (*79*). These are made from the commercial platinum inlay electrodes (Beckman No. 39273) mentioned previously. The glass envelope of the commercial electrode is first cut to about one half its length using a glass saw. By carefully cutting around the glass envelope, the inner wire which makes contact with the platinum is left intact and can then be cut with pliers to extend $\frac{1}{2}$ in. or so beyond the cut-off tubing. Next a hole is drilled to a depth of $\frac{1}{2}$–$\frac{3}{4}$-in. in a board which is clamped in a fixed position on a drill press. The electrode is placed in the hole in an upright position and should fit snugly enough to ensure good vertical alignment. Without moving the board, the drill is removed from the chuck and replaced with a $\frac{1}{4}$-in.-diameter brass rod about 3–6 in. in length. This serves as the rotator shaft. The drill press is brought down until the brass rod is ca. $\frac{1}{2}$–$\frac{3}{4}$ in. from the end of the electrode and is thus automatically centered in the proper position. Several turns of the internal connector wire are wound tightly around the center rod and may be soldered or glued if desired. With the rod now in position, one of a number of epoxy resins or potting compounds can be poured in the tube. The resin is filled to the top of the glass tube and allowed to set while the entire unit is kept in the drill press. (Some of the older commercial electrodes have a sealing compound inside which is easily removed after the sawing operation by rinsing with acetone.)

Although this type of electrode has an outer glass rim or shroud of only 3–4 mm, it gives excellent results over a range of speeds from 2 to 40 rps. The area is quite satisfactory for the usual studies. Although one may hesitate at first to dismember a relatively expensive commercial unit, the cost is quite low when one considers the time and effort usually spent in achieving a good platinum to glass seal. It is the experience in the writer's laboratory that this RDE performs better, in every sense, than a "homemade" platinum RDE.

WORKING ELECTRODES

Disks of carbon paste are easily made by precision drilling of a well into a Teflon rod. The rod is then attached in a suitable manner to the shaft (for instance, via a threaded hole on the other end, as mentioned above). Alternatively, a metal shaft can be set in plastic and a hole for the carbon paste drilled in the other end. Electrodes of this type have been described by Prater and Adams (81).

When a brass or other metallic rod is the center shaft of a RDE, electrical contact, is normally made via a brush arrangement. Ordinary multistrand copper household wire, wetted with mercury, has been found quite satisfactory. If the metal rod is mated to the motor shaft through a chuck or other metallic connector, there may be electrical pickup from the motor. In this case a Teflon sleeve or other insulator can be interposed between the shaft and motor. A good grade of chuck or collett or other centering device should be used to eliminate wobble.

Since the primary utility of the RDE is the variation of i_L with rotation rate, a controllable-speed motor needs to be used for all but routine analytical work. Many laboratory units employ a motor with integral tachometer generator. The output signal from the tachometer generator is used as a control signal in an operational amplifier feedback loop. There are commercially available a wide variety of motor-speed-control devices which are satisfactory. Calibration of rotation rates can be accomplished with a stroboscope or via photocell pickups. Straight gear changing of speeds is cumbersome over a wide range but quite accurate.

The motor and electrode should be mounted on a rigid support, preferably detached from the cell holder and other hardware. A convenient unit shown in Fig. 9-6, where the motor is clamped to a horizontal metal plate which is supported by vertical risers (e.g., Uni-Strut) bolted to the wall or side of a laboratory table. The cell is then brought up under the electrode via an extension jack or other suitable device. (It should be observed that all the RDE devices discussed are for moderate rotation rates—up to 60–80 rps. Where very high speeds are employed, precision engineering is required. In addition, cavitation of the solution is a major problem.)

The ring-disk electrode of Fig. 4-6 is indeed more difficult to fabricate. Since one desires an electrode with a very small insulation spacing between inner disk and the ring, the problem really is a challenge in precise centering and machining. (Probably it should be pointed out that these fabrication problems are a trial for most chemists, but not very difficult for skilled machinists.) Bruckenstein has prepared some ring disks with very small insulation spacing by first silver-soldering a heavy platinum foil on an accurately machined stainless-steel shaft. The steel shaft provided proper rigidity for the electrode. After machining the edges of the platinum, the insulation ring is prepared by "painting" on several layers of epoxy resin to give an oversized insulation layer. This is machined down and a stainless-steel tube of proper dimension fitted tightly around the resin. Platinum is soldered again to this for rigidity and machined. Now the end of the unit plus a length up the shaft is cast in resin, care being taken

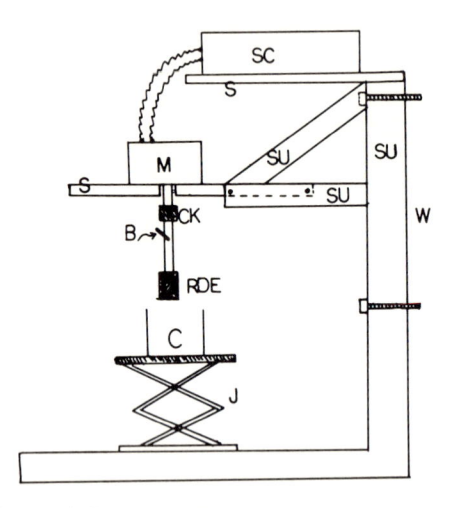

Fig. 9-6. Typical rotated disk electrode apparatus. Sc, speed-control unit for motor; M, motor; S, shelf supports; CK, chuck or other shaft coupler; RDE, rotated disk electrode; C, cell; J, laboratory jack; SU, support members (typically Unistrut); W, laboratory wall or bench upright; B, brush.

to prevent bubble formation, and then the entire end cut down on a lathe to expose the platinum surfaces (51). Further detailed instructions of the construction of disk electrodes can be found in the original literature.

H. Unusual Electrode Configurations

There exist a considerable number of solid electrodes of highly specialized design. First, the unusual requirements of biological and physiological research have led to a number of special electrode systems, especially for measuring oxygen in vivo. The design and performance of such electrodes is beyond the scope of this work and is well documented in several sources (52–55).

An unusual platinum electrode with output response like that of the DME was designed by Lyalikov and Karmazin (56). A platinum needle was surrounded by an open-ended glass tube and immersed in the polarographic solution. Nitrogen was passed through the glass tube and the platinum was alternately immersed in the solution by the escaping nitrogen bubbles. Lyalikov and co-workers used the electrode for the reduction of several metal ion systems. A similar bubbling electrode was recently described by Cozzi and Desideri (57). The advantages of such electrodes are difficult to discover. The mere fact that the current is made to oscillate analogous to a DME is hardly a desirable feature. Although the diffusion layer is periodically renewed, it is only in this way that the bubbling electrode differs from an ordinary platinum surface.

Other workers have advocated periodic renewal of the diffusion layer (retaining measurement of transient current response), and a recent apparatus by Roffia and Vianello gives excellent results in this fashion (80). Here the electrode is moved abruptly back and forth in the solution, and, during a rest cycle, the current–time behavior is recorded.

These electrodes in which the diffusion layer is periodically altered (by bubbles, pulsed movement, etc.) differ from the regular vibrating electrodes in which steady convective mass transport is operative. The periodic renewal is a hybrid style of measurement—falling between the extremes of i–t measurements under strict diffusion conditions and steady-state current observations with forced convection. Either of the latter techniques are, in general, easier to apply.

Skobets and co-workers described a platinum wire which was caused to move linearly through the solution (58). An ingenious use of a wire electrode in paper chromatography was described by Langer. Here a amalgamated gold wire was pressed directly against a paper-chromatograph strip for direct polarography of the components on the strip (59). Ewald and Lim described a platinum-wire electrode and cell for high-pressure voltammetry studies up to 3000 atm (60).

One of the most unusual applications of solid electrodes was the design of an "artificial nose"—a galvanic cell arrangement of graphite and platinum electrodes to study the role of oxidation–reduction in olfaction. This voltammetric noise is reported to be much more sensitive to short chain alcohols than the human counterpart (61).

The various "Solion" devices which have been described in the instrumentation literature for several years are based on mass transport and controlled electrolysis at solid electrode surfaces. These devices can be used as integrators, amplifiers, elapsed-time indicators, etc. Information on the design and application is available from Self-Organizing Systems, Inc., Dallas, Texas.

9-3. REFERENCE HALF-CELLS

All the reference half-cells used in polarography can be coupled to any type of working electrode. There are no specific types peculiar to solid electrode voltammetry. Some of the electrolytic cells mentioned previously are particularly suited for use with small side-arm reference cells, which are described briefly in this section. Unless otherwise designated, the term reference electrode will mean the saturated calomel electrode (SCE) in this discussion. Since the great bulk of polarographic $E_{1/2}$'s are recorded vs. SCE, only those with a flair for bookkeeping will find it advantageous to use other forms of calomel electrodes (1 N or 0.1 N). The silver–silver chloride (Ag/AgCl) and lead amalgam–lead sulfate [Pb(Hg)/PbSO$_4$] find limited usage.

Figure 9-7 shows the form of side-arm SCE found particularly useful by the writer. The glass dimensions (Fig. 9-7A) can be altered to suit individual requirements. The external connection is a heavy-gage platinum wire sealed in the bottom and originally extending about 1 in. beyond the glass. To this is soldered a piece of heavy, bare copper wire. The platinum is then carefully bent around the base of the tube and the copper connection run up the outside of the tube (Fig. 9-7B). Finally, the copper is bent horizontal to provide a connector point for battery-clip connections. The base connection is enclosed with sealing wax for protection. The upright copper wire is firmly held against the tube by a small wrapping of electrical tape.

The inside of the electrode should be absolutely dry before the mercury is added. A dry acetone rinse should be used with the acetone removed by a vacuum line. About ½ in. of instrument-grade mercury is added until the platinum wire is well covered. The calomel paste is made by grinding powdered mercurous chloride with a few drops of mercury in a mortar and pestle until the mass becomes light gray throughout. The contents of the mortar are then pasted with a few milliliters of saturated potassium chloride. The calomel paste is slid gently down on top of the mercury using a spatula. The paste layer should be about ½–¾ in. in height.

Before the half-cell is filled, the salt-bridge top is completed. This can be made of agar–agar in the usual fashion. A somewhat simpler plug can be made with filter paper. A small triangular piece of filter paper (about ½ in. on the sides) is wet with the KCl solution and rolled between

REFERENCE HALF-CELLS

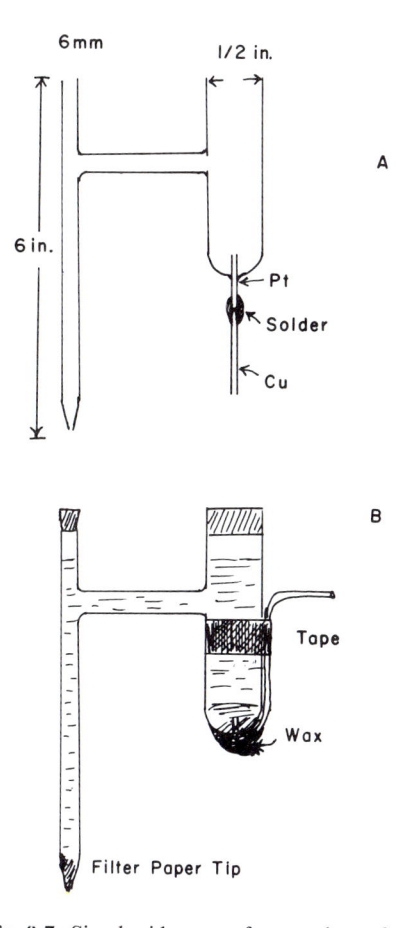

Fig. 9-7. Simple side-arm reference electrodes.

the fingers to a conical shape. The point is then forced down to the end of the tube and tamped tight from the inside with a long length of heavy-gage wire. There should be no free space between the end of the plug and the glass tip, or high-resistance air bubbles will be trapped in this space. With a little practice such a plug can be made which has low resistance but very little leakage. The entire half-cell and salt bridge is then filled with saturated potassium chloride. Small corks are added to complete the half-cell. It is not wise to use tight and permanent stoppers since the SCE (as prepared above) will require occassional filling. A "rubber policeman" on top of the side arm is convenient for refilling operations.

Especially since $E_{1/2}$'s at solid electrodes are seldom more reproducible than ±2–3 mV, it is unnecessary to use special-grade mercurous chloride and potassium chloride in the electrode preparation. Instrument-grade mercury should always be used however, since ordinary mercury may contain gross amounts of metal impurities. Probably the most common cause of nonreproducible potentials with the SCE is moisture around the mercury–platinum connection. Such an electrode may function properly for long periods but suddenly give erratic potential fluctuations. If suspected of being wet, the entire electrode should be emptied and rebuilt The appearance of crystals of KCl in the salt-bridge tip causes no trouble unless the resistance increases. A SCE prepared as described above should have a resistance between 500 and 900 Ω as measured with an ac bridge. The value of 0.246 V (positive with respect to the hydrogen electrode) is usually assigned to the SCE.

If the presence of chloride ions in the polarographic solution is objectionable, the SCE can be immersed in an auxiliary solution, with a potassium nitrate salt bridge connecting to the test solution. This procedure is bulky at best, and it is often easier to substitute an alternative reference electrode. The $Pb(Hg)/PbSO_4$ electrode is constructed formally like the SCE. The lead amalgam (approximately saturated) can be prepared by warming a few milliliters of mercury in an evaporating dish with 15–20 ml of 1 M sulfuric acid. Powdered lead is stirred in until the mercury becomes mushy. The amalgam is then washed thoroughly with distilled water. Often the amalgam will be partially solidified by this cooling and can be stirred together with a small amount of mercury until again pasty. This amalgam paste can be poured directly into the empty half-cell, but it is preferable to add a small amount of mercury first, to establish true contact with the platinum wire. On top of the amalgam layer (about ¾ in. deep) is poured a paste of lead sulfate wetted with 1 M sulfuric acid. The rest of the electrode and salt bridge is filled with 1 M sulfuric saturated with lead sulfate. A filter-paper plug or agar–agar may be used for the bridge tip.

The potential of the $Pb(Hg)/PbSO_4$ electrode naturally depends on the amalgam concentration as well as the sulfate or sulfuric acid concentration. An electrode of exact potential can, of course, be prepared by the rather tedious process of controlling the amalgam concentration. It is far easier, and equally satisfactory for polarographic purposes, to prepare the approximately saturated amalgam and, after it has equilibrated, measure its potential vs. an SCE or other standard. The nominal potential of the $Pb(Hg)/PbSO_4$ prepared as described is −0.27 V vs. hydrogen, or about 0.52 V negative with respect to SCE. The Pb(Hg)/PbSO$_4$ electrode has been widely used by Cooke et al. in coulometric work (62).

Both internal and external Ag/AgCl electrodes have been used from time to time. Cooke preferred the Ag/AgCl electrode in his work with rotating amalgam electrodes, since the cell could be inverted for cleaning purposes without altering the reference potential (63). The electrode was made by electrodepositing an adherent deposit of AgCl on a helix of 16-gage pure silver wire. The wire was immersed in a solution of 1 M potassium chloride. The reference half-cell was attached permanently to the main vessel through a sintered glass connection. Further details of construction are to be found in the original literature. An excellent review of silver chloride electrode and their preparation is given by Janz and Taniguchi (64).

Among others, Gaylor et al. found an internal Ag/AgCl reference very advantageous (42). The wax-impregnated graphite rod was used as a support around which was wound the Ag/AgCl wire. The wire was prepared by winding about 10 cm of 20-gage silver wire loosely around the carbon rod. The silver chloride coating was prepared by anodizing in 5 N HCl for 10 min using a 1.5-V source. This arrangement is especially useful for stirred solutions and with essentially nonaqueous media. Accurate positioning of the reference electrode is simplified.

The rather casual directions given for the preparation of the reference electrodes should not be taken to indicate that careless work is satisfactory. Very stringent requirements are indeed imposed on the preparation of reference electrodes for certain physicochemical measurements, e.g., where emf's are measured to 0.1 mV or better. However, the inherent accuracy of $E_{1/2}$ observation in solid electrode work does not require this effort.

A variety of reference electrodes have been used in nonaqueous solvents. For the common solvents such as acetonitrile and dimethylformamide, the ordinary aqueous SCE is recommended (65). Some difficulties such as precipitation of potassium perchlorate at the salt-bridge tip when perchlorates are used as supporting electrolytes are to be expected but can be circumvented without undue trouble.

9-4. INSTRUMENTATION

A. Instrumentation for Voltammetry

The instrumentation for solid electrode voltammetry does not differ fundamentally from that employed with the DME. However, there are very few commercial polarographs which are at all satisfactory for extensive solid electrode work. In addition, it can be seen from Chapter 5 that there is a wide range of voltammetric measurements one may wish to carry out. It is, therefore, worthwhile to consider the design and construction of simple recording polarographs specifically oriented toward solid electrode applications. (A rudimentary manual polarograph has been described in Chapter 1, Section I. Detailed treatments of manual polarographs are given in the monographs on polarography.)

An approximately ideal recording polarograph for solid electrode work would have the following characteristics:

1. A variety of sweep rates from ca. 0.2 to 20 V/min. (Higher frequencies are useful but 20 V/min is about maximum for pen and ink recording.)
2. Capabilities for single- and cyclic-sweep operation.
3. Potentiostatic control (controlled-potential polarography).

There are two classes of polarographic circuits. First, the conventional design which employs two electrodes is shown in Fig. 9-8. The polarograph is really only a voltage divider which varies the applied voltage to a cell composed of the working electrode, WE, and the reference half-cell, REF The iR drop across the variable, precision resistance, R_s, which is in series with the polarographic cell, is fed to the pen axis of a strip-chart recorder to indicate the polarographic current. (In the Leeds and Northrup Electrochemograph, a unique recording system eliminates the variable dropping resistor for current measurement, but Fig. 9-8 is representative of the general design of many recording polarographs.) The synchronous motor M varies the adjustable tap on the voltage divider at a rate which is correlated with the recorder chart drive. It is quite clear that instruments using a two-electrode system record at best only current vs. *applied voltage* polarograms.

Such devices do not suffice for critical examinations of theoretical equations in voltammetry since the theory is developed on the basis that one observes current vs. *potential of the working electrode*. This viewpoint was not discovered recently. Kolthoff and Lingane pointed out that the potential of the WE should be monitored independently (*1*). An excellent paper some time ago by Furness emphasized the utility of three-electrode polarographic measurements (*66*). Müller distinguished carefully between current–voltage curves (two-electrodes) and current–potential curves, which are taken with three electrodes (*3*).

The true potential of the WE in Fig. 9-8 is indeed quite different from

INSTRUMENTATION 293

the applied voltage as read from the voltage divider position. This is evident from the following considerations. Assuming the voltage divider is a linearly wound potentiometer or slide wire, the applied voltage varies linearly with shaft rotation or slider position only if no current is drawn from the voltage divider. Otherwise the so-called "loading error" causes the divider voltage to be nonlinear with respect to slider position (an

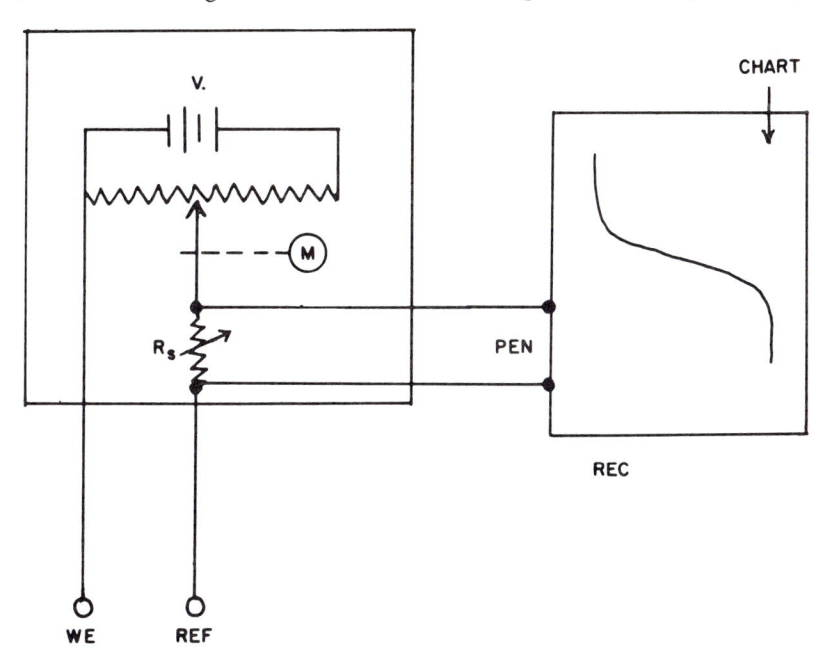

Fig. 9-8. Conventional recording polarograph using two electrodes. V, Linear voltage divider, battery plus precision potentiometer or slide wire; M, synchronous motor; R_s, standard dropping resistor; REC, strip-chart recorder; WE, working electrode; REF, reference electrode.

effect which cannot be correlated directly with fixed recorder chart speed). This zero current drain is an obvious impossibility in a polarographic circuit, where current necessarily flows through the network R_s, WE-REF, which is in parallel with the voltage divider. However, this nonlinearity problem is a minor one. By keeping the polarographic current small with respect to the current which flows through the divider resistance, the loading error can be reduced to negligible proportions. Thus the resistance for a "polarographic bridge" is always kept low (50–100 Ω), so that milliampere currents flow through the voltage divider resistance while the polarographic currents in the parallel path are normally never greater than a few microamperes.

By far the more important problem is that the potential of the WE must be different from the applied voltage whenever polarographic current flows by the amount of iR drop through R_s and the cell resistance. Now, in those instruments where R_s is varied for current sensitivity, the value iR_s is a quantity which, although usually small, varies from one polarogram to the next. In aqueous solutions with small cell resistances (i.e., less than 1000 Ω) and polarographic currents of a few microamperes, the iR_{cell} correction obviously is only a few millivolts. In nonaqueous media this value can become quite large.

At first glance these iR corrections (or differences between the potential of the WE and the applied voltage) seem to be minor ones. Although it is convenient and desirable to have an instrument to eliminate them, the appropriate corrections are relatively easy to understand and apply. Unless unusually high cell resistances are encountered, iR corrections in DME polarography are no particular problem.

In voltammetry at a stationary electrode the effects are just a bit more subtle. Assume that a typical peak polarogram is being run at a linear diffusion electrode. As current begins to flow, the WE potential lags behind the applied voltage because of the associated iR drops. This effect is a maximum in the region of the ascending portion of the peak polarogram but then effectively reverses as the polarographic current decreases after the peak. In any event the *real potential sweep rate* at the WE is far from linear and less than that recorded as applied voltage sweep. Now i_p is directly proportional to $V^{1/2}$. The $V^{1/2}$ of importance is that sweep rate which is effective at the WE. Thus, with the slower and distorted sweep rate, i_p will be less than predicted theoretically. This is why an uncorrected plot of i_p vs. concentration almost always falls below the line predicted by theory. As the concentration increases, currents increase, real sweep rates decrease, and i_p is less than predicted.

Further, suppose a current–time curve is to be run at a stationary electrode to check for linear diffusion dependence. The E_{app} is preset to be at least 50–100 mV past the peak potential. Here, upon sudden initiation of the applied voltage, a large current needs be passed in the first few seconds of electrolysis. Now, even with small solution resistances, the value of iR_{cell} will be considerable. If, indeed, it is large enough to drop the potential of the WE to a value less than the peak potential, then the instantaneous current will not be the maximum value to be expected.

INSTRUMENTATION

Further, as current decreases, the potential of WE varies and the initial conditions for linear diffusion vary. Constancy of the product ($it^{1/2}$) can hardly be expected under these conditions.

Again, iR corrections can be applied. The corrections for true peak currents given by Delahay and Stiehl work quite well (67). The point, however, is that it seems rather unnecessary to continue a "horse and buggy" approach to voltammetry, since relatively simple instrumentation is available to record current–*potential* curves directly.

Basically the desired approach is to employ a three-electrode system as seen in Fig. 9-9. Here the auxiliary electrode is one whose voltammetric

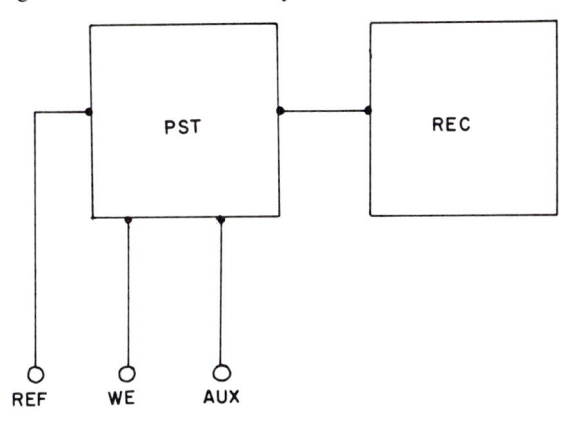

Fig. 9-9. Controlled-potential polarograph using three-electrode system. PST, Potentiostat; REF, reference electrode; WE, working electrode; AUX, auxiliary electrode; REC, recorder.

characteristics are relatively unimportant. The potentiostat, or potential controller, merely supplies whatever current is necessary to the WE-AUX circuit so that the potential of the WE is held at any desired value with respect to REF. The potential of the WE vs. REF is sensed with a high input impedance device so that practically no current is drawn in its measurement. As subunits, or integral with the potentiostat, are a voltage sweeping device and an output section for a recorder. In actual fact, if another REF electrode were added to the two-electrode circuit of Fig. 9-8, then one has the equivalent of Fig. 9-9. Now, a pair of hands replaces the potentiostat and adjusts the slide wire to whatever value is necessary to maintain the potential of the WE at the desired value vs. the "new" REF. In this case the potential between WE and the new REF should be observed with a vacuum-tube voltmeter. The AUX merely happens to be a calomel or other reference electrode—in any event, its potential is no longer important; it is being used only to complete the current path.

Potentiostatic control at polarographic current levels is best accomplished through the use of operational amplifiers. The name operational amplifier (OA) is given to high-gain amplifiers used with a variety of negative feedback configurations to carry out various mathematical operations in computers, etc. They are available as compact, plug-in units and are extremely versatile for laboratory instrumentation. Specific adaptations of OA's to electroanalytical instrumentation were made some time ago by Booman (*68*) and DeFord (*69*). In the last few years a wealth of information has appeared in the literature which makes it redundant to reproduce specific circuitry here. Highly finished polarographic instruments have been described by Kelly and co-workers (*70,71*). Excellent discussions of operational amplifiers are contained in the texts by Bair (*72*) and Malmstadt et al. (*73*). A complete symposium on operational amplifiers in analytical instrumentation was recently published (*74*). This compilation provides details on almost any desired circuit for electroanalytical applications and contains a wide spectrum of references.

With the wide range of information available, one is faced with a real problem of deciding how complex or versatile an instrument to construct. The modular concept which adds in OA units as desired for specific operations is certainly desirable but can be carried to an extreme. It is clear that any unit which has repetitive capabilities (i.e., cyclic sweep) can be used for single-sweep operation and also for fixed-potential operation. Since all three of these operations are highly desirable, the minimum capabilities should, it would seem, include cyclic voltage sweeping. Additional modules for a-c voltammetry or special output functions such as differentiation or integration of response signals are additional capabilities which may be desired by some workers. In the writer's laboratory some extremely simple OA circuits have been used with only the capabilities noted above (cyclic, single-sweep, and fixed potential). These instruments contain no unique circuitry nor is any claim intended for originality of design. They are specifically oriented toward solid electrode applications and have proved reliable over several years of continuous operation (*75,76*).

Much of the utility of cyclic voltammetry is lost unless $X\text{-}Y$ recording is used. A large number of $X\text{-}Y$ recorders are now available. Many incorporate an easy conversion to X-time recording. Often cyclic voltammetry is used for an overall examination of an electrode reaction and one switches to single-sweep operation or chronopotentiometry for precise evaluation of kinetic parameters. It may not be necessary to have the highest grade of X-Y recorder and there are some excellent units available at relatively low prices.

It is probably safe that an X-Y recorder with a 1-sec pen response (or greater) rapidly becomes one of the most useful instruments in a laboratory engaged in electrochemical research. If cyclic voltammetry is to be practiced at high frequencies, a commercial signal generator is used as the scanner and the output is presented on an oscilloscope.

B. Equipment for Chronopotentiometry

It is a rather straightforward matter to add a module which converts a controlled-potential polarograph to constant-current output for chronopotentiometry. Simple battery (or regulated power supply) resistor combinations serve equally well for most solid electrode applications. After the last section this opinion may sound inconsistent to the reader and be adjudged heresay by many colleagues. However, consider the ordinary usage in solid electrode chronopotentiometry. The constant currents employed range most frequently from a few microamperes to a very few milliamperes. For pen and ink recording, times less than 1 sec are difficult to measure and times longer than 50–60 sec involve natural convection difficulties. Heavy-duty "B" batteries or simple regulated power supplies will provide milliampere currents for these times with stabilities which are excellent. Ordinary toggle switches provide sufficiently rapid response for most operations, although more rapid mercury-wetted relays may be desired. A simple circuit which fulfills the needs indicated above is shown in Fig. 9-10.

The supply voltage, PS, can be four or five large radio "B" batteries or a 200- to 300-V regulated power supply (the Heathkit regulated supply, Model EUW-15, has been found to be excellent for this purpose). If batteries are used, S_1 can be used when the unit is to be inoperative for long periods. The current levels are selected by S_2, which switches in series resistors selected to give currents from ca. 20 μA to a few milliamperes. Moderate adjustment of these currents can be provided by wire-wound radio pots marked "Coarse" and "Fine," although these are refinements which can be left out of the circuit. S_4 initiates the constant-current pulse by switching the current from a fixed dummy load DL (ca. 1000 Ω) to the cell. For reverse-current chronopotentiometry, the current-reversal switch S_3 is used. This should be a nonshorting toggle switch. Current can be measured as the iR drop across R_s, which is a

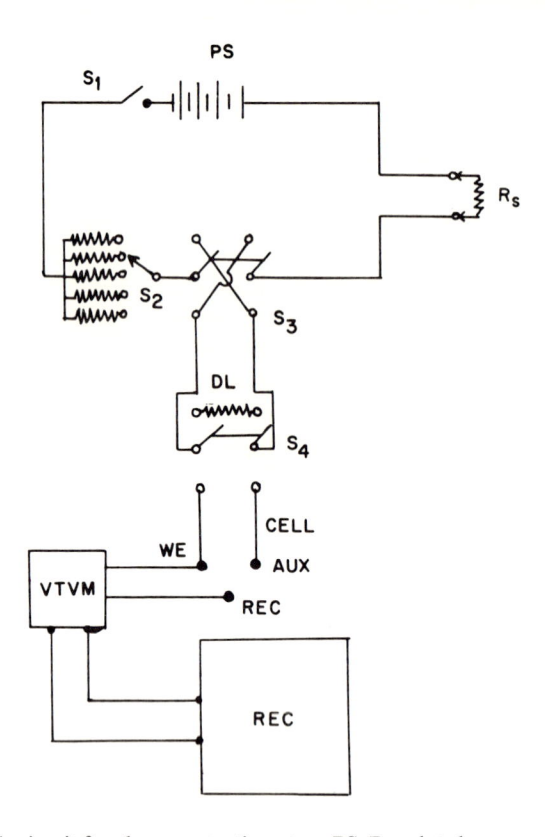

Fig. 9-10. Simple circuit for chronopotentiometry. PS, Regulated power supply or four "B" batteries; S_1, on–off switch, SPST; S_2, current-range selector switch; S_3, current reversing switch, DPDT; S_4, selector switch, "Operate-Dummy Load"; R_s, decade box—current-measuring resistance; WE, working electrode; AUX, auxiliary electrode; REF, reference electrode; VTVM, vacuum-tube voltmeter or pH meter with recorder output; REC, strip-chart recorder, 1-sec pen, 8 in./min chart speed or greater.

standard decade box. The vacuum-tube voltmeter can be any unit with an output to a recorder—a Leeds and Northrup Model 7664 pH meter has been widely used for this purpose. The recorder is a 1-sec-pen-speed instrument with a chart speed of about 8 in./min. Alternatively, an X-Y recorder with the X axis on the time base can be used. Most time bases on X-Y recorders require checking for linearity of sweep and it may be necessary to use an external time sweep to achieve the desired linearity.

Unless high currents are employed or very short transition times are to be observed, it is difficult to visualize why any circuit more involved than that of Fig. 9-10 would be desired. Although it may be satisfying to have currents of 10.00, 20.00, 40.00, etc., microamperes selected by push-button switches, or current reversal at selected voltage gates, all these refinements are accomplished at the expense of frequent calibrations. It is, however, true that if one desires or needs programmed currents or very fast switching situations, this breadboard type of circuitry will not suffice. The unit shown in Fig. 9-10 is normally built on a large piece of pegboard in the writer's laboratory. This is fastened temporarily to the back of a lab bench and can be moved as desired. All inputs and outputs are via bananna jacks and plugs which keeps the circuitry free to be moved.

C. Commercial Recording Polarographs

It is safe to say that all recording polarographs on the market perform in excellent fashion for the purposes for which most of them were designed—routine polarography at the DME. Except in a few instruments, variable-voltage sweep rate is lacking, and since most use strip-chart recorders, the best cyclic voltammetry capabilities are missing. Probably the largest drawback is the lack of controlled-potential output. With the exception of one or two recent instruments, all commercial instruments are two-electrode polarographs. It is unfortunate for those interested in electrochemical research that even routine polarography is a relatively unused analytical technique in industrial laboratories. A greater potential market over the years would undoubtedly have resulted in controlled-potential polarographs concurrent with the development of operational-amplifier circuitry during the past 5 years. It is not difficult to modify most commercial polarographs to include variable sweep rates and even cyclic features (although to be highly satisfactory this implies substitution of an X-Y recorder for the usual strip-chart system). This may be feasible for those wishing to extend the range of their electrochemical investigations.

The comments above are in no way intended as a criticism of commercial polarographic instrumentation. It is true, with most instruments having scan rates of 200 mV/min or greater, that one can run stationary electrode polarograms quite satisfactorily. Taking note of iR corrections, $it^{1/2}$ curves can be obtained. But, in general, the commercial instruments are simply not oriented toward the electroanalytical techniques most applicable to the solid electrode methodology discussed in this monograph.

REFERENCES

1. I. M. Kolthoff and J. J. Lingane, *Polarograph* Vol. 1, Wiley (Interscience), New York, 2nd ed., 1953.
2. J. Heyrovsky, *Polarographic*, Springer, Berlin, 1941.
3. O. H. Müller, *The Polarographic Method of Analysis*, Mack Printing Co., Easton, Pa., 2nd ed., 1951.
4. L. J. Meites, *Polarographic Techniques*, Wiley (Interscience), New York, 1951.
5. G. W. C. Milner, *The Principles and Applications of Polarography and Other Electroanalytical Processes*, Longmans, London, 1957.
6. J. J. Lingane, *Electroanalytical Chemistry*, Wiley (Interscience), New York, 1953.
7. P. Delahay, *New Instrumental Methods in Electrochemistry*, Wiley (Interscience), New York, 1954.
8. S. S. Lord and L. B. Rogers, *Anal. Chem.*, **26**, 284 (1954).
9. D. B. Julian and W. R. Ruby, *J. Am. Chem. Soc.*, **72**, 4719 (1950).
10. J. F. Hedenberg and H. Freiser, *Anal. Chem.*, **25**, 1355 (1953).
11. F. T. Eggertsen and F. T. Weiss, *Anal. Chem.*, **28**, 1008 (1956).
12. I. M. Kolthoff and J. Jordan, *Anal. Chem.*, **24**, 1071 (1952).
13. I. M. Kolthoff and J. Jordan, *J. Am. Chem. Soc.*, **74**, 4801 (1952).
14. F. Baumann and I. Shain, *Anal. Chem.*, **29**, 303 (1957).
15. M. M. Nicholson, *J. Am. Chem. Soc.*, **76**, 2539 (1954).
16. M. M. Nicholson, *J. Am. Chem. Soc.*, **79**, 7 (1957).
17. M. M. Nicholson, *Anal. Chem.*, **27**, 1364 (1955).
18. T. Pavlopoulos and J. D. Strickland, *J. Electrochem. Soc.*, **104**, 116 (1957).
19. T. Berzins and P. Delahay, *J. Am. Chem. Soc.*, **75**, 555 (1953).
20. H. A. Laitinen and I. M. Kolthoff, *J. Phys. Chem.*, **45**, 1079 (1949); also Ref. *1*, p. 415.
21. M. J. Allen and V. J. Powell, *Trans. Faraday Soc.*, **50**, 1244 (1954).
22. I. M. Kolthoff and J. Jordan, *J. Am. Chem. Soc.*, **75**, 1571 (1953).
23. I. M. Kolthoff, J. Jordan, and A. Heyndrickx, *Anal. Chem.*, **25**, 884 (1953).
24. I. M. Kolthoff and N. Tamaka, *Anal. Chem.*, **26**, 632 (1954).
25. T. Tsukamoto, T. Kambara, and I. Tachi, *Proc. 1st Intern. Polarog. Congr., Prague, 1951*.
26. R. N. Adams, C. N. Reilley, and N. H. Furman, *Anal. Chem.*, **25**, 1160 (1953).
27. D. G. Ferrett and C. S. G. Phillips, *Trans. Faraday Soc.*, **51**, 390 (1955).
28. E. R. Nightengale, *Anal. Chim. Acta*, **16**, 493 (1957).
29. S. Bruckenstein and T. Nagai, *Anal. Chem.*, **33**, 1201 (1961).
30. Obtained through E. H. Sargent Co., Chicago, Ill.
31. Servo Corporation of America, Hicksville, N.Y.
32. A perforated masonite board obtainable from Country Workshop, Newark, N.J.
33. I. M. Kolthoff and J. Jordan, *J. Am. Chem. Soc.*, **76**, 3843 (1954).
34. H. P. Silverman, Ph.D. Thesis, Stanford Univ., Stanford, Calif., 1957.
35. E. D. Harris and A. J. Lindsay, *Analyst*, **76**, 647, 650 (1951).
36. A. J. Lindsay, *Anal. Chim. Acta*, **13**, 200 (1955).
37. E. R. Roberts and J. S. Meek, *Analyst*, **77**, 43 (1952).
38. H. Lund, *Acta Chem. Scand.*, **11**, 491, 1323 (1957).
39. W. Dirscherl and K. Otto, *Leybold Polarograph. Ber.*, **1**, 49 (1953).
40. F. Oehme and D. Noack, *Chem. Tech. (Berlin)*, **7**, 270 (1955).
41. V. F. Gaylor, P. J. Elving, and A. L. Conrad, *Anal. Chem.*, **25**, 1078 (1953).

REFERENCES

42. V. F. Gaylor, A. L. Conrad, and J. H. Landerl, *Anal. Chem.*, **29**, 224, 228 (1957).
43. P. J. Elving and A. Krivis, *Anal. Chem.*, **30**, 1645, 1648 (1958).
44. J. B. Morris, Ph.D. thesis, Pennsylvania State Univ., University Park, 1956.
45. J. B. Morris and J. M. Schempf, *Anal. Chem.*, **31**, 286 (1959).
46. P. J. Elving and D. L. Smith, *Anal. Chem.*, **32**, 1849 (1960).
47. G. A. Ward, *Talanta*, **10**, 261 (1963).
48. E. S. Jacobs, *Anal. Chem.*, **35**, 2112 (1963).
49. S. Azim and A. C. Riddiford, *Anal. Chem.*, **34**, 1023 (1962).
50. Z. Galus and R. N. Adams, *J. Phys. Chem.*, **67**, 866 (1963).
51. S. Bruckenstein, personal communication, 1966.
52. P. W. Davies, Chapter in *Physical Techniques in Biological Research*, Vol 4 (W. L. Nastuk, ed.), Academic Press, New York, 1962.
53. N. T. S. Evans and P. F. D. Naylor, *J. Polarog. Soc.*, **1960**, 2, 9, 16, 22, 25.
54. B. Hagihara, *Biochim. Biophys. Acta*, **46**, 134 (1961).
55. J. D. Eye, L. H. Reuter, and K. Keshaven, *Water Sewage Works*, **108**, 231 (1961).
56. Y. S. Lyalikov and V. I. Karmazin, *Zavodsk. Lab.*, **14**, 138 (1948).
57. D. Cozzi and P. G. Desideri, *J. Electroanal. Chem.*, **1**, 301 (1960).
58. E. M. Skobets, I. D. Panchenko, and V. D. Ryabokon, *Zavodsk. Lab.*, **14**, 1307 (1948).
59. A. Langer, *Anal. Chem.*, **28**, 426 (1956).
60. A. H. Ewald and S. C. Lim, *J. Phys. Chem.*, **61**, 1443 (1957).
61. H. L. Rosano and S. Q. Sheps, *American Chemical Society, 145th Meeting, New York, Sept. 1963*, Abstract 4I, see also, *Chem. Eng. News*, **41** (38), 51 (1963).
62. W. D. Cooke, C. N. Reilley, and N. H. Furman, *Anal. Chem.*, **24**, 205 (1952).
63. W. D. Cooke, *Anal. Chem.*, **25**, 215 (1953).
64. G. L. Janz and H. Taniguchi, *Chem. Rev.*, **53**, 397 (1953).
65. R. C. Larson, R. T. Iwamoto, and R. N. Adams, *Anal. Chim. Acta*, **25**, 371 (1961).
66. W. Furness, *Analyst*, **77**, 345 (1952).
67. P. Delahay and G. L. Stiehl, *J. Phys. Colloid Chem.*, **55**, 570 (1951).
68. G. L. Booman, *Anal. Chem.*, **29**, 213 (1957).
69. D. D. DeFord, Div. of Anal. Chem., *American Chemical Society, 133rd Meeting, San Francisco, 1958*, Abstracts.
70. M. T. Kelly, H. C. Jones, and D. J. Fisher, *Anal. Chem.*, **31**, 1475 (1959).
71. M. T. Kelly, D. J. Fisher, and H. C. Jones, *Anal. Chem.*, **32**, 1262 (1960).
72. E. J. Bair, *Introduction to Chemical Instrumentation*, McGraw-Hill, New York, 1962.
73. H. V. Malmstadt, C. G. Enke and E. C. Toren, *Electronics for Scientists*, Benjamin, New York, 1962.
74. Operational Amplifiers Symposium, *Anal. Chem.*, **35**, 1770 (1963).
75. Z. Galus, H. Y. Lee, and R. N. Adams, *J. Electroanal. Chem.*, **5**, 17 (1963).
76. J. R. Alden, J. Q. Chambers, and R. N. Adams, *J. Electroanal. Chem.*, **5**, 152 (1963).
77. L. S. Marcoux, K. B. Prater, B. G. Prater, and R. N. Adams, *Anal. Chem.*, **37**, 1446 (1965).
78. W. R. Ruby and C. G. Tremmel, *J. Electroanal. Chem.*, **12**, 216 (1966).
79. L. S. Marcoux and R. N. Adams, *Anal. Chem.*, **39**, 1898 (1967).
80. S. Roffia and E. Vianello, *J. Electroanal. Chem.*, **12**, 112 (1966).
81. K. B. Prater and R. N. Adams, *Anal. Chem.*, **38**, 153 (1966).

10 APPLICATIONS TO ORGANIC COMPOUNDS

10-1.	Basic Patterns for Anodic Oxidation of Aromatic Compounds	305
10-2.	Aromatic Hydrocarbons	308
10-3.	Primary Aromatic Amines	327
10-4.	Secondary Aromatic Amines	345
10-5.	Tertiary Aromatic Amines	351
10-6.	Aromatic Diamines	356
10-7.	Aromatic Hydroxy Compounds	363
10-8.	Sulfur Compounds	369
10-9.	Miscellaneous Aromatic and Heterocyclic Systems	370
10-10.	Aliphatic Hydrocarbons	372
10-11.	Aliphatic Acids	372
10-12.	Aliphatic Alcohols and Aldehydes	375
10-13.	Aliphatic Amines and Amides	375
10-14.	Aliphatic Halides	377
10-15.	Reduction Processes	377
References		378

The main purpose of all the previous sections in this monograph is to provide background for a survey and discussion of electrochemical reactions of organic materials at solid electrodes. The special interest in organic solid electrode studies is justified by the fact that there exists no general or unified theory of anodic oxidation pathways for organic substances. Such interest is increased by the growth of research on fuel cells and other electrochemical energy-conversion systems. In addition, the detailed electrochemistry of aromatic and heterocyclic molecules may well provide important correlations for biological redox systems.

Inorganic compounds are of limited interest. In no manner does this imply that inorganic redox systems are unimportant. Rather, inorganic materials can be studied (ordinarily via reduction) more conveniently and usually with greater reliability at mercury electrodes. The overall picture of electron-transfer mechanisms in inorganic redox systems has been developed in elegant fashion by Taube, Marcus, Halpern, Vlcék, and Sutin, among others.

A generalized theoretical treatment of the reduction of conjugated hydrocarbons, including electrochemical reduction, has been presented by Hoijtink (*1*). Peover and co-workers have initiated a definitive study of the general behavior of aromatic systems (*2–4*).

There is a large body of information on the mechanisms of *chemical* oxidations of organic species. Much of this has been summarized in two recent monographs by Waters (*5*) and Stewart (*6*). A few books present surveys of anodic oxidations. Fichter's *Organische Elektrochemie* (now out of print) is a treasure of old but valuable data (*7*). It has been said that there are two types of organic electrochemists—"those who have read Fichter and those who have not." Surely the latter are the underprivileged group. Allen's *Organic Electrode Processes* is concerned mainly with preparative electrolyses (*8*).

The excellent reviews of organic polarography initiated by Wawzonek (*9*) and continued by Pietrzyk (*10*) provide coverage of recent work on anodic oxidations. An outstanding contribution by Sherlock Swann, a pioneer in organic electrochemical synthesis, is providing a thorough bibliography of electroorganic publications (*11*). Clark's *Oxidation-Reduction Potentials of Organic Systems* involves classical potentiometry almost entirely but contains much information of concern to organic oxidation processes (*12*). An excellent review of the anodic oxidation of organic compounds which succeeds in bringing Fichter's book up to date has been presented by Weinberg and Weinberg (*82*).

Experimental procedures used to study anodic processes have been given in the previous chapters and are not repeated here. Details of technique or interpretation are mentioned only where they have special relevance to the particular study. The literature survey for this chapter extends through November 1967. It is selective and almost certainly contains inadvertent omissions.

A few reductions are of interest, but this chapter concerns oxidations almost entirely. It is difficult to devise a totally satisfactory classification since there are not as many functional groups as in dropping mercury polarography. The organization chosen is a mixture of functional grouping and type of compound. The intent is only to have a scheme which will enable one to find compounds of interest as easily as possible. The compounds in the survey are examined in the framework of the general pattern of anodic oxidation presented next.

10-1. BASIC PATTERNS FOR ANODIC OXIDATION OF AROMATIC COMPOUNDS

It will be very useful to outline the general anodic behavior of organic compounds. No attempt is made to give a general theory. This would require a detailed description of the electron-transfer mechanisms and all preceding processes. Generally only the events concurrent with and following electron transfer are included here. In fact, it is the special purpose of this chapter to show that the electrochemistry of organic oxidations is *mainly the chemistry of follow-up reactions*.

Restricting attention to aromatic compounds, one can begin with a formulation of the charge transfer itself. Heterogeneous processes which precede it, such as adsorption, as well as homogeneous chemical reactions, are ignored. Adsorption phenomena are extremely important but not yet well characterized. Preceding reactions, especially acid–base dissociations, are prevalent but seldom dominate the overall electrochemistry of aromatic systems.

Representing any aromatic species by Ar, the initial electron transfer can be written

$$Ar \rightleftharpoons Ar(\delta^+) + n_1 e$$

We will develop this discussion by considering Ar to be an aromatic hydrocarbon, but Ar can, in a most general sense, be an already-substituted aromatic species. It can be shown by numerous examples that very frequently (perhaps in all cases) $n_1 = 1$ and a monocation radical, $Ar^{\cdot+}$, is the first result of the electron transfer. In any event, $Ar(\delta^+)$ is an electron-deficient species. The entire course of the overall electrochemical reaction depends on the properties of $Ar(\delta^+)$. There are clearly two situations to consider.

A. $Ar(\delta^+)$ is Stable

First, $Ar(\delta^+)$ can be a stable species within the time interval of the particular electrochemical experiment. It should be remembered that this time interval extends from a few seconds or less in the case of a fast sweep polarogram, to minutes or longer for a controlled-potential electrolysis. Hence the stability of $Ar(\delta^+)$ is a relative parameter. It needs be qualified further with regard to solvent system [in general, stability of $Ar(\delta^+)$ in aqueous or hydroxylic systems vs. nonaqueous or aprotic solvents]. *Within this latitude it is useful to classify certain* $Ar(\delta^+)$*'s as stable.*

There are three possibilities which can give rise to this degree of stability.

1. High Degree of Charge Delocalization in $Ar(\delta^+)$

This condition applies to unsubstituted hydrocarbons with large pi systems. The charge delocalization can be spread out to a degree where there are no longer positions of sufficient reactivity advantageous to nucleophilic attack. [$Ar(\delta^+)$, being electron-deficient, will necessarily undergo some type of reaction with electron-rich or nucleophilic species.] A few examples of such highly delocalized, stable $Ar(\delta^+)$'s have been found recently in hydrocarbon oxidations.

2. Reactive Sites in $Ar(\delta^+)$ Blocked

This situation, which is far more prevalent in hydrocarbons, involves blocking of some or all of the reactive sites in $Ar(\delta^+)$ by relatively electro-inactive substituents. Groups like methyl and phenyl serve this purpose as opposed to functional groups such as —OH, —NH$_2$, —SH, etc. Stability conditions 1 and 2 are easily predicted via simple HMO calculations, and examples of experimental verifications will be treated later.

3. $Ar(\delta^+)$ Stabilized by Specific Functional Groups

Here, in many cases, Ar is a disubstituted aromatic. The stabilization is, in a sense, charge delocalization but is particularly endowed by the functional groups which strongly interact with the basic pi structure. The general class of phenylenediamines and various methoxy-stabilized $Ar(\delta^+)$'s are examples of this behavior. In this class of compounds both mono-cations and dications may be stable. Although classified as stable, it must be recognized that the lifetimes of these species may be markedly dependent on pH and other solution conditions. Thus, while the mono- and dications of various phenylenediamines are quite stable at certain pH's, at other acidities rapid hydrolyses of the amino functions occur. Hence, with functional groups substituents, the stability classification of $Ar(\delta^+)$ is frequently arbitrary and often is best described in the next category.

B. $Ar(\delta^+)$ Undergoes Extensive Chemical Follow-Up Reactions

This is the common behavior of $Ar(\delta^+)$ and makes the electrochemistry of anodic oxidations difficult but fascinating to study. As noted in Chapters 5 and 8, these follow-up reactions can be subdivided into two classes.

1. EC Processes

In an EC process, the product of the follow-up chemical reaction is formed in the oxidized state. The chemical reaction provides no "extra" current over that in the primary electron transfer. In the overall reaction

$$\text{Ar} \rightleftharpoons \text{Ar}(\delta^+) + n_1 e \qquad \text{E}$$

$$\text{Ar}(\delta^+) + \text{Z} \xrightarrow{k_f} \text{P (oxidized)} \qquad \text{C}$$

it will be remembered that chemical removal of $\text{Ar}(\delta^+)$ has an effect on the potentials and height of a peak polarogram (see Chapter 5). The oxidized product, P, may be observed by sweep reversal and its concentration, and the magnitude of k_f, can often be determined (Chapters 4, 5, and 8). P may be further oxidizable at higher potentials. But, at the potential of the primary charge transfer, a net passage of only n_1 electrons ever occurs. This distinguishes the simple EC reaction from the more complex ECE.

The most common EC processes are solvolysis reactions after electro-oxidation. The oxidation of some azines and certain nucleophilic displacement reactions appear to also follow an EC path.

2. ECE Processes

The ECE process is discussed in detail in Chapter 8, Section 8-5. For an aromatic oxidation one can write the general scheme

$$\text{Ar} \longrightarrow \text{Ar}(\delta^+) + n_1 e \qquad \text{at } E_1$$

$$\text{Ar}(\delta^+) + \text{Z} \xrightarrow{k_f} \text{C}$$

$$\text{C} \longrightarrow \text{D} + n_2 e \qquad \text{at } E_2$$

[For consistency with previous notation, only the symbols A and B have been replaced by Ar and $\text{Ar}(\delta^+)$ respectively.] The distinguishing characteristic of the ECE reaction is that the intermediate chemical product, C, is formed in the *reduced state*. However, it is not necessary that it be more easily oxidized than the starting Ar, i.e., that $E_2 \leq E_1$. However, this is common and will be the case if the nucleophile Z is strongly electron-donating to the overall pi system represented by C. Z may be electron-withdrawing, and C will be more difficult to oxidize with $E_2 > E_1$. In this case, if $|E_2 - E_1|$ is sufficiently large, the process may be studied as an EC reaction by restricting the potential to only slightly more anodic than E_1.

The interesting case in which Z is another molecule of Ar(δ^+)—a cation radical dimerization—is frequently met.

$$\text{Ar} \rightleftharpoons \text{Ar}(\delta^+) + n_1 e \qquad \text{at } E_1$$

$$\text{Ar}(\delta^+) + \text{Ar}(\delta^+) \xrightarrow{k_2} \text{C}$$

$$\text{C} \rightleftharpoons \text{D} + n_2 e \qquad \text{at } E_2$$

In most cases HMO calculations readily predict that the dimer C will be more easily oxidized than Ar. Intuitively, since the pi system is essentially "doubled," one expects such a result. On the other hand, with substituted Ar's, the dimer may be twisted from a planar configuration, and calculations predict C to be more difficult to oxidize. Examples of this exist in the couplings of aryl amine cations to give o-benzidines.

3. E-E Processes

In certain instances Ar(δ^+) appears to undergo a further, direct electron transfer without intervening chemical reaction. There are frequent examples of what appear to be overall $2e$ oxidations. Experimentally they cannot be separated into discrete $1e$ steps and the question as to whether they indeed follow such a path becomes largely academic. In the scheme of anodic oxidations presented here, such stable entities as dications are perfectly acceptable. The point to be made is that quite generally the monocations have recognizable lifetimes.

Experimental examples and reliable theoretical predictions are available only for aromatic species, but the remarks of the previous section can be extended to any organic system. The initial product, most generally designated A(δ^+), will always be an electron-deficient entity.

The remaining sections of this chapter provide a survey of organic compounds.

10-2. AROMATIC HYDROCARBONS

A. Polyacenes (Unsubstituted)

It is ideal to begin the survey with unsubstituted aromatic hydrocarbons because their oxidations illustrate very well the extremes in stability of the species Ar(δ^+). Owing to the limited solubility and relatively high oxidation potentials, all hydrocarbons have been studied in nonaqueous

AROMATIC HYDROCARBONS 309

media—usually acetonitrile. Extensive tables of $E_{1/2}$'s are not reproduced for hydrocarbons. For this information the reader is referred to the original literature—especially Ref. *2-4, 13-19, 21,* and *30*. It should be noted that some of these studies employ the Ag$^+$/Ag reference electrode in acetonitrile. To convert from the 0.1 N Ag$^+$/Ag electrode to an aqueous SCE, 0.300 V should be added to the listed values (*25*). It should also be remembered that any potential data for systems with fast follow-up reactions will be slightly dependent on sweep (or rotation) rates and concentrations. Reproducibility in potential measurements of hydrocarbons greater than ±10–20 mV should not be expected unless all significant parameters are controlled (*2,3,19*).

Uncontrolled potential oxidations of hydrocarbons, yielding a mixture of quinones, acids, and other products, are reported in Fichter (*7*). The first careful voltammetry was done by Lund (*13*). Lund's work indicated an initial two-electron (2e) loss. Several studies correlating oxidation $E_{1/2}$'s with HMO parameters tacitly assumed a 1e process (*14-18*). However, Parkanyi and Zahradnik's HMO–$E_{1/2}$ study showed current levels in agreement with a 2e process (*19*). Stanienda distinguished 1e oxidations in a study of diphenylpolyenes (*20*). Preparative-scale-controlled potential electrolyses (*70*) and anodic substitution studies (*21*) were divided in evidence between the 1e or 2e process. The 1e abstraction was indicated for certain hydrocarbons of interest in electrochemiluminescence (*22*) and well-resolved EPR spectra were obtained for a few electrochemically oxidized hydrocarbons (*23,24*). Despite all this work uncertainty continued concerning the basic nature of this electron transfer. Very recently, from three independent groups, unequivocal information on the number of electrons transferred has appeared. The first of these, by Peover and co-workers, was done in acetonitrile and mainly utilized quantitative relationships of cyclic voltammetry (*2-4*). Phelps et al. used cyclic voltammetry and EPR in methylene chloride to illustrate cation radical stability (*28*). Even more stable 1e products were obtained in the solvent nitrobenzene by Marcoux et al. Rotated disk electrodes and EPR were used to evaluate the number of electrons transferred (*29*).

Each of these studies showed relatively stable monocation radicals formed in quite a few cases. [The general species Ar(δ^+) is clearly Ar$\cdot^+$ here.] Surprisingly, even for some completely unsubstituted hydrocarbons, there were indications of stable Ar$\cdot^+$'s. A high degree of charge delocalization is the only logical explanation of such stability. It is useful at this point to see how well one might predict this behavior via chemical reactivity concepts of simple HMO theory.

There are several excellent accounts of HMO calculations designed for chemists interested in applications to organic chemistry (*30–32*). It is assumed that the reader is qualitatively familiar with the calculations of MO energies, coefficients, free valences, localization energies, etc., for simple molecules. Quite extensive tabulations of these parameters are available for more complex species (*33*).

An important reactivity parameter for the species Ar·+ is the unpaired electron density, ρ, at the various positions in the radical ion. It is reasonable to assume that positions with large values of ρ are advantageous centers for reactions. Conversely, the smaller is ρ and the more spread out over a large number of atoms (corresponding to a high degree of "charge" delocalization), the more likely it is that the radical ion will be relatively stable. In forming a cation radical, the electron is lost from the highest filled molecular orbital (HFMO). In simple HMO theory the value of ρ at a given atom position is given by the square of the coefficient, c, of the corresponding atomic orbital in the HFMO. (This unpaired electron density is not to be confused with total pi electron density, q_r, or with spin density.)

Table 10-1 summarizes the ρ values at various positions for most of the unsubstituted hydrocarbons used in the anodic studies mentioned above. The numbering system corresponds to that used by Coulson and Streitweiser, from which all the numerical data were taken (*33*). The ρ values are merely the squares of the indicated coefficients in the HFMO (e.g., for anthracene the coefficients listed for molecular orbital 7 in Coulson and Streitweiser). The ρ values for positions not ordinarily involved in chemical reactions (i.e., positions 11, 12, 13, and 14 for anthracene) are not tabulated. For general comparisons, two other reactivity parameters, free valences (FV), and localization energies (L^+) are listed. These values pertain to the parent hydrocarbon.

Before using Table 10-1 for any predictions, two points must be very clear. First, only gross trends will be revealed. Second, the exact nature of the follow-up reactions of Ar·+'s has not been established. Hence it is not at all clear that ρ is a proper reactivity parameter. Only in a few instances where additive reagents (pyridine, alcohol, etc.) purposely were introduced into controlled-potential oxidations have products been isolated. All attempts so far to identify products from the voltammetric scale oxidations of "pure" hydrocarbons have failed. It is known from RDE measurements that the order of the follow-up reaction is at least greater than unity with respect to hydrocarbon (*34*). Polymer-like insulating films are often formed on the electrodes even in fast sweep methods and are a great source of difficulty in larger-scale controlled-potential oxidations. These facts strongly suggest that the general follow-up reaction is a coupling process leading to dimers and polymers. The use of the unpaired electron density in Ar·+ as a reactivity parameter appears reasonable.

On the above basis, if we use the ρ values of Table 10-1 to predict the most stable Ar·+, we should not be looking for a compound with very small unpaired density "everywhere." The summation of ρ over *all* positions in a given radical ion will necessarily be unity. Rather, one should look for the compounds with ρ values small and well distributed over the possible reactive sites. Compounds with unpaired electron density concentrated at only a few sites will certainly tend to be reactive—regardless of the nature of the reaction.

Using these criteria, of the compounds in Table 10-1, perylene should form the most stable cation radical. Even at the 3, 4, 9, and 10 positions the value of ρ is less than that of any other Ar·+ listed, and the overall density is spread well over a fairly large structure. This conclusion is unquestionably verified by experiment. Of the *unsubstituted* hydrocarbons oxidized to date, the perylene cation radical (Per·+) is by far the most stable. The electrochemical stability criteria are (1) the ratio of $i_{p,c}/i_{p,a}$ in cyclic voltammetry, (2) constancy of the quantity $i_p/V^{1/2}C$ in single-sweep voltammetry or (3) the ratio $i_L/\omega^{1/2}C$ in RDE measurements, and (4) observance of an identifiable EPR spectrum by in situ electrolysis.

In nitrobenzene as solvent (Ar·+'s appear to be considerably more stable in nitrobenzene than the more common solvent acetonitrile), the cyclic voltammetry of perylene shows $i_{p,c}/i_{p,a} = 1.02$, even at a slow scan rate of 5.05 V/min. The $i_p/V^{1/2}C$ value is constant. The ratio $i_L/\omega^{1/2}C$ at the RDE is constant to rotation rates as low as $\omega = 16$ rad/sec (*29*). In nitrobenzene, the EPR spectrum of Per·+ is observed easily. Perylene was not studied by Peover et al. or by Bard and co-workers.

The ρ values for anthracene and tetracene bear out the well-known fact that the 9, 10 and 5, 6, 11, and 12 positions, respectively, are highly reactive sites. One would hardly expect either cation radical (A·+ or T·+) to be very stable. Anthracene shows no indications of limiting to a 1*e* process at high sweep or rotation rates, even in nitrobenzene. Tetracene has a tendency to approach 1*e* behavior at high rotation rates at the RDE in nitrobenzene. This is in accord with Table 10-1, which suggests that T·+ might be slightly more stable than A·+. Phelps et al. found no evidence that T·+ was stable in methylene chloride up to about 40 V/min (*28*).

Peover and White indicated that 1,2-benzanthracene approached a 1*e*

TABLE 10-1

Reactivity Parameters for Aromatic Hydrocarbons

Compound[a]	Positions[b]	ρ^c	FV	L^{+d}
Anthracene				
	1(4,5,8)	0.0967	0.459	2.23
	2(3,6,7)	0.0484	0.409	2.42
	9(10)	0.1936	0.520	2.01
Tetracene				
	1(4,7,10)	0.0562	0.461	2.20
	2(3,8,9)	0.0335	0.411	2.39
	12(5,6,11)	0.1475	0.529	1.93
1,2-Benzanthracene				
	1	0.0000	0.440	2.73
	2	0.0412	0.409	2.43
	3	0.0090	0.403	2.49
	4	0.0256	0.450	2.31
	5	0.0829	0.455	2.25
	6	0.0888	0.455	2.26
	7	0.1980	0.514	2.04
	8	0.1049	0.458	2.24
	9	0.0376	0.407	2.44
	10	0.0557	0.408	2.43
	11	0.0900	0.457	2.26
	12	0.1543	0.502	2.10
Phenanthrene				
	1(8)	0.1156	0.450	2.32
	2(7)	0.0018	0.402	2.50
	3(6)	0.0992	0.408	2.45
	4(5)	0.0543	0.440	2.37
	9(10)	0.1772	0.451	2.30
Chrysene				
	1(7)	0.0876	0.451	2.30
	2(8)	0.0144	0.402	2.49
	3(9)	0.0533	0.408	2.45
	4(10)	0.0581	0.442	2.35
	5(11)	0.0548	0.440	2.35
	6(12)	0.1490	0.457	2.25

AROMATIC HYDROCARBONS   313

Compound[a]	Positions[b]	ρ[c]	FV	L^{+}[d]
Pyrene				
	1(3,6,8)	0.1414	0.468	2.19
	2(7)	0.0000	0.393	2.55
	10(4,5,9)	0.0876	0.452	2.27
Perylene				
	1(6,7,12)	0.0835	0.460	2.20
	2(5,8,11)	0.0130	0.396	2.51
	3(4,9,10)	0.1076	0.473	2.14

[a] The "internal" positions are not numbered on the diagrams. Planes and points of symmetry are as indicated.
[b] The first number corresponds to the listed coefficient in Coulson and Streitweiser; those in parentheses are equivalent by symmetry.
[c] $\rho = c^2$, where c is the coefficient of the atomic orbital.
[d] Localization energy for formation of a residual cation in units of β.

process in acetonitrile at high sweep rates, as shown by $i_{p,c}/i_{p,a} \to 1$ and the limiting value of $i_p/V^{1/2}C^3$. (Presumably, this cation radical might be more stable in nitrobenzene or methylene chloride.) This result is somewhat surprising in view of the fairly high ρ values at the 7 and 12 positions. However, the unpaired density is well spread out, and also some degree of added steric hindrance to coupling reactions may aid the stability.

Pyrene cation is highly reactive at the 1, 3, 6, and 8 positions, and, although it has not been studied thoroughly, there seems to be little evidence for stability of Py·+. Chrysene cation radical would seem to be more stable—only at the 6 and 12 positions is there moderate unpaired electron density. The results of Peover and White are in accord with this. Phenanthracene has not been studied, but there appears to be no doubt that Phen·+ would be quite reactive.

It appears that the ρ values do provide a useful guide to the stabilities of the Ar·+'s. It will be observed that the free valence and localization energies (for the parent hydrocarbons) parallel the ρ values. At a given position with large ρ, there is a correspondingly high free valence and low localization energy. Linear relations between free valences and free-radical reaction rates are well established (32). It is not clear that localization energies apply to the supposed type of reaction here. Although it is conceivable that similar conclusions on Ar·+ stability could be reached by considering the latter two parameters, the use of unpaired electron density—calculated directly for the radical ion—seems more sound.

Although the predictions appear quite successful, one should not become overconfident of their validity. Consider the ρ values for p-terphenyl

Position	$\rho = c^2$
1 (13)	0.0543
2 (6, 14, 18)	0.0600
3 (5, 15, 17)	0.0077
4 (16)	0.0882
7 (10)	0.1232
12 (8, 9, 11)	0.0488

Except for the 7 and 10 positions, which are not open for reaction in the usual sense, the unpaired density is very small. The p-terphenyl cation radical should be at least as stable and perhaps more stable than that of perylene. Experimentally, this is not the case. Hansen et al. studied p-terphenyl briefly among a series of nitro-p-terphenyls (37). They reported that $E_p = 1.78$ V vs. SCE in acetonitrile. The cyclic voltammetry at 3 V/min indicated total irreversibility. Further cyclic studies have shown no indication of a stable 1e product up to 40 V/min in acetonitrile (39). (Owing to the high anodic potential it is not possible to examine p-terphenyl in nitrobenzene.) A satisfactory explanation of the behavior of p-terphenyl is not available. It may be that the follow-up reactions are completely different—perhaps involving cleavage of the polyphenyl linkage.

B. Phenyl-Substituted Polyacenes

In this section we consider some of the simple polyacenes in which certain positions are substituted by "nonelectroactive" groups. This implies a group which, in itself, does not impart electroactivity to a hydrocarbon. A phenyl substituent fits this description, whereas —NH$_2$, —OH, —OCH$_3$, etc., ordinarily make a relatively nonelectroactive hydrocarbon such as benzene easily oxidizable. The latter functionally substituted hydrocarbons are discussed elsewhere as amines, phenols, etc. Halogen and hetero-substituted hydrocarbons are treated in the next sections.

It is evident from the previous material that stable Ar·+'s should be formed if the reactive sites are blocked. This is indeed the case and 9,10-diphenylanthracene (9,10-DPA) produces 9,10-DPA·+, which is stable by all the electrochemical and EPR criteria in acetonitrile, methylene chloride, and nitrobenzene (3,28,29).

Tetracene cation was seen to be quite unstable, owing to the reactive 5, 6, 11, and 12 positions. However, rubrene, which is 5,6,11,12-tetraphenyltetracene, gives an extremely stable radical cation (28,29). In DMF the rubrene cation is far less stable (38).

Pyrene has the 1, 3, 6, and 8 positions reactive (also moderate reactivity at the 4, 5, 9, and 10 positions), and blocking in 1,3,6,8-tetraphenylpyrene gives a completely stable cation radical in methylene chloride and nitrobenzene (28,29).

Methyl groups are also effective substituents. Thus 9,10-dimethylanthracene gives a moderately stable 9,10-DMA·+ in acetonitrile (2). The stability is enhanced somewhat in nitrobenzene, as evidenced by a ratio $i_{p,c}/i_{p,a}$ of 0.84 at the slow sweep rate of 5 V/min (29). It is interesting that qualitatively, 9,10-DPA·+ is more stable under all conditions than the 9,10-DMA·+ counterpart. It would appear that, in addition to blocking extra charge delocalization in the phenyl derivative adds stabilization. This is emphasized by comparing the behavior of 9-phenyl- and 9-methylanthracene, where one reactive site is left open. In this case the 9-phenyl cation shows considerable stability ($i_{p,c}/i_{p,a}$ = 0.45 at 5 V/min), but the 9-methylanthracene exhibits totally irreversible oxidation (29).

Marcoux applied the blocking concept to a molecule which previously had not been studied electrochemically. The compound selected was 1,2,3,6,7,8-hexahydropyrene:

which can be considered as naphthalene alkylated in the 1, 4, 5, and 8 positions. Since the reactive α position of the "naphthalene" is blocked, one could expect a stable cation radical. Also, calculations show that the energy of the HFMO for hexahydropyrene is higher than that for naphthalene, so the former would be more easily oxidized. Experimentally, hexahydropyrene is considerably easier to oxidize and $i_{p,c}/i_{p,a}$ and $i_p/V^{1/2}C$ both indicate a relatively stable cation. A strong EPR signal was obtained on in situ electrolysis but could not be resolved satisfactorily (*34*).

Cauquis and co-workers have studied the oxidation of a series of arylamino-9-anthracenes of the general structure

where R was H or —C_6H_5 and R' a variety of substituents. This blocking allowed EPR observation of stable cation radicals (*36*).

The most extensive investigation of blocking effects was carried out by Zweig and co-workers in conjunction with cation radical stabilities of electrochemiluminescent hydrocarbons (*35*). Most of the blocking group involved electron-donor substituents (methoxy, dimethylamino, etc.), hence they are not necessarily "electroinactive" in the sense described above. Zweig et al. showed clearly how donor substituents raised the energy of the HFMO and hence lowered the oxidation potential for a given hydrocarbon. Moreover, the enhancement of this effect was illustrated when the donor substituents were introduced at positions where the parent molecule had high electron density. For a few molecules such as 9,10-bis(2,6-dimethoxyphenyl)anthracene and 1,6-bisdimethylaminopyrene, cation radical lifetimes of ca. 15 sec in DMF were observed via cyclic voltammetry.

From the foregoing data there is no question but that the anodic oxidation of aromatic hydrocarbons proceeds in many cases via a 1*e* intermediate. When reactive sites are blocked, the 1*e* product may be very stable. The unequivocal identification of reasonably stable cation radicals in the oxidation of some of the hydrocarbons suggests that *an initial 1e abstraction is common to all these systems*. When the electrochemistry indicates a greater number of electrons removed, it can be concluded that rapid follow-up reactions of Ar·+ with the environment take place. These are ordinarily ECE-type reactions, and further electrons are transferred. The environment includes solvent, hydrocarbon species, or added components.

Very little is known about fast follow-up reactions of Ar·+ with the solvent, but the stability of Ar·+ is obviously markedly affected. From the limited data available, Ar·+'s are increasingly stable in the order acetonitrile < methylene chloride < nitrobenzene. These are the most common solvents used to date. The $i_p/V^{1/2}C$ data of Fig. 10-1 illustrate the

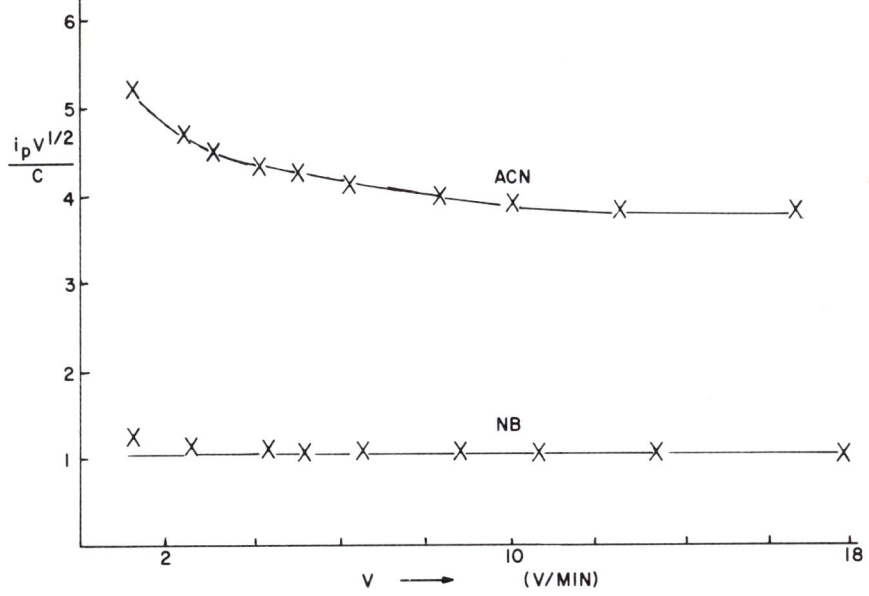

Fig. 10-1. Relative stabilities of 9-phenylanthracene cation radical in acetonitrile vs. nitrobenzene. ACN, acetonitrile; NB, nitrobenzene.

considerable increase in stability of the 9-phenylanthracene cation radical in nitrobenzene as compared to acetonitrile. Variations of $E_{1/2}$ with a greater variety of solvents for the stable 9,10-DPA·+ have been studied by Peover (4,26). Actually, with the more stable Ar·+'s relatively slow solvent interactions probably account for the decay paths. With the shorter-lived Ar·+'s from unsubstituted hydrocarbons there have been no indications that the follow-up reactions are with the solvent.*

As mentioned previously, although product identification has not been successful as yet, it appears that follow-up processes of the highly reactive

* Recent studies show 9,10-dichloroanthracene from oxidation of anthracene in methylene chloride (A. J. Bard, personal communication, 1968).

Ar·+'s lead to polymeric hydrocarbon entities. This is undoubtedly favored by the strong tendency of aromatic hydrocarbons to adsorb on the electrode surface (27), resulting in a high concentration of hydrocarbon species for reaction. Whether these are reactions between two Ar·+ species or Ar·+ with parent Ar (or other possibilities) is not known. In any event, if coupling occurs, the resulting dimeric species will ordinarily be at least as easily oxidized as the starting material. Hence an ECE reaction develops and further polymer-like reactions result in a "cascade" of electrons transferred. The overall extent of these reactions is apparently controlled by the polymer covering the electrode with a relatively insulating film and stopping further electrooxidation. In some cases this filming is minor, but still the limiting values of $i_p/V^{1/2}$ or $i_L/\omega^{1/2}$ (i.e., for short chemical reaction times) indicate a net flow of several electrons and hence rapid ECE processes. These ECE reactions are obviously competitive with the third type of Ar·+ interaction, which is with added components in solution. The latter comprises the area of anodic substitution reactions, which is the last topic discussed under aromatic hydrocarbons.

C. Halogen-Substituted Hydrocarbons

Lund originally examined 9, bromo- and 9,10-dibromoanthracene in acetonitrile (13). Parkanyi and Zahradnik made a more extensive investigation of the $E_{1/2}$'s of haloaromatics (19). The bromo derivatives are all

TABLE 10-2

$E_{1/2}$ Values for Halogen-Substituted Hydrocarbons[a]

Compound	$E_{1/2}$, V vs. SCE[b]
1-Br-naphthalene	1.85
2-Br-naphthalene	1.90
1-Cl-anthracene	1.40
2-Cl-anthracene	1.31
9-Br-anthracene	1.33
	(1.28)[c], second wave (1.78)[c]
9,10-di-Br-anthracene	(1.45)[c], second wave (1.77)[c]
9-Br-phenanthrene	1.79
1-Br-pyrene	1.49
6-Br-chrysene	1.60
4-Br-diphenyl	1.95

[a] In acetonitrile, 0.5 M NaClO$_4$, vibrating Pt electrode.
[b] All values recalculated from Ag/Ag$^+$ reference to aqueous SCE by adding 0.300 V.
[c] From Lund (13).

AROMATIC HYDROCARBONS

oxidized at more positive potentials than the parent hydrocarbons. The data of Table 10-2 are mainly from Parkanyi and Zahradnik. Limiting-current data for bromo derivatives at the vibrating platinum electrode seem consistent with a 2e oxidation (19) as originally proposed by Lund (13). Cyclic studies have not been made, and there is no evidence in the literature for or against 1e intermediates.

D. Hetero-Substituted Hydrocarbons

Hoping to probe further the relations between hydrocarbon molecular orbital parameters and their electrochemical properties, Marcoux investigated the anodic voltammetry of azines (34). By introducing one or more nitrogen atoms in the ring system, the electronic properties are seriously altered but the general structure can be retained. The relations between electronic properties and reductive electrochemistry of azines have been reviewed thoroughly (40–43).

The effect of substituting a nitrogen in the aromatic framework is to considerably lower the energy of the HFMO, thus making it considerably more difficult to oxidize. Contrary to expectations, it is still possible to oxidize both mono- and diazines in acetonitrile under the hydrocarbon conditions. Table 10-3 lists the $E_{p/2}$ and $i_p/V^{1/2}C$ data for the compounds studied (34).

The results are surprisingly different from those of the hydrocarbons. All the azines and diazines in Table 10-3 gave *totally irreversible oxidations*. Figure 10-2 for acridine is typical of the cyclic voltammetry. The value of

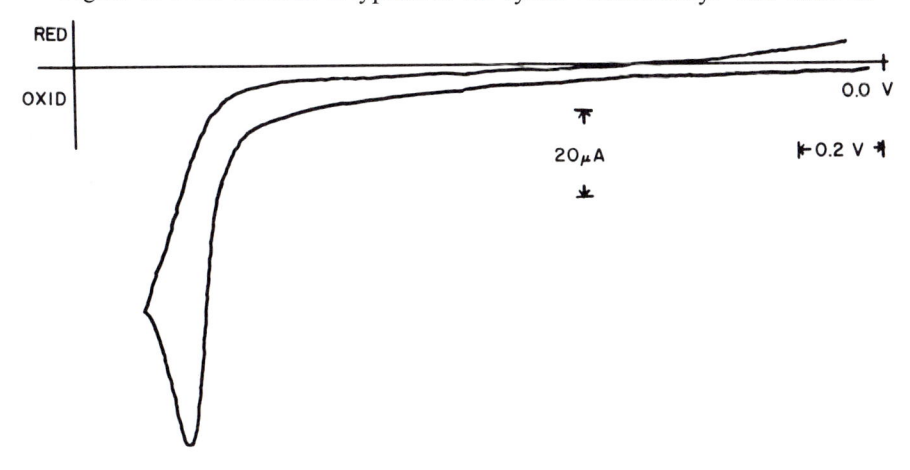

Fig. 10-2. Cyclic polarogram of acridine oxidation. In acetonitrile, 0.1 *M* TEAP platinum button electrode.

TABLE 10-3
Anodic Oxidation Behavior of Selected Azines and Diazines[a]

Compound	Structure	$E_{p/2}$, V vs. SCE[b]	$(i_p/V^{1/2}C) \times 10^{-4}$
Monoazines			
Quinoline[c]		1.97	2.18
Isoquinoline[c]		1.84	1.93
3-Methylisoquinoline		1.67	2.43
Acridine		1.58	3.16
5-6-Benzoquinoline		1.69	3.01
7,8-Benzoquinoline		1.72	3.10
Phenanthridine		1.80	2.20
3,4-Benzacridine		1.73	2.34
Diazines			
Quinoxaline		2.19	3.1
Phenazine		1.91	3.30 (3.92)

AROMATIC HYDROCARBONS 321

Compound	Structure	$E_{p/2}$, V vs. SCE[b]	$(i_p/V^{1/2}C) \times 10^{-4}$
3,4-Benzo-cinnoline		1.72	2.67
5,6,7,8-Dibenzo-quinoxaline		1.85	3.4 (3.78)
1,2,3,4-Dibenzo-phenazine		1.82	2.58 (2.42)

[a] In acetonitrile, 0.1 M tetraethylammonium perchlorate, Pt electrode.
[b] Measured at 5.05 V/min sweep rate.
[c] Cyclic polarogram indicates some impurities present.

$i_p/V^{1/2}$ was constant from 1 to 30 V/min, and when this value was reduced to unit concentration as $i_p/V^{1/2}C$, listed in Table 10-3, it was consistent with a 1e oxidation. (This was shown by comparison with a known 1e oxidation for a similar system, 5,10-dihydro-5,10-dimethylphenazine, which is discussed a bit later. Also, the diazines show reversible reductions in acetonitrile at platinum, leading to characterized anion radicals. Hence for the diazines the $i_p/V^{1/2}C$ for oxidations could be compared with that for the known 1e reduction. The reduction values are listed in parentheses in Table 10-3.)

The azines and diazines all apparently undergo a 1e oxidation with a rapid follow-up reaction. The cyclic voltammetry shows that the products of these follow-up reactions are completely electroinactive in the anodic region. This is entirely different from their hydrocarbon analogues, where the follow-up reactions are ECE.

Using acridine as a test system, attempts were made to identify the nature of the follow-up reactions of azines and diazines. Although no products are formed which are electroactive in the anodic region, a reducible species at ca. −0.5 V is formed. Almost identical behavior is given by 9-phenylacridine, and hence the 9 position is more or less eliminated as the site of the follow-up reaction. Another highly hindered acridine, 1,2,4,5,7,8-hexamethylacridine, seems to follow the same pathway, although there are some indications of an initial cation radical formed in this case. A product has been isolated from the acridine oxidation and tentatively identified via mass-spectral analysis as an acridine type of dimer (39).

A few other azine studies are in the literature. Turner and Elving used pyridine as a solvent for oxidations at pyrolitic graphite electrodes. They suggested that the background oxidation at ca. +1.4–1.6 V resulted in formation of a N-pyridylpyridinium adduct (44). Loveland and Dimeler briefly examined pyridine, pyrrole, and thiophene in acetonitrile (16). Again, in conjunction with electrochemiluminescence work, Zweig's group examined the oxidation of a large number of aryl-substituted isobenzofurans and isoindoles. Cation radical lifetimes of ca. 20 sec in DMF were found in the case of some of the substituted isoindoles (38). A few other heterocyclic studies are mentioned in the review by Tomilov and Fioshin (53).

An ideally behaved phenazine compound is 5,10-dihydro-5,10-dimethylphenazine (DMPZ):

DMPZ

This compound undergoes two successive, reversible 1e oxidation–reduction steps in acetonitrile and propylene carbonate. It behaves as a perfectly reversible system when subjected to all the electroanalytical diagnostic tests. It can be used with confidence as a model system to test other multistage electrochemical reactions. In other solvents (aqueous, DMF, etc.) the second stage of oxidation shows fast follow-up reactions (45).

The phenothiazines and their stable cation radicals have been studied in acetonitrile (46,47). Owing to the importance of the N-substituted derivatives as tranquilizer drugs, their redox properties were examined in aqueous media. All the tranquilizers oxidize fairly easily at ca. +0.5 V in aqueous or alcoholic acid. Large-scale oxidations at gold electrodes indicated that sulfoxides were formed (48). Stable radical species in aqueous solution from the oxidation of chloropromazine and other tranquilizers were identified via EPR (*49,50*). Merkle and Discher reinvestigated the oxidation of chloropromazine derivatives. They showed conclusively that the usual oxidation to sulfoxide obtained in 1 N sulfuric acid could be split into two 1e stages in 12 N sulfuric acid. The first stage, forming the stable cation radical, occurs at ca. +0.6–0.7 V, and the second stage to sulfoxide appears at ca. +1.0 V. They obtained excellent results using controlled-potential millicoulometry by successively oxidizing first to the radical cation, then further to the sulfoxide in 12 N sulfuric acid. The latter coulometry results were consistent with direct 2e oxidation in 1 N acid (*54*).

Allen and Powell studied a variety of indole alkaloids including reserpine-like compounds (*51*). Only 6-methoxy-substituted indoles were oxidizable in an acidic medium. Based only on an apparent 1e change in acid medium, it was proposed that the electrode reaction involved introduction of a hydroxyl group at the indole 5 position. Cyclic studies of other indole derivatives, particularly compounds related to serotonin, indicate possible 1e initial oxidation followed by fast follow-up reactions, perhaps to dimers. The anodic electrochemistry of indoles is very difficult, owing to serious filming of the electrode surface, but certain methylated derivatives behave reasonably (*52*).

E. Hydrocarbon Anodic Substitution Reactions

Anodic substitution reactions comprise an important area of preparative organic electrochemistry. These substitution processes include acetoxylation, alkoxylation, and cyanation of aromatics as well as halogen substitutions and nitrations. In the past two years important advances have been made. For years substitution reactions have been considered only in the framework of Kolbe synthesis. In the Kolbe electrosynthesis, acetoxy radicals decarboxylate and couple to aliphatic hydrocarbons. In the acetoxylation substitution process, the acetoxy radical was supposed to attack the substrate hydrocarbon, forming the acetate.

A correct interpretation of substitution reactions came from studying the voltammetry of the hydrocarbons. Again, almost concurrently, several groups arrived at similar conclusions. Ross et al. observed that naphthalene drastically decreased the production of ethane in the usual Kolbe reaction (*55*). Mango and Bonner studied the acetoxylation of diphenylethylenes and found anomalies from the usual Kolbe process (*56*). These studies were at uncontrolled potential. Salzberg and coworkers showed clearly that the *i–E* curves for acetic acid–acetate ion (Kolbe mixture) were lowered in potential upon addition of aromatic hydrocarbons (*57,58*). The most comprehensive work is that of Eberson and Nyberg.

Eberson first suggested that in a system composed of acetic acid, acetate ion, and aromatic compound, the aromatic would be far more easily oxidized than acetate ion (*59*). Voltammetry in acetic acid–0.5 *M* sodium acetate medium verified these predictions (*60*). Table 10-4 shows some of the data acquired by Eberson and Nyberg.

TABLE 10-4
$E_{1/2}$ Values of Some Aromatic Hydrocarbons in Acetic Acid–0.5 *M* NaAc

Aromatic compound	$E_{1/2}$, V vs. SCE
Anthracene	1.20
trans-Stilbene	1.51
Anisole	1.67
Phenathrene	1.68
Naphthalene	1.72

Both anisole and naphthalene are favorite substrates for acetoxylation and give good yields of the respective *o*- and *p*-acetoxyanisole and 1-acetoxynaphthalene. Yet, as pointed out by Eberson and Nyberg, these compounds are oxidized at potentials well below that required for the Kolbe reaction. The latter never gives high yields below +2.0 V. SCE. Thus it is clear the aromatic hydrocarbon is oxidized to an Ar(δ^+) intermediate which then undergoes reaction with acetate ion as a nucleophile. This mechanism was substantiated and amplified for both ring and side-chain acetoxylations in later papers utilizing controlled-potential electrolysis and careful product analysis (*61,62*).

Eberson and Nyberg favor a two-electron oxidation of the aromatic with a concerted reaction with acetate ion to give a Wheland-type intermediate as

(Benzene is used here just as an example of a general aromatic system—actually a substituted or larger aromatic hydrocarbon must be used to allow oxidation at reasonable potentials.)

Eberson and Nyberg suggest a possibility of consecutive 1e transfers as

$$ArH \xrightarrow{-1e} ArH^{\cdot+} \quad (a)$$

$$ArH^{\cdot+} \xrightarrow{OAc^-} Ar\underset{OAc}{\overset{H}{\diagup}}\cdot \xrightarrow{-1e} Ar\underset{OAc}{\overset{H}{\diagup}}{}^+ \quad (b)$$

$$Ar\underset{OAc}{\overset{H}{\diagup}}{}^+ \longrightarrow Ar-OAc + H^+ \quad (c)$$

This is the third type of reaction postulated earlier for a short-lived cation radical, i.e., interaction with an added component of the environment—in this case the nucleophile acetate ion. In view of the earlier discussions, this mechanism seems preferable to the writer. It is difficult electrochemically to differentiate between the initial 1e or 2e processes in this case. One normally would use a rapid sweep or RDE technique to show the presence of the ECE reaction [step (b) above] and that results limiting toward a 1e reaction obtain at short time gates (fast sweep or rapid rotation rates). This will not work in the present case since there is an additional ECE complication. In general, the acetoxy products themselves are oxidized at the same potentials or even lower than the initial hydrocarbons. Thus the limiting behavior would not be a valid indication of the initial reaction. Unequivocal proof of the presence or absence of an initial 1e step must await further experiments. The question is not entirely academic since, if the initial cation radical forms, rational electrochemical approaches to preventing its other "natural" decomposition reactions would lead to improved acetoxylation results.

A variety of other substitution processes follow the route given above; i.e., a reaction of $Ar(\delta+)$ definitely occurs with added reagent. These include side-chain acetoxylation (61,62) and cyanation. Koyama and co-workers showed a mixture of 2- and 4-cyanoanisole formed on oxidation of anisole in methanol–sodium cyanide (63). Parker and Burgert showed conclusively that this reaction proceeds by interaction of an anisole cation intermediate and not via cyano radicals (64).

Methoxylation of aromatic and heterocyclic compounds has been studied widely but usually without controlled potential (65–69). Although methoxy radicals have been suggested as active intermediates, some of the results by Weinberg et al. appear to parallel the situation for acetoxylation *(61,68,69)*.

Finally, two specific studies of anthracene oxidation should be considered from the viewpoint of interaction of Ar(δ^+) with added reactants. The first of these is the original study by Lund. In this the oxidation of anthracene in acetonitrile containing pyridine resulted in isolation of the 9,10-dihydroanthranyldipyridinium diperchlorate *(13)*. This set the stage for the continuing concept of two-electron oxidations of aromatic systems. There is no question but that a net 2e transfer occurs with pyridine present. It is equally clear that the number of electrons observed in the absence of pyridine (acetonitrile only) is variable and experimentally usually greater than two. The interpretation given previously of an initial 1e transfer followed by rapid ECE reactions accounts for the variable "electron count" depending on the experimental technique. With pyridine present, it is perfectly reasonable to conclude that the "natural" ECE reactions of anthracene are displaced by very favorable reactions with pyridine to give a stabilized dipyrido adduct and hence an effective 2e change *(3,29)*. It is most likely that this involves an initial 1e oxidation (E), followed by reaction with pyridine (C), and this intermediate loses one more electron (E). Addition of one more pyridine gives a stable stopping point. Hydrocarbon oxidations stopping at two-electrons are special cases and not the general situation.

One other oxidation study of anthracene by Friend and Ohnesorge led to an isolable product *(70)*. In acetonitrile containing ca. 0.5 M absolute ethanol they obtained 10,10'-bianthronyl. This presumably is formed by coupling of two anthracene cation radicals to 9,9'-bianthryl, which is subsequently oxidized to the bianthronyl. The exact nature of the ethanol interaction is not known. However, again, in the absence of ethanol no bianthronyl-type products are found *(34)*.

F. Comments on Correlations of Oxidation Potentials and Molecular Orbital Parameters

A significant portion of the anodic electrochemistry of aromatic hydrocarbons has concerned correlations of $E_{1/2}$'s with HMO parameters *(14–20,30,31)*. Additional examples will be encountered later. It is necessary to consider a particular fallacy of this type of correlation and evaluate its limitations.

The usual assumption of all the HMO correlations is that a reversible

$E_{1/2}$ is linearly related to the energy of a HFMO (or, for reductions, to the energy of the lowest empty MO). As we have seen so far, only a very few strategically substituted aromatic hydrocarbons give completely stable cation radicals. Most of the hydrocarbon oxidations involve fast and extensive chemical follow-up reactions and are overall totally irreversible. Yet the $E_{1/2}$–HMO correlations are highly satisfactory. Hence the alleged requirement of reversible $E_{1/2}$ values is completely unnecessary. What is needed is that the $E_{1/2}$'s of a series of related compounds behave similarly and give values moderately close to the reversible ones. These conditions are ordinarily met in most studies. The general nature and rate of the follow-up reactions are similar for a series of compounds. The perturbation on the reversible $E_{1/2}$'s is small and usually present in all compounds examined. However, as already cautioned by Peover, care should be taken where it is recognized that $E_{1/2}$'s will be markedly dependent on sweep rate, concentration, etc. (2,3). In fact, one experimental approach can be recommended without question. Where follow-up reactions predominate, $E_{1/2}$'s should be determined via fast-sweep techniques or, even better, using the rotated disk electrode. In this way, the effect of fast follow-up reactions on observed $E_{1/2}$'s can be minimized (34).

The remarks above should not cause any stigma to be attached to $E_{1/2}$–HMO correlations. Although, in anodic oxidations, they rarely measure up to the lofty ideals of theory, experimentally they work admirably well.

10-3. PRIMARY AROMATIC AMINES

A. Aniline

The first voltammetric data on aniline oxidation were obtained by Lord and Rogers in their pioneering study of recorded polarograms at solid electrodes (71). Adams and co-workers determined the formal potential over a wide pH range using chronopotentiometry at platinum electrodes (72). The $E_{1/4}$ vs. pH data are shown in Fig. 10-3 (the $E_{1/4}$–pH curves for N-methyl- and N,N-dimethylaniline are also given in Fig. 10-3). The curve shows an expected break at about pH 4.7 which is close to the known pK_a of anilinium ion. Beyond pH 9, the $E^{o'}$ appears pH-independent. Kuwana and Adams found both $E_{1/4}$ and E_p values at the mercury chloride film anode to be in general accord with the data of Fig. 10-3 (73).

The electrochemical oxidation of aniline produces a species $Ar(\delta^+)$ in which the amino group supplies relatively little stabilization over that of the parent hydrocarbon. (The parent benzene is, of course, very difficult to

328   APPLICATIONS TO ORGANIC COMPOUNDS

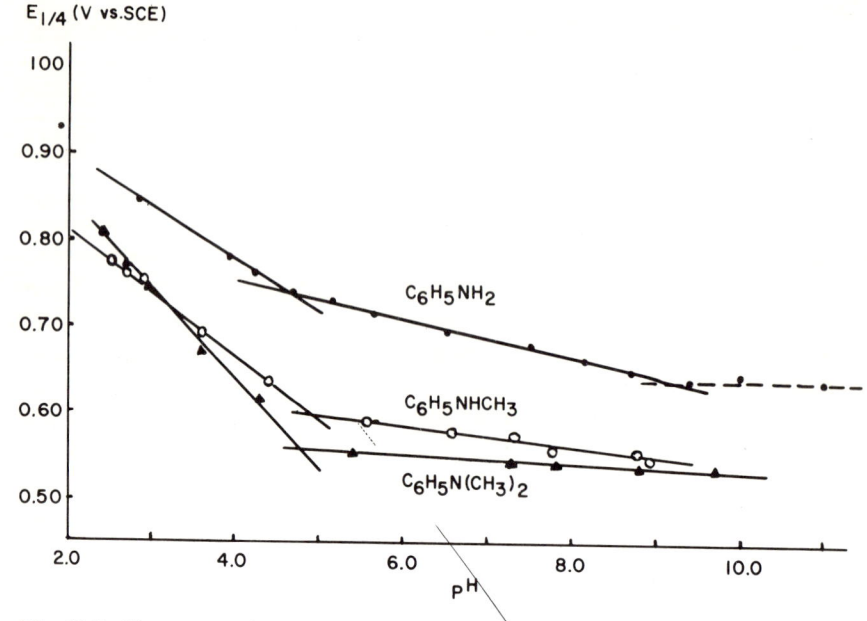

$E_{1/4}$ (V vs.SCE)

Fig. 10-3. Chronopotentiometric $E_{1/4}$–pH behavior of anilines. In aqueous buffers, platinum-foil electrode.

oxidize. If its cation radical is formed, indeed, it is incapable of detection by ordinary electrochemical techniques.)

The anodic reaction of aniline presumably involves a 1e oxidation to produce the cation radical:

$$\text{C}_6\text{H}_5\text{NH}_2 \longrightarrow \text{C}_6\text{H}_5\text{NH}_2^{\cdot+} + e \qquad (10\text{-}1)$$

The principal resonance structures of the radical cation are

(1)   (2)   (3)

The products of aniline oxidation indicate rapid follow-up reactions involving structures (1) and (2), giving dimeric and polymeric products. Few, if any, ortho coupling products appear.

A number of investigations showed that continued electrolysis of acidic aniline solutions produced a dark green precipitate on the anode. These studies are reviewed by Mohilner et al., who identified the anode precipitate via IR and chemical tests as emeraldine (74). Emeraldine is an octamer of the so-called "aniline black" series, whose structure has been given as

Emeraldine

Nigraniline, a closely related member of the series, is slightly more oxidized (three quinoidal rings) then emeraldine. Such compounds obviously arise via a cascade of head to tail condensations and further electrooxidations. From a careful Tafel analysis at varying aniline concentrations, Mohilner deduced that the rate-controlling charge transfer was second-order in aniline and involved two electrons, pointing to the aniline cation radical as the initial reactant. All the other ECE steps are rapid with respect to the initial electron transfer. The first likely product (from a single stage of head to tail condensation) would be p-aminodiphenylamine (or N-phenyl-p-phenylenediamine):

It was easily verified that such a step was consistent with the ECE series—under the same conditions the oxidation of p-aminodiphenylamine is ca. 0.4 V less anodic than that of aniline. Further, emeraldine can also be obtained by sustained anodic oxidation of the intermediate p-aminodiphenylamine.

It should be emphasized that the mechanism proposed by Mohilner et al. probably is valid only for the relatively strong acid conditions used (3.7 M sulfuric acid). No thorough study of the controlled-potential electrolysis products has been made over a wide pH range. Under certain pH conditions one might expect to find competition between the emeraldine-type condensation and hydrolysis of some of the intervening partially oxidized imino functions leading to different products. Recent studies by Leedy have shown the hydrolysis rate of the oxidized form of p-aminodiphenylamine in moderate acid (50% acetone–50% 1 M perchloric acid) to be too slow to measure by reverse-current chronopotentiometry (75). This sets an upper limit for the rate at about 10^{-3} sec^{-1}.

Bacon has demonstrated that p-aminodiphenylamine is the predominate product of aniline oxidation from pH 0 to 6.5 during the time scale of

330 APPLICATIONS TO ORGANIC COMPOUNDS

cyclic voltammetry (*80*). *Significant amounts of benzidine are also formed in the more acid range.* This is well illustrated in Fig. 10-4. There is beautiful matching of the cyclic polarograms of aniline, *p*-aminodiphenylamine, benzidine, and a mixture of all three. The benzidine, which must arise via tail to tail coupling, is not formed beyond about pH 4.

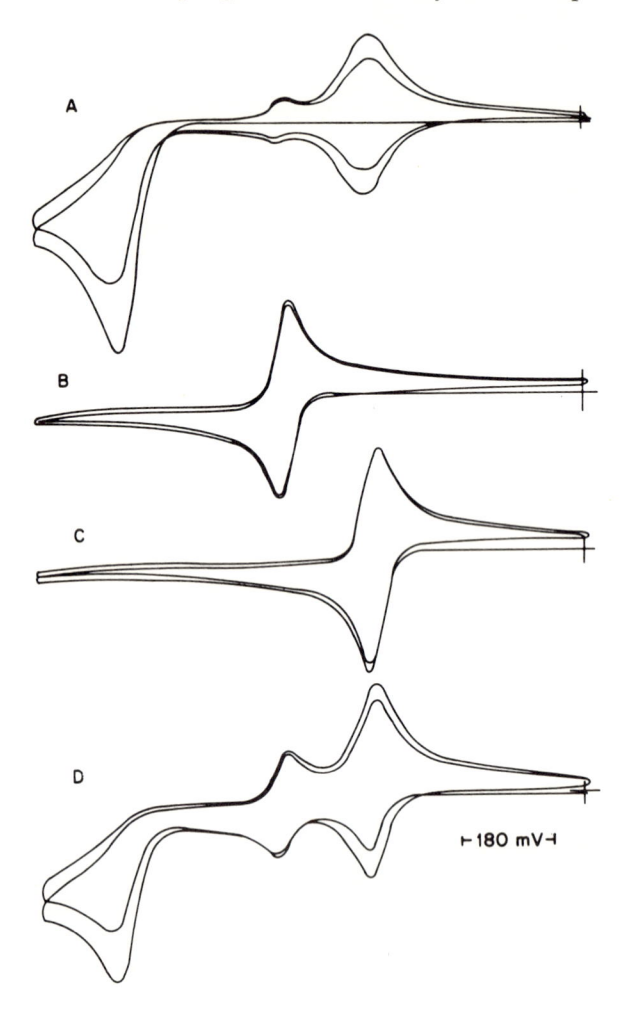

Fig. 10-4. Matching cyclic voltammetry for aniline oxidation. A, Aniline alone; B, benzidine alone; C, *p*-aminodiphenylamine alone; D, mixture of A, B, and C; all at pH 2.3, carbon paste electrode, scan rate 8 V/min.

Assuming an initial 1e oxidation, the first step in the formation of p-aminodiphenylamine is represented by Eq. (10-1). Using the resonance forms (1) and (2), the head to tail coupling can occur as

$$\text{Ph-NH}_2^{\oplus} + \text{(cyclohexadienyl-NH}_2^{\oplus}) \longrightarrow \text{Ph-NH-C}_6\text{H}_4\text{-NH}_2 + 2\text{H}^+ \quad (10\text{-}2)$$

Since the p-aminodiphenylamine is much more easily oxidized than the aniline, at the applied potential a further 2e oxidation occurs:

$$\text{Ph-NH-C}_6\text{H}_4\text{-NH}_2 \longrightarrow \text{Ph-N=C}_6\text{H}_4\text{=NH} + 2\text{H}^+ + 2e \quad (10\text{-}3)$$

The benzidine formation is similarly formulated via two units of structure (2). A further 2e oxidation of the benzidine to give the dication

$$\text{H}_2\overset{+}{\text{N}}=\text{C}_6\text{H}_4\text{-C}_6\text{H}_4=\overset{+}{\text{N}}\text{H}_2$$

occurs in this case. Both the benzidine and p-aminodiphenylamine pathways represent an overall ECE process.

It is interesting that considerable amounts of benzidine are observed in the cyclic experiments, since none of the previous reports of aniline oxidation have mentioned benzidine formation (7,74). Bacon determined the relative amounts of benzidine vs. p-aminodiphenylamine formed as a function of acidity by two methods. First, a potentiostatic oxidation at the main aniline oxidation peak was carried out for ca. 15 sec and then a cathodic sweep initiated. The relative peak currents for the oxidized forms of benzidine and p-aminodiphenylamine were easily obtained (although difficult to evaluate exactly because of lack of proper base lines to measure each i_p). Second, reverse-current chronopotentiometry was applied and the relative cathodic τ's for the oxidized forms measured. These data are summarized in Fig. 10-5, which shows practically no benzidine formation higher than pH 4 but some 30–40% in the 2–6 M acid region. This finding sheds some doubt on the characterization of the emeraldine-type compounds (formed by sustained controlled-potential electrolysis in strong acid solution) as being entirely head to tail polymers. In fact, the original analysis of the "aniline black"-type compounds was hardly sophisticated enough to discern the internal linkages in the polymer.

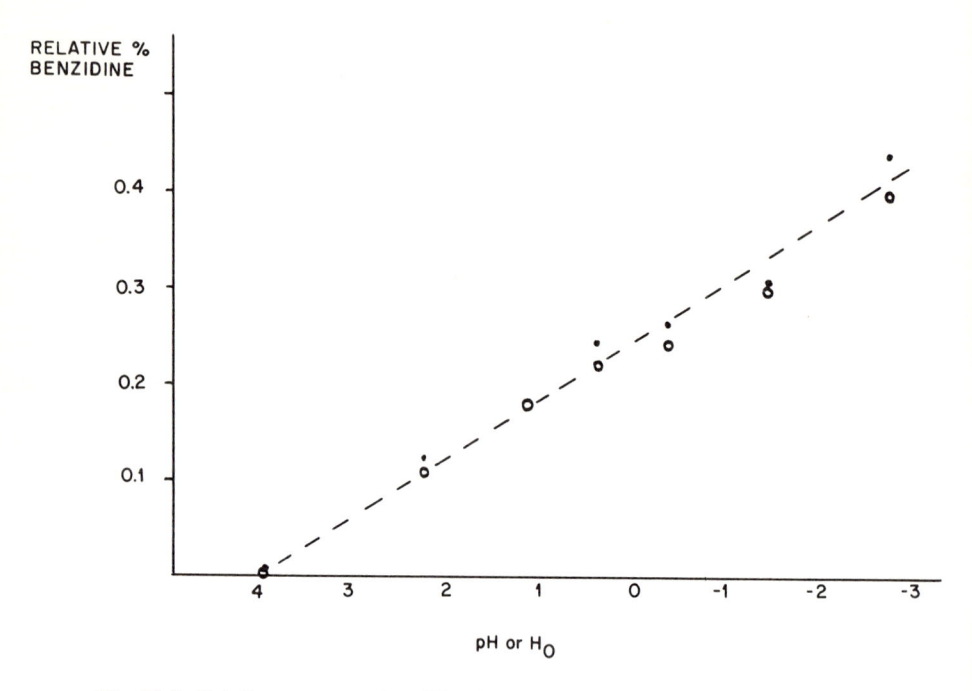

Fig. 10-5. Relative percentage benzidine formation in aniline oxidation. , From cyclic voltammetry; , from reverse-current chronopotentiometry.

It is probably best to describe emeraldine as a relatively reproducible entity with eight "aniline units" in the overall octamer.

Finally, it must be observed that the 1e initial oxidation of aniline is not proved unequivocally by any of the experimental results mentioned above. Both benzidine and p-aminodiphenylamine formation can be accounted for by an initial 2e oxidation to an aniline dication. This species can then react with an unoxidized molecule of aniline to give the same product as a coupling of two cation radicals. This type of reaction is less likely, especially in the case of benzidine formation. To provide the dication with a likely reactant partner, one needs postulate the unoxidized aniline species as the charge-separated resonance structure (4).

(4)

Such structures are certainly less favorable reaction intermediates than the cation radicals. However, in the cyclic voltammetry there appears to be no absolutely conclusive experimental way to differentiate between a 1e vs. 2e initial oxidation in the presence of the fast follow-up reaction.

Voorhies and Davis obtained interesting results with aniline at pH 2 adsorbed on carbon black electrodes (76). Lund surveyed aniline in acetonitrile (13) and Salzberg and Leung (58) give some $E_{1/2}$ values in connection with preparative studies. Most recently Wawzonek and McIntyre investigated aniline and substituted anilines in acetonitrile with pyridine present (81). The latter results are discussed next with substituted anilines.

B. Ring-Substituted Anilines

Many of the available data on ring-substituted anilines consist of $E_{1/2}$ measurements or large-scale preparative work. Fox et al. examined a variety of anilines (and dimethylanilines) at carbon electrodes for the purpose of correlating $E_{1/2}$ with linear free-energy functions (77). A similar study of phenols and anilines was done by Suatoni et al. (78) Both studies used slow voltage sweep rates at stationary WIGE's, but the relative potentials are reliable.

1. Toluidines

As expected from the electron-donating properties of the methyl group, the isomeric toluidines are slightly more easily oxidized than the parent anilines. The $E_{1/2}$'s shift to lesser values with increasing pH. Table 10-5

TABLE 10-5
$E_{1/2}$ of Toluidines (V vs. SCE)

	pH 1.0	pH 5.6
Ortho	0.80	0.60
Meta	0.83	0.61
Para	0.78	0.54

summarizes some of the earlier data from Lord and Rogers (71) at pH 1.0 together with that of Suatoni et al. (78) at apparent pH 5.6 (acetic acid–sodium acetate, 50:50 aqueous–i-propanol buffer). The para isomer is most easily oxidized but there is little hope of any $E_{1/2}$ separation for analytical purposes. The corresponding anisidines and phenetidines (methoxy and ethoxy anilines, respectively) show similar $E_{1/2}$ relationships (78).

2. Chloroanilines

All three chloroanilines are ca. 0.1 V more difficult to oxidize than the parent with $E_{1/2}$'s at pH 5.6 of 0.67 (para), 0.74 (ortho), and 0.77 (meta) V vs. SCE (78). Introduction of a second chlorine in 2,5-dichloroaniline raises $E_{1/2}$ to 0.80 V.

3. Nitroanilines

The strongly electron-withdrawing nitro function makes the $E_{1/2}$ of the nitroanilines quite anodic. In fact, with *p*-nitroaniline, the wave is ordinarily merged with background at platinum electrodes. The oxidation is possible at carbon paste and WIGE's. Olson et al. gave the $E_{p/2}$–pH dependence figures for *p*-nitroaniline at carbon paste electrodes given in Table 10-6 (79). The $E_{1/2}$'s of a few other substituted anilines are given in the original literature (77,78).

TABLE 10-6
Anodic $E_{p/2}$ of *p*-Nitroaniline

Background electrolyte	pH	$E_{p/2}$ V vs. SCE
1 M NH$_3$, 1 M NH$_4$Cl	9.2	0.78
1 M NaClO$_4$	7.0	0.95
1 M HAc, 1 M NaAc	4.3	0.97

Bacon also examined the cyclic voltammetry of a series of *p*-substituted anilines in acidic aqueous media. These were found to also undergo a head to tail coupling giving the corresponding 4′-substituted *p*-aminodiphenylamine but in a partially oxidized form:

$$X-\langle\bigcirc\rangle-N=\langle\bigcirc\rangle=NH$$

Authentic samples of the substituted diphenylamines, where X was —OCH$_3$, —OC$_2$H$_5$, —Cl, and —COOH, were prepared. The cyclic voltammetry of these samples matched perfectly the follow-up products from the corresponding anilines.

In contrast to the aniline situation, the *p*-substituted derivatives oxidize via an EC mechanism. Controlled potential coulometry of the —OCH$_3$,

—OC_2H_5, and —Cl derivatives show one electron per mole. In the aniline case, the intermediate coupling reaction releases protons with the product in the reduced form and a further electron transfer occurs. With the 4-substituted aniline, the leaving group in the coupling is presumably an anion (i.e., chloride), and the diphenylamine is already formed in an oxidized state—an EC reaction. In addition to the compounds mentioned above, the *p*-nitro- and *p*-cyanoanilines behave similarly, but the cyclic voltammetry is not as well defined (*80*). The *p*-methyl (*p*-toluidine) does not react in this fashion, and the mechanism of its oxidation remains unclear. The results of Bacon agree perfectly with the quantitative coulometry on *p*-phenetidine by Santhanam and Krishnan (*83*). However, their interpretation of the oxidation product as the 4,4′-diethoxyhydrazobenzene seems untenable in light of the present results.

Wawzonek and McIntyre obtained very different results when *p*-substituted anilines were oxidized in acetonitrile in the presence of pyridine. In this case N—N coupling occurs to give hydrazobenzenes which are further oxidized to the substituted azobenzenes. These products were verified by voltammetry and large-scale electrolyses with product identification (*81*). Gough and Peover studied a variety of substituted anilines in acetonitrile (*2*). The aqueous vs. acetonitrile results emphasize that

TABLE 10-7
$E_{p/2}$ Data for Some Ring-Substituted Anilines (*80*)

Compound	Background electrolyte	$E_{p/2}$ V vs. SCE[a]
m-Chloroaniline	pH 2.4[b]	0.89
p-Chloroaniline	pH 1.9	0.82
p-Iodoaniline	pH 2.4	0.79
2,5-Dichloroaniline	pH 2.4	0.94
o-Anisidine	pH 2.4	0.68
p-Anisidine	pH 2.4	0.58
2-Methyl-*p*-anisidine	pH 2.4	0.58
p-Aminobenzoic acid	pH 2.4	0.83
p-Aminobenzonitrile	pH 1.9	0.93
o-Toluidine	pH 2.4	0.80
m-Toluidine	pH 2.4	0.74
p-Toluidine	pH 2.4	0.77
2,4,6-Trimethylaniline	pH 2.4	0.65

[a] At carbon paste–Nujol electrode; scan rate 8 V/min potentials subject to slight shifts with scan rate.
[b] Britton and Robinson buffers unless otherwise noted.

much remains to be studied regarding the influence of solvents, supporting electrolytes, and other additives on the course of anodic oxidations.

Some half-peak potentials for the oxidation of various ring substituted anilines at carbon paste electrodes are given in Table 10-7.

C. Aminophenols

The aminophenols are considered here rather than as phenols because of the importance of their oxidized forms (quinoneimines) in a variety of aromatic diamine oxidations.

Fox and Waters obtained unequivocal EPR identification of the cation radical of *p*-aminophenol (and related para-substituted amines). This was an aqueous oxidation with Ce(IV). A high-flow-rate EPR technique was required, owing to the short radical lifetime (*84*). No evidence is known for the electrochemical 1*e* product. Instead, a 2*e* oxidation of *p*-aminophenol (PAP) is observed leading to the quinoneimine (QI):

$$\text{PAP} \rightleftharpoons \text{QI} + 2H^+ + 2e$$

The QI has reasonable stability but, depending on pH, readily undergoes hydrolysis to *p*-benzoquinone (Q):

$$\text{QI} + H_2O \; (OH^-, H_3O^+) \xrightarrow{k_h} Q + NH_3 \; (NH_4^+)$$

The total electrochemical reaction is a simple EC process and one which has caused PAP to be studied widely. It has, in fact, become a standard test system for evaluating the applicability of electroanalytical techniques to EC processes. The hydrolysis rate of oxidized PAP has been measured by reverse-current chronopotentiometry, cyclic voltammetry, potentiostatic, and other methods [for a review of the techniques see Hawley and Adams (*85*)]. Most of these studies were not interested in the hydrolysis of quinoneimine per se but rather were satisfied with the hydrolysis rate being consistent with proving out the electroanalytical technique. Actually the rates and mechanism of quinoneimine hydrolyses are of particular significance. First, the hydrolysis of a quinoneimine is often a decisive step in the follow-up reactions of diamine oxidations. Second, the imino function of a quinoneimine is a type of Schiff base. The hydrolytic reactions of Schiff bases are of importance in enzymatic and diverse biological reactions (86).

Feiser, in his classical paper on the potentiometric titration of unstable redox systems, set the groundwork for the more modern electrochemical measurements of hydrolysis rates of quinoneimines and diimines (EC processes in general). In fact, this type of "discontinuous potentiometric titration," which was practiced in various degrees by Feiser (94), Clark (12), and others, is worth contrasting with, say, cyclic voltammetry. In the discontinuous titration method, to the reductant are added (in separate experiments) various amounts of an oxidant. The oxidant is mixed rapidly and, with this originally fixed ratio of Ox/Red, the potential is followed vs. time by the classical (zero current) potentiometric method. Now, chemical follow-up reactions which alter the concentration of Ox or Red (in the case of the quinoneimine hydrolysis, the chemical transformation of Ox) are related to the potential variation with time. With judicious insight into the nature of the follow-up reactions, Feiser actually measured remarkedly good hydrolysis rates for p-benzoquinoneimine and other compounds. At the time, the reactions were not known for certain to be hydrolyses. The fast electroanalytical techniques, especially cyclic voltammetry, usually identify the individual chemical intermediates via their characteristic voltammetric curves. In addition, much more precise rate measurements are obtained with the newer techniques.

Malachesky apparently first pointed out that the hydrolysis mechanism of a quinoneimine can be formulated in a fashion similar to that for Schiff bases. Jencks, Cordes, and others have given comprehensive explanations of the hydrolysis of Schiff bases (86–88). The hydrolysis rates often reach limiting values at both high and low pH's and thus show a bell-shaped profile of hydrolysis rate vs. pH. The original literature should be consulted for details of the mechanism.

A typical bell-shaped pH–hydrolysis rate (k_{obs}) profile for 3-methylbenzoquinoneimine (from the 2e oxidation of 3-methyl-p-aminophenol) obtained by Malachesky is seen in Fig. 10-6 (90). The value of k_{obs} approaches a limiting value of 5×10^{-2} sec^{-1}. The 3-methyl compound was used rather than the unsubstituted p-aminophenol to eliminate undesirable 1, 4 addition reactions at high pH.

Much earlier Knobloch showed that the hydrolysis rate of the quinoneimine of vitamin K_5 (2-methyl-4-aminonaphthol) was independent of pH

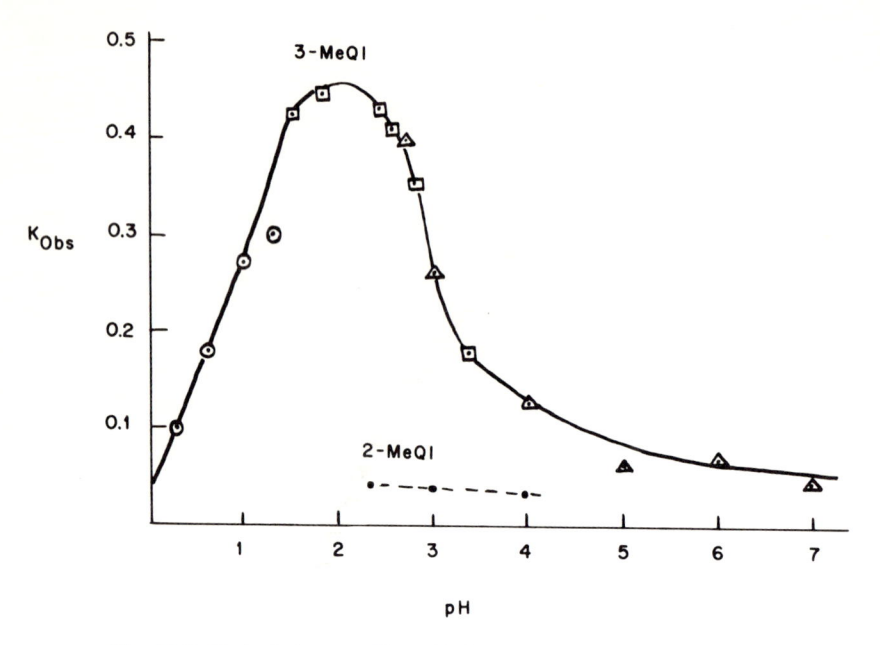

Fig. 10-6. Hydrolysis-rate pH profile for 3-methylbenzoquinoneimine.

at both extremes of pH. This rate is considerably slower (limiting values ca. 7×10^{-3} sec^{-1}) and was measured by DME polarography (*89*). Bell-shaped profiles have been obtained for the hydrolysis of several other quinoneimines, and the similarity of these and Schiff base hydrolyses seems clear.

From the electrochemistry viewpoint it is unfortunate that quinoneimine hydrolyses are this complex. Complete data on k_h for various quinoneimines would be very useful in assessing the relative role of, for instance, quinone formation in various electrooxidations. Unfortunately, each quinoneimine has its own k_h vs. pH dependence, and the individual characteristics are dependent on the pK_a's, etc. Only a few pH profiles have been measured. Without rather complete profiles, it is difficult to compare relative hydrolysis rates of different quinoneimines. It is not necessarily valid, of course, to compare the k_h values of two quinoneimines at the same pH unless their pH profiles are very similar. Nevertheless, the data which follow are useful. It represents the best information presently available on quinoneimine hydrolyses pertinent to anodic oxidation studies.

PRIMARY AROMATIC AMINES   339

The early chronopotentiometric data on k_h for p-benzoquinoneimine were by Testa and Reinmuth but were measured at 30°C. All the data listed in Table 10-8 are at 25°C. The various techniques give internally

TABLE 10-8
Hydrolysis Rate of Quinoneimine of p-Aminophenol
in Acidic Media (25°C)

Molarity of H_2SO_4	k_h, sec^{-1}	Technique[a]	Ref.
1.02	0.0074	TLCR	91
0.51[b]	0.019	RCC	85
0.20	0.052	RRDE	92
0.051	0.100	RCC, CV, PS	85
pH 2.4 buffer	0.152	RCC	85
pH 4.0 buffer	0.129	RCC	85
Higher pH[c]			

[a] RCC, reverse-current chronopotentiometry; CV, cyclic voltammetry; PS, potentiostatic step; RRDE, rotated ring-disk electrode; TLCR, thin-layer chronopotentiometry.

[b] For data at 30°C given by Testa and Reinmuth, see the summary in Ref. 85.

[c] For data at higher pH which may not be indicative of pure hydrolysis, see Ref. 94.

consistent values, and the listing for 0.051 M sulfuric acid is a rounded value. For details on the precision of the measurements, which is quite good, the original literature should be checked. The rate constants are all pseudo-first-order in units of sec^{-1}.

It would appear that the p-benzoquinoneimine k_h peaks at about pH 2.5, but not enough reliable measurements are available to define the curve well. Voorhies and Davis found an apparent k_h for p-benzoquinoneimine adsorbed on carbon black which was considerably smaller than the above values (76).

A few values exist for ring-substituted p-aminophenols. The quinoneimines of the 2,6-dihalo-p-aminophenols hydrolyze somewhat faster than the unsubstituted compound at equal acidities. (Note that the numbering system for p-aminophenols starts with the phenolic end as the 1-position.) The 2,6-dichloro and 2,6-dibromo derivatives have been measured at 2.0 M and 0.05 M sulfuric acid concentrations (85) and are given in Table 10-9.

TABLE 10-9

Quinoneimine of	Medium	k_h, sec^{-1}
2,6-Dichloro-p-aminophenol	2.0 M H$_2$SO$_4$	0.016
	0.05 M H$_2$SO$_4$	0.18
2,6-Dibromo-p-aminophenol	2.0 M H$_2$SO$_4$	0.019
	0.05 M H$_2$SO$_4$	0.18

The complete pH profile for the 3-methyl-p-aminophenol is given in Fig. 10-6. There is a striking difference between it and the 2-methyl derivative. The quinoneimine of the 3-methyl compound hydrolyzes at least 10 times faster than that of the 2-methyl in the acidic region. Owing to the difficulty of electrochemically measuring k_h for the 2-methyl, a complete pH profile is not available. A few values in the pH range 2–4 are included in Fig. 10-6. By raising the temperature to 43°C, Malachesky measured 0.074 sec^{-1} in 0.05 M sulfuric acid (*90*). (It should be noted that Ref. *85*, p. 385, inadvertently reverses the nomenclature for 2-methyl and 3-methyl-p-aminophenol.)

It would be of considerable interest to have complete data on the N-methyl derivatives

from oxidation of N-methyl-p-aminophenol ("Metol") and N,N-dimethyl-p-aminophenol. At present the only qualitative data indicate that k_h for the N-methyl and N,N-dimethyl is considerably slower than that of the unsubstituted imine in moderately acid solutions. If one were to predict the behavior of the quinoneimines from some Schiff base analogues studied by Koehler et al. (*88*), in moderate acid solution, the relative hydrolysis rates would be in the order —NH$_2$ > —NHCH$_3$ > —N(CH$_3$)$_2$. This is qualitatively borne out by the cyclic polarograms of the three aminophenols at pH 2.3 seen in Fig. 10-7. The initial oxidation peaks for the p-aminophenol (PAP), N-methyl-p-aminophenol (N-Me-PAP), and the dimethyl compound (N,N-DiMe-PAP) are marked "Ox" in each case. The practically reversible reduction of each corresponding quinoneimine is marked "R." In the case of PAP, this peak is small and the reduction of the hydrolyzed quinone, marked "Q," indicates relatively rapid hydrolysis. The corresponding "Q" peak for N-Me-PAP is just barely

PRIMARY AROMATIC AMINES 341

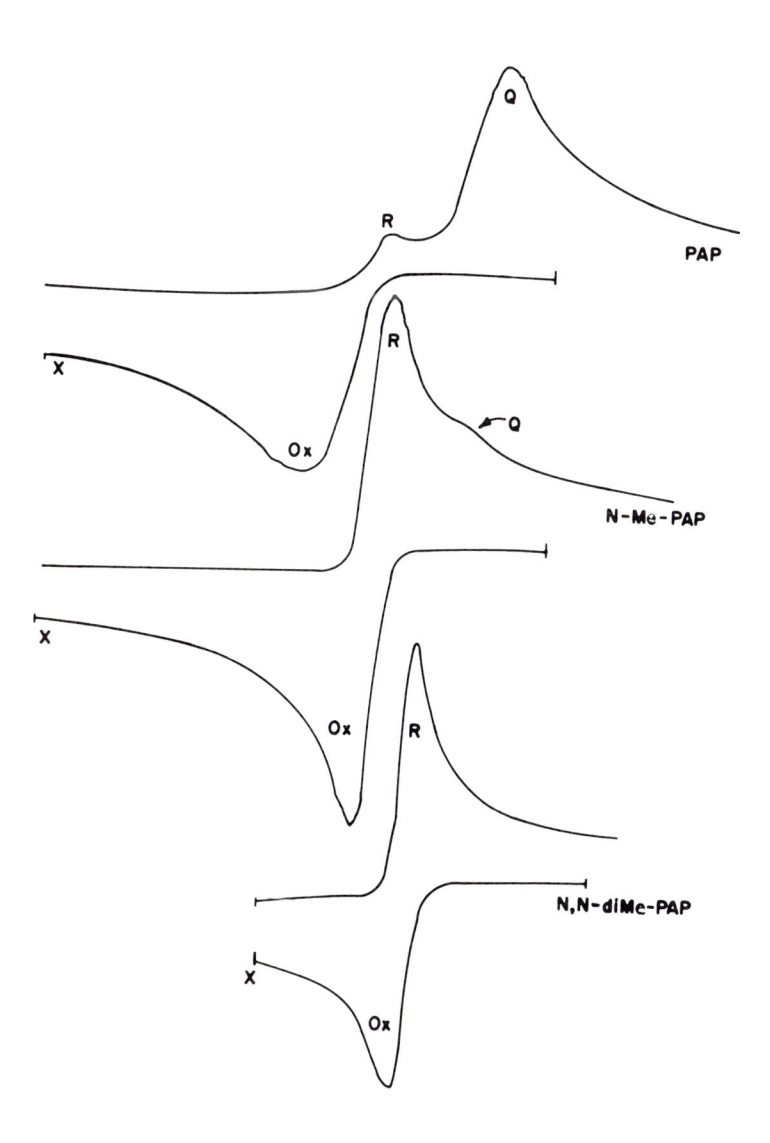

Fig. 10-7. Qualitative comparison of hydrolysis rates of N-methyl-substituted quinone-imines. PAP, oxidation of *p*-aminophenol; N-Me-PAP, oxidation N-methyl-*p*-aminophenol; N,N-DiMePAP, oxidation of N,N-dimethyl-*p*-aminophenol.

evident. The "Q" peak for the N,N-DiMe-PAP is simply not seen in the time span of the cyclic experiment (actually, the forward sweep was "held' at the anodic points marked "X" in each case to increase the time gate for the hydrolysis).

In basic solution, the Schiff base analogy shows that the hydrolysis of the—N(CH$_3$)$_2$ derivative should increase rapidly, whereas the unsubstituted and monomethyl should reach pH-independent limiting values (see Ref. 88, Fig. 1). Quantitative electrochemical data for the quinoneimines are difficult to obtain because of the slowness of hydrolysis at low pH and competing 1,4 reactions at high pH. However, it is known from a recent study of the overall electrooxidation of N,N-dimethyl-p-aminophenol by Marcus and Hawley that k_h is slow in an acid medium and increases markedly at pH 7-8 (93).

Some very interesting data were obtained on N-phenyl-p-aminophenols recently by Leedy. Three compounds studied were

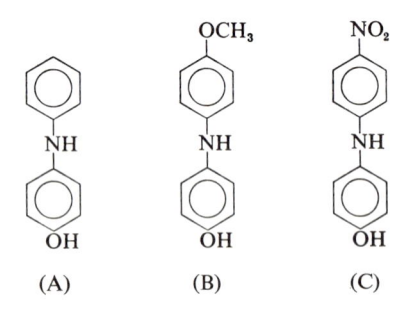

The N-phenyl-p-aminophenols (note that these compounds can also be considered as substituted diphenylamines) were oxidized in 50% acetone–50% 1 M perchloric acid. The acetone is necessary for solubility reasons. The hydrolysis rates of their quinoneimines are tabulated in Table 10-10.

TABLE 10-10

Hydrolysis Rates of Quinoneimines of Substituted N-Phenyl-p-Aminophenols in 50% Acetone–50% 1 M HClO$_4$

(A)	0.037
(B)	0.010
(C)	0.064

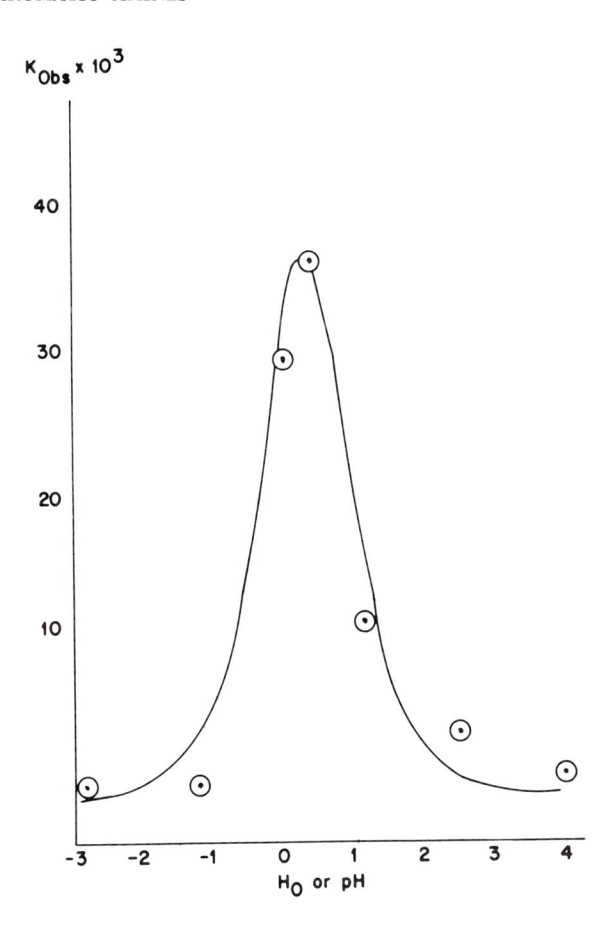

Fig. 10-8. Hydrolysis-rate pH profile for N-phenyl-p-benzoquinoneimine. Solid line, theory calculated for Schiff's base hydrolysis analogy; circles, experimental k_{obs}.

The hydrolyses of (A) and (B) were measured by both RCC and CV, but only RCC was applicable to compound (C). Results from both techniques were in satisfactory agreement.

Leedy also determined the complete pH profile for N-phenyl-p-benzoquinoneimine (from compound A) in 50% acetone-50% aqueous acid an buffers. This is shown in Fig. 10-8. The solid line is the calculated dependence based on the Schiff base hydrolysis analogy, and the experimental points again verify this behavior. It also is in agreement with the previous suggestion that the hydrolysis of the mono N-methyl-p-benzoquinoneimine should show some sort of bell-shaped pH dependence. Further studies of similar compounds are forthcoming and should add materially to the understanding of benzoquinoneimine hydrolysis rates in electrochemical systems (75).

The o-benzoquinoneimines from the oxidation of the o-aminophenols undergo rapid hydrolyses, but the overall reactions are complex, owing to other follow-up processes. However, the quinoneimine of 4,5-dimethyl-o-aminophenol is blocked with respect to such complications, and valid hydrolysis rates have been obtained by Petrie and Adams for the range 2 M perchloric acid through pH 8 buffers (115).

Some half-peak potentials for the oxidation of various p-aminophenols at carbon paste electrodes are given in Table 10-11.

TABLE 10-11
$E_{p/2}$ Values for Substituted p-Aminophenols[a]

Compound	Background electrolyte	$E_{p/2}$, V vs. SCE
p-Aminophenol	2 M H$_2$SO$_4$	0.56
	0.51 M H$_2$SO$_4$, 0.5 M Na$_2$SO$_4$	0.53
	0.10 M H$_2$SO$_4$	0.46
	pH 2.3 (B and R buffer)	0.42
3-Methyl-p-aminophenol	pH 1.8 (H$_2$SO$_4$)	0.44
	pH 4.5	0.22
	pH 6.0 (B and R buffer)	0.09
	pH 9.0 (B and R buffer)	−0.09
2,6-Dichloro-p-aminophenol	2 M H$_2$SO$_4$	0.54
2,6-Dibromo-p-aminophenol	0.51 M H$_2$SO$_4$	0.47
N-Methyl-p-aminophenol	pH 1.6 (HClO$_4$)	0.40
	pH 2.3 (B and R buffer)	0.37
	pH 9.0 (B and R buffer)	−0.10
N,N-Dimethyl-p-aminophenol	pH 1.5 (HClO$_4$)	0.42
	pH 2.3 (B and R buffer)	0.38

[a] At carbon paste–Nujol electrodes; scan rates varying between 2 and 10 V/min; potentials subject to slight shifts with scan rate.

D. Naphthylamines and Large Hydrocarbon Primary Amines

Studies of these compounds have been limited to analytical and $E_{1/2}$ data (15,19,72). In general, the oxidation-products film electrode surfaces heavily, even in nonaqueous solvents. However, Adams et al. showed that quantitative chronopotentiometry was applicable to a series of naphthyl amine sulfonic acids of interest as dyestuff intermediates (72). Practically nothing is known about the mode of oxidation of these larger amines. In spite of their poor electrochemical characteristics, excellent HMO–$E_{1/2}$ correlations were obtained by Parkanyi and Zahradnik (19).

10-4. SECONDARY AROMATIC AMINES

A. N-Alkylanilines

N-methylaniline (MA) is about the only compound of this category which has been investigated in any depth. The $E_{1/4}$–pH behavior of MA was given in Fig. 10-3 and it is seen to be considerably more easily oxidized than aniline over the entire pH range. Galus and Adams investigated the oxidation mechanism at platinum and carbon paste electrodes and found the reactions identical (95). The overall process involves oxidation of MA and tail to tail coupling to the N,N'-dimethylbenzidine (DMB). This is more easily oxidized than MA, so an ECE process ensues, forming the fully oxidized benzidine diquinoid (DMBOx).

In addition, the DMBOx reacts quite rapidly with excess MA present to form a new semireversible redox system at even lower potential. In fact, chemically prepared DMBOx can be titrated potentiometrically with MA and the stoichiometry corresponds to a compound with the general composition DMBOx(MA)$_2$. This product was not identified but may well be a double 1,4 addition product of MA to the benzidine diquinoid structure. In the electrode reaction it is formed in the reduced state and undergoes subsequent electrooxidation. Galus used rotated disk techniques to establish an order of unity for MA in the initial oxidation step. There is some question as to whether one or two electrons are involved in this step. In its overall characteristics, the oxidation of MA is very similar to that of N,N-dimethylaniline, which is discussed later.

Panketh reported the polarographic oxidation potential at platinum of N-sec-butylaniline (96). Weinberg and Weinberg's review indicates a few other studies of mono-N-alkylanilines (82).

B. Diphenylamines (N-Phenylanilines)

Although diphenylamines can be considered as N-phenylanilines, they form a class of compounds well known by the former name. On the other hand, simple unsubstituted diphenylamines apparently undergo coupling reactions typical of the anilines, so the latter classification has some usefulness.

One can group diphenylamines into two broad categories. First, those which are unsubstituted in the *p*-phenyl groups oxidize and undergo fast and extensive chemical follow-up reactions. The second class, ordinarily with para substituents, form relatively stable diquinoid structures whose further reactions are much more slow.

Eggertsen and Weiss studied the $E_{1/2}$'s of several diphenylamine compounds in connection with antioxidant behavior (97). Several other isolated reports of $E_{1/2}$ data exist (72,96). Most of the mechanism studies on the unsubstituted diphenylamines has been done in acetonitrile. The oxidation is quite complex and a complete understanding is not yet available.

Assuming a 1*e* oxidation, HMO calculations show an appreciable unpaired electron density at both the *o*- and *p*-phenyl positions. Thus, in addition to para-para coupling to give ordinary benzidines, one can expect *o*-benzidines as well as mixed ortho-para types. In addition, appreciable N-N coupling to the tetraphenylhydrazines is prevalent. Nelson showed that cyclic voltammetry of diphenylamine in acetonitrile gives a well-defined primary oxidation ($E_{p/2} = +0.91$ V vs. SCE) (98). Two follow-up redox systems develop. One can clearly be identified as due to N,N'-diphenylbenzidine (the usual para-para coupling product). The other, suspected to be tetraphenylhydrazine, is not identifiable via cyclic voltammetry, since the $E_{p/2}$ of the authentic product is at $+0.79$ V and is thus "buried" in the same potential region as the N,N'-diphenylbenzidine. In situ electrolysis in a Cary spectrophotometer indicates that both products are formed but absolute identification of the hydrazine is difficult. It appears that the N,N'-diphenylbenzidine and tetraphenylhydrazine couplings account for about 50% of the electrooxidation products. Presumably ortho coupling products make up the rest. (A known photo-oxidation product, the carbazole, is definitely not formed from electrochemical oxidation in acetonitrile.) Dvorak et al. recently studied diphenylamine at a rotated platinum electrode in acetonitrile. They concluded that oxidized N,N'-diphenylbenzidine was the major product but they also noted another small component which was not identified (99).

In 1 *M* sulfuric acid, assuming only tetraphenylhydrazine formed, Heusler and Schurig estimated a $t_{1/2}$ for the coupling reactions of 3 sec. They employed a double-ring rotating electrode (100).

Nelson studied a large variety of para-substituted diphenylamines. With the mono-para-substituted compounds, relatively rapid follow-up reactions were evident leading to unidentified products. The half-peak potentials of the main oxidation step for some of these compounds are given below.

The di-para-substituted diphenylamines gave, in general, relatively stable cation radicals. These, too, however, decayed slowly, and the final oxidation products are not certain. The $E_{p/2}$ values are listed for reference in Tables 10-12 and 10-13.

TABLE 10-12
$E_{p/2}$ Data for Mono-Para-Substituted Diphenylamines[a]

Substituted diphenylamine R_1R_2NH		
R_2	R_2	$E_{p/2}$ V vs. SCE
$C_6H_4NH_2$	C_6H_5	0.47
C_6H_4OH	C_6H_5	0.60
$C_6H_4OCH_3$	C_6H_5	0.72
$C_6H_4OC_2H_5$	C_6H_5	0.88
C_6H_4NO	C_6H_5	0.98
$C_6H_4NO_2$	C_6H_5	1.45

[a] In acetonitrile, 0.1 M tetraethylammonium perchlorate; scan rate 16.7 V/min.

TABLE 10-13
$E_{p/2}$ Data for Di-Para-Substituted Diphenylamines[a]

Substituted diphenylamine R_1R_2NH		
R_1	R_2	$E_{p/2}$, V vs. SCE
$C_6H_4OCH_3$	$C_6H_4OCH_3$	0.58
$C_6H_4CH_3$	$C_6H_4CH_3$	0.77
$C_6H_4N(CH_3)_2$	$C_6H_4N(CH_3)_2$	0.09
C_6H_4Cl	C_6H_4Cl	1.04

[a] In acetonitrile, 0.1 M tetraethylammonium perchlorate; scan rate 16.7 V/min.

The dimethylamino compound is known as leuco Bindschedlers' Green and its cation radical is quite stable. Dvorak et al. also studied the 4,4'-ditolyl, 4,4'-diamino, and 4,4'-dimethylamino derivatives in acetonitrile but without positive identification of the final products (99).

More quantitative data and product identification on the para-substituted diphenylamines has come from aqueous–acetone solution studies. The hydrolysis rates of the N-phenyl-p-aminophenols (hydroxydiphenylamines) were mentioned previously. The overall EC reaction is

$$\text{(hydroxydiphenylamine)} \longrightarrow \text{(quinone imine)} + 2H^+ + 2e$$

$$\text{(quinone imine)} \xrightarrow[k_h]{H_3O^+} \text{(benzoquinone)} + \text{(substituted anilinium)}$$

To prove this process, Leedy isolated and identified both hydrolysis fragments. A portion of an exhaustively electrolyzed sample (coulometry showed $n_T = 2.0 \pm 0.2$ electrons/mole) was adjusted to pH 8–9 and checked for p-benzosemiquinone radical. A five-line EPR spectrum with $a_H = 2.35$ G showed that unequivocally quinone was present.

Another portion was analyzed for the substituted aniline by adjusting the pH, diazotizing, and coupling with N-(1-naphthyl)ethylenediamine. The resulting absorption spectra were checked with those obtained from authentic samples.

Using the techniques above and comparisons of cyclic voltammetry, Leedy examined the overall behavior of the following para-substituted diphenylamines in 50% acetone–50% 1 M HClO$_4$:

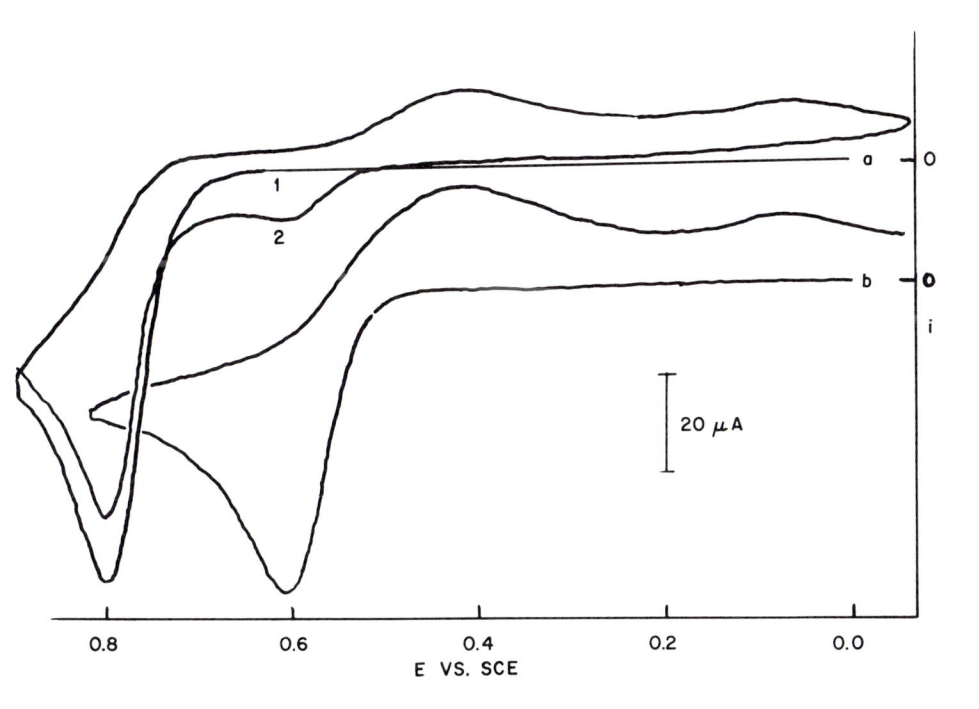

Fig. 10-9. Cyclic polarograms of 4-nitro-4'-methoxydiphenylamine. a, Cyclic polarogram of 4-nitro-4'-methoxydiphenylamine: 1, first anodic sweep; 2, second anodic sweep. b, Comparison cyclic polarogram of N-(p-nitrophenyl)-p-aminophenol; in 50% acetone, 50% 1 M HClO$_4$.

These are the corresponding methyl ethers of the N-phenyl-p-aminophenols or may be considered as substituted N-phenyl-p-anisidines. Figure 10-9a shows the cyclic voltammetry of the nitro derivative which is typical of the series of compounds. On the first anodic sweep, the initial oxidation is observed with $E_{p/2} = 0.76$ V. Upon sweep reversal, no reverse current is observed for the conjugate oxidant. Instead, two broad peaks at ca. 0.42 and 0.08 V are seen. On the second and all following anodic sweeps, a new anodic peak at $E_{p/2} \simeq 0.56$ V is observed. The new redox systems can be clearly identified as belonging to the N-(p-nitrophenyl)-p-aminophenol–quinoneimine couple and hydroquinone–quinone from its subsequent hydrolysis. A comparison of the cyclic voltammetry of N-(p-nitro-phenyl)-p-aminophenol under identical conditions is given as Fig. 10-9b. Rapid ejection of alkoxy groups upon electrooxidation is well established and the overall process can be written

$$\text{R-C}_6\text{H}_3(\text{NH})\text{-C}_6\text{H}_4\text{-OCH}_3 \longrightarrow \text{[quinoneimine-OCH}_3\text{]}^+ + \text{H}^+ + 2e$$

$$\text{[quinoneimine-OCH}_3\text{]}^+ \xrightarrow[\text{fast}]{\text{H}_2\text{O}} \text{quinoneimine=O} + \text{CH}_3\text{OH} + \text{H}^+$$

$$\text{quinoneimine=O} \xrightarrow{\text{H}_3\text{O}^+, \, k_{obs}} \text{quinone} + \text{R-C}_6\text{H}_4\text{-NH}_3^+$$

This mechanism is consistent with reverse-current chronopotentiometry and coulometric studies. In addition, product identification of the quinone and substituted anilines was obtained as mentioned previously. If the alkoxy ejection is rapid, the hydrolysis rate of the quinoneimine should be identical when measured from electrooxidation of either the substituted phenol or the corresponding anisidine. This was found to be exactly the case. The above example illustrates how quinoneimine hydrolysis plays a major role in the overall electrode reaction. Several other substituted diphenylamines were examined in this fashion by Leedy (75).

10-5. TERTIARY AROMATIC AMINES

A. Dimethylanilines

The oxidation pathway of N,N-dimethylaniline (DMA) has been studied by a wide variety of techniques, including tritium tracer identification of products on a micro scale. The details are well covered in the literature (*101–103*) and need not be repeated here. [It should be pointed out that the large-scale electrooxidation of DMA yielding TMB was clearly established in 1922 by Fichter and Rothenberger (*105*).] The $E_{1/4}$–pH behavior of DMA is shown in Fig. 10-3.

The overall ECE process is

$$DMA \rightleftharpoons DMA^{\cdot +} + e \quad (E)$$

$$2\,DMA^{\cdot +} \xrightarrow{k} TMB\,(N,N'\text{-Tetramethylbenzidine}) + 2H^+ \quad (C)$$

$$TMB \rightleftharpoons TMBOx + 2e \quad (E)$$

The early studies suggested an initial 2e charge transfer with the chemical coupling occurring between a DMA²⁺ and an unoxidized molecule of DMA. However, for reasons indicated later, the cation radical coupling appears more likely (*104*). The cation radical has not been detected as such in the electrochemical experiments and the coupling reaction is very fast.

A few minor complications exist in addition to the above reactions. Some degree of o-benzidine formation was detected. Also, TMBOx undergoes relatively slow interaction with excess DMA to give an unknown redox system of lower potential. Hydrolysis of the dimethylimino function can be expected, but this seems to not play any detectable role in the time span of cyclic voltammetry. The TMB oxidation can proceed via two 1e stages, and the cation radical TMB·+ is well known.

When carbon paste electrodes are used for organic mechanism studies there is always the possibility that electroactive species or products may be extracted into the paste liquid. Chambers and Lee investigated this situation for DMA using N,N-dimethyl-^{14}C-aniline (106). Indeed, it was found that extraction of DMA occurred whenever a significant fraction of DMA existed in the free-base form. The radioactivity retained in the electrode showed the same potential dependence as that typical of the adsorption of uncharged molecules on metal electrode surfaces. The TMBOx formed in DMA oxidation was not retained. Surprisingly, when DMA was purposely mixed with the bulk of the carbon paste and the electrode then used in suitable buffer, the cyclic voltammetry was identical to that obtained with an ordinary electrode. Similar results were obtained with other amines and diamines. The extraction phenomenon presumably always is present whenever uncharged (extractable to the organic phase) molecules are involved. Further studies of the significance of this process are needed. The overall conclusions of mechanism studies to date do not appear to be compromised by the extraction problem. It does emphasize, however, that as many techniques as possible, independent of the electrochemistry, should be used in such studies.

DMA has received considerable attention in several nonaqueous solvents. In acetonitrile, TMB is the major product and the subsequent TMB oxidation is clearly a two-stage process (98,99). TMB is also formed in anhydrous acetic acid (107).

The importance of the solvent and electrolyte composition is forcibly demonstrated by the recent studies of Weinberg and co-workers. In methanol-containing potassium hydroxide Weinberg and Brown (108) found two products from DMA oxidation: N-methoxymethyl-N-methylaniline (5) and N,N-bis(methoxymethyl)aniline (6).

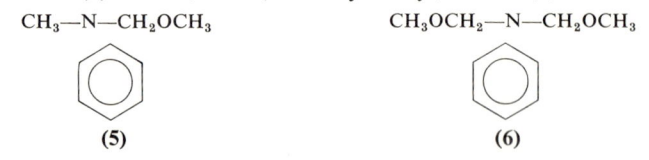

CH$_3$—N—CH$_2$OCH$_3$ CH$_3$OCH$_2$—N—CH$_2$OCH$_3$

(5)   (6)

In contrast, Weinberg and Reddy showed, using methanol with ammonium nitrate as supporting electrolyte, that the mono nitrate of TMB was the product (109). The importance of adsorption of the DMA and the DMA·+

is emphasized by Weinberg and Reddy. In the KOH–methanol, base-assisted deprotonation of adsorbed DMA·+ followed by methanolysis leads to the methoxylated product. In the acidic (ammonium nitrate) methanol, the adsorbed DMA·+ undergoes the TMB coupling reaction.

Fichter and Schönmann illustrated good yields for thiocyanation of DMA using a graphite anode and an acidic, alcoholic solution of ammonium thiocyanate. Between 60 and 70% yields of the *p*-SCN derivative were obtained (*110*).

B. Substituted Dimethylanilines

Zweig and co-workers studied the oxidation of substituted dimethylanilines in a comprehensive investigation of the cumulative effects of the —N(CH$_3$)$_2$ group in pi systems. Along with charge-transfer and proton-resonance chemical shifts, the oxidation-potential correlations helped establish a reliable set of heteroatom parameters for HMO calculations. These parameters gave consistent results for both HMO energy levels and electron (charge) distributions. The voltammetry was done in acetonitrile at a RPE. These heteroatom parameters are important for HMO calculations pertinent to the electrochemistry of aromatic amines and the original paper should be consulted for details (*111*). [DMA was included in a HMO–$E_{1/2}$ correlation of amines by Dvorak et al. (*99*).]

Latta and Taft studied the EPR spectra of para-substituted DMA cation radicals in detail (*112*). The radical ions were produced by in situ electrolysis and most were reasonably stable. The *p*-bromo and *p*-benzoate derivatives gave spectra identical with that of TMB·+ showing group elimination and para coupling. Nelson et al. studied the follow-up reactions of para-substituted DMA radical cations in acetonitrile (*98,113*).

C. Triphenylamines

The triphenylamines (TPA's) provide an excellent example of the extremes of stability of their 1*e* oxidation intermediates depending on substituent effects. If the TPA is completely para-substituted, very stable mono cation radicals are formed. At the other extreme, with completely unsubstituted TPA, or only partial para-substituted TPA's, rapid coupling reactions produce tetraphenylbenzidines (TPB's) which, at the applied potentials, are further oxidized to diquinoid forms. The overall process, analogous to that for N,N-dimethylaniline, is

$$\text{TPA} \underset{}{\overset{-e}{\rightleftharpoons}} \text{TPA}^{\cdot+} \longrightarrow \quad (E)$$

$$2 \; \text{Ph}_2\text{N-C}_6\text{H}_6\cdot \overset{k_b}{\longrightarrow} \text{TPB} + 2\text{H}^+ \quad (C)$$

$$\text{TPB} \underset{}{\overset{-2e}{\rightleftharpoons}} \text{TPB}^{2+} \quad (E)$$

The overall oxidation of TPB goes readily through two 1e stages, and the TPB·+ cation radicals are easily observed in many cases. A wide variety of TPA oxidations were studied by Seo et al. Details of the electrochemical, EPR, and other techniques used to characterize the overall reaction, together with $E_{p/2}$ data, can be found in the original literature (104).

As mentioned, the stabilities of the initial TPA·+'s vary tremendously with para substitution. If all three para positions are blocked, the TPA·+'s are very stable and the cyclic voltammetry shows only a single, highly reversible redox system with $i_{p,c}/i_{p,a} = 1.00$. p-Methoxy groups exert an unusual stabilizing influence on the TPA·+'s. Thus the cation radical of trianisylamine is extremely stable not only in acetonitrile but also in 50% acetone–water buffers from pH 2–6. Even a single p-methoxy group produces a quite stable cation radical. In the absence of methoxy substituents, a single "open" para position gives rise to rapid coupling reactions. Nitro groups are highly "destabilizing" and coupling occurs.

Nelson examined, in more quantitative fashion, a wide variety of triphenylamines, starting with tri-para-substituted compounds and selectively eliminating substituents and moving them to ortho and meta positions (114).

For the tri-para-substituted compounds, the stability of the corresponding TPA·+ can be expressed in terms of the characteristic ratios $i_p/V^{1/2}C$, $it^{1/2}/C$, and $i\tau^{1/2}/C$ for peak voltammetry, chronoamperometry, and chronopotentiometry, respectively. These data are summarized in Table 10-14. The ratios are all consistent with a simple 1e transfer with no follow-up chemical reactions.

TABLE 10-14
Electrochemical Characteristics of Tri-Para-Substituted Triphenylamines

Substituted triphenylamine, $R_1R_2R_3N$						
R_1	R_2	R_3	$E_{p/2}{}^a$	$i_p/V^{1/2}C^b$	$it^{1/2}/C^c$	$i\tau^{1/2}/C^d$
C_6H_4OMe	C_6H_4OMe	C_6H_4OMe	0.52	28.0	45.5	73.0
C_6H_4Me	C_6H_4Me	C_6H_4Me	0.75	29.4	49.0	76.0
C_6H_4F	C_6H_4F	C_6H_4F	0.95	28.5	51.0	84.0
C_6H_4Cl	C_6H_4Cl	C_6H_4Cl	1.04	30.5	52.5	81.5
C_6H_4Br	C_6H_4Br	C_6H_4Br	1.05	30.0	49.5	79.5
C_6H_4COOMe	C_6H_4COOMe	C_6H_4COOMe	1.26	26.5	45.0	71.5
C_6H_4OMe	C_6H_4OMe	$C_6H_4NO_2$	0.86	27.2	46.0	72.5
C_6H_4Me	C_6H_4Me	$C_6H_4NO_2$	1.03	27.0	48.0	75.5
$C_6H_4NO_2$	$C_6H_4NO_2$	$C_6H_4NO_2{}^e$				

[a] In volts vs. SCE, solvent acetonitrile, 0.1 M TEAP, platinum electrode.
[b] i_p, peak current; V, scan rate (1.25–16.7 V/min).
[c] From chronoamperometric measurements.
[d] From chronopotentiometric measurements.
[e] No solvent could be found to dissolve this compound.

TABLE 10-15
Coupling Rates of Mono-Para-Substituted Triphenylamines

Substituted triphenylamine	$E_{p/2}{}^a$	k^b
RPh_2N		
C_6H_4OMe	0.76	ca. 1.0^c
C_6H_4Ph	0.89	$4.63 \pm 0.6 \times 10^1$
C_6H_4Me	0.88	$1.46 \pm 0.12 \times 10^2$
C_6H_4Cl	0.99	$9.6 \pm 0.5 \times 10^2$
C_6H_5(TPA)	—	$2.4 \pm 0.5 \times 10^3$
C_6H_4CN	1.14	$6.8 \pm 1.2 \times 10^3$
$C_6H_4NO_2$	1.17	$1.3 \pm 0.1 \times 10^4$

[a] In volts vs. SCE, measured at a scan rate of 16.7 V/min.
[b] Bimolecular rate constant in liters per mole per second; concentration of triphenylamines was varied from 1.0×10^{-4} M to 2.0×10^{-3} M, solvent acetonitrile, 0.1 M TEAP in all cases.
[c] Difficult to measure accurately by electrochemical techniques due to slowness.

As para substituents are removed, the mono cations are reactive and couple to form benzidines. As a measure of this reactivity, Nelson measured the observed bimolecular rate constant k_b by standard potentiostatic techniques for an ECE process. The rates were evaluated via a working curve provided by digital simulation techniques (*114*). (Some rate constants were also determined by rotated disk measurements.) Some of these rates are given in Table 10-15 for mono-para-substituted TPA's. In all these cases, other coupling reactions cannot be excluded, but there appears ample evidence that the *p*-benzidine reaction predominates. (For the tri-*o*-methoxy TPA it was possible to provide unequivocal evidence for *p*-benzidine formation. The cyclic voltammetry and EPR spectrum of an authentic sample matched those obtained from the parent TPA.) Other ortho and meta derivatives were studied and a fairly comprehensive picture of the anodic oxidation of TPA's, consistent with simple HMO predictions, is available. The original literature should be consulted for further details (*114*).

10-6. AROMATIC DIAMINES

A. Unsubstituted Phenylenediamines

The potentiometric studies of Michaelis et al. (*116*), Feiser (*94*), and others initiated the anodic electrochemistry of *p*-phenylenediamine (PPD), and it has continued via solid electrode voltammetric techniques (*49,117–120*). Unfortunately, these studies have supplied only parts of the story. A qualitative interpretation can be given, but at this time a complete picture of the PPD oxidation over the entire pH range is not available. The details are obscured by several problems:

1. Precursor Reactions

The two amino functions of PPD have acidic dissociations corresponding to $pK_{a_1} = 2.8$ and $pK_{a_2} = 6.2$. At pH 1–2, PPD exists principally as the diprotonated species H_2PPD^{2+}, and even at ca. pH 4.5 about 90% $HPPD^+$ is present. Chronopotentiometry shows that a plot of $i_0\tau^{1/2}$ vs. i_0 decreases with increasing i_0 in accordance with an acid-dissociation precursor reaction. This was investigated quantitatively by Mark and Anson (*121*) and Hawley (*122*). [The recombination rates found by Mark and Anson appear to be too low for diffusional control (*122*).]

2. Hydrolysis Reactions

In the acidic pH range, two successive stages of hydrolysis are present. First, the diimine is hydrolyzed to the quinoneimine and then the latter to benzoquinone.

3. Nucleophilic Addition Reactions

The greatest difficulty is that the diquinoid structures formed above (diimine, quinoneimine, and quinone) are all capable of undergoing 1,4 addition reactions with various nucleophiles. The most logical reactant is the excess of parent PPD. (The reader is referred to Chapter 8, Section 8-5 where the general picture of the hydrolyses and 1,4 additions of PPD were introduced.)

The precursor and acid–base relationships are eliminated to simplify the equations below summarizing the overall PPD process.

Primary electron transfer;

$$\text{PAP} \rightleftharpoons \text{PDI} + 2e + 2H^+$$

The 1e intermediate semiquinone is quite stable and EPR confirmation of it is observed from about pH 2 to pH 6.

Hydrolyses:

$$\text{PDI} \xrightarrow[H_2O]{k_1} \text{QI} \xrightarrow[H_2O]{k_2} Q$$

Not even a partial pH profile is available for the hydrolysis of PDI (k_1), but a few important facts are known. In moderately dilute acid, k_1 is very rapid with respect to k_2 for the quinoneimine, QI. This is shown by the fact that the observed hydrolysis rate constant k_h is identical when measured from the oxidation of either *p*-aminophenol (PAP) or PPD. These results, measured by RCC at two acidities and two different electrode surfaces, are given in Table 10-16 (*122*). At pH 2.4, the hydrolysis rates of PAP and PPD seem to be of the same order of magnitude.

TABLE 10-16
Comparison of Observed Hydrolysis Rates

		k_h observed, sec^{-1}	
Medium	Source[a]	At CPE[b]	At Pt[c]
0.29 M HClO$_4$	PAP	0.032	0.030
	PPD	0.032	0.030
0.051 M H$_2$SO$_4$	PAP	0.102	—
	PPD	0.102	0.100

[a] Compound oxidized for hydrolysis-rate study, at 25°C; PAP, p-aminophenol; PPD, p-phenylenediamine.
[b] Carbon paste electrode (Nujol).
[c] Platinum button electrode.

Since the p-aminophenol k_h is known to be about 0.15 sec^{-1} (see Section 10-3C), that for PPD can be assumed to be about 0.1 sec^{-1} in this pH range.

At high pH's the hydrolysis rates are known from the extensive deamination studies of Tong (*123*). At pH 8 the values of k_h for PAP and PPD are ca. 1.9×10^{-4} sec^{-1} and 4.3×10^{-4} sec^{-1}. [Hydrolysis data for a number of N,N-dialkyl-p-phenylenediamines are contained in this and continuing studies by Tong et al. but all are at alkaline pH's (*123–125*).]

4. 1,4 Addition (Coupling) Reactions

The possibility of 1,4 coupling reactions in the oxidation of PPD were outlined in Feiser's study (*94*). There is no question as to their occurrence in the PPD oxidation, but their rates and variation with pH are practically unknown.

The 1,4 addition (Michael) reaction of a nucleophile with a quinoidal system is well established. Its predominate role in the electrochemical oxidation of catechols and catecholamines has been reported (*126*). Recently Piekarski and Adams made a thorough electrochemical study of the kinetics of the 1,4 addition of substituted anilines and amino acids to o-benzoquinone (*127*).

For the PPD case the reactions can be generalized as

The formulation (NH, O), (NH$_2$, OH), etc., indicates that the reacting quinoid can be the diimine, the quinoneimine, or the quinone, and the resulting product will be a substituted diamine, aminophenol, or hydroquinone. Which of these reactions will occur depends, first, on the rates of the two hydrolysis reactions which produce the quinoidal compounds. Hence the 1,4 additions vary with pH because of the pH variation of the precursor hydrolyses. Assuming all three quinoidal forms are available, one can make "ball-park" predictions of the relative 1,4 addition rates. The 1,4 addition has been shown to involve reaction of the free-base form of the attacking nucleophile (here PPD) on the partially positively charged 4 position of the quinoid. Thus one would predict the relative rates to be in the order

$$\text{PDI} < \text{Q} < \text{HPDI} < \text{H}_2\text{PDI}$$

The 1,4 addition to PDI should be somewhat slower than that to Q, since oxygen is more electronegative than nitrogen. A partial positive charge at the 4 site is more facilitated in quinone than PDI. The protonated diimine, HPDI, on the other hand, should 1,4-add more rapidly than Q because the protonated imine function can be considered slightly more electronegative than oxygen. The HPDI and the diprotonated H$_2$PDI are only important in moderately strong acid solutions. Hawley was able to measure the rate of 1,4 addition of HCl to *p*-benzoquinoneimine. In 6 *M* HCl, the rate was 0.016 sec^{-1} and 0.008 sec^{-1} in a mixture of equal volumes of 6 *M* HClO$_4$ and 6 *M* HCl. The addition to quinone in 6*M* HCl was too slow to measure by standard electrochemical techniques (*122*).

5. Further Electron Transfers

Since the attacking nucleophile in the 1,4 addition reactions of PPD is electron-donating, the resulting products (formed in their reduced states) are, in general, as easy or easier to oxidize than the starting PPD. They are further oxidized at the applied potential and a net ECE reaction develops. At this point, "double" 1,4 additions can occur. These are well known in chemical interactions of quinones and amines.

Until more specific data are available on the hydrolysis and 1,4 addition rates and their variations with pH, it is hopeless to specify the PPD oxidation in more detail. It should be noted that gross differences in the behavior of the PPD system will be observed depending on the "time gate" of the electrochemical experiment. The hydrolyses and 1,4 additions play a major role at almost all pH's in long-time coulometry but may be somewhat bypassed with very fast techniques.

The $E_{1/2}$–pH behavior of the isomeric phenylenediamines has been determined at both platinum (*117*) and carbon paste electrodes (*119*). Analysis of mixtures via voltammetry is difficult due to follow-up reactions which have been partially interpreted (*119*).

The ortho isomer (OPD) films very badly at almost all pH ranges, and it is very difficult to derive any reasonable information from the cyclic voltammetry. Elving and Krivis suggested that OPD underwent an initial 2*e* oxidation followed by coupling to give 2,3-diaminophenazine and further oxidation (*118*). Air or electrochemically oxidized OPD solutions show identical fluorescent spectra with that of 2,3-diaminophenazine (*128*).

The meta compound (MPD) oxidizes most difficultly of the three isomers, and nothing is known about its electrooxidation products in aqueous media. Presumably polymeric coupling products are formed.

PPD in acetonitrile shows two distinct 1*e* waves (*2,99*) and the cation radical PPD·+ has been well characterized in this medium (*129*). Neither OPD nor MPD appears to have been studied in nonaqueous media.

B. Ring-Substituted Phenylenediamines

Only a few ring-substituted *p*-phenylenediamines appear to have been studied. A few $E_{1/2}$'s for such compounds are given by Eggertsen and Weiss (*97*). Toluenediamines were studied chronopotentiometrically by Voorhies and Parsons (*138*).

C. N-Alkyl Phenylenediamines

Owing to their usage as antioxidants and as photographic developers, the N-alkyl-*p*-phenylenediamines have received widespread attention. This is especially true of the N,N-dialkyl-PPD's. The $E_{1/2}$'s of a large number of such compounds were examined by the Eastman Kodak group (*130,131*). Other studies contain further $E_{1/2}$ data (*73,96,97,120, 134–137*).

In aqueous media, following oxidation to the diimine, the N,N-dialkyl function is most readily hydrolyzed, especially at high pH's. Data on these alkaline hydrolyses are available, as mentioned previously (*123–125*). In acidic media, practically nothing is known about the hydrolysis rates, and the details of the oxidation of these compounds remains about as speculative as for the unsubstituted PPD's.

The N,N-dimethyl-PPD, whose cation radical and fully oxidized form are both frequently called Wurster's Red, has been studied frequently (*73,132,134*). Complicating coupling reactions obscure the n value in long-term coulometry, but Christensen and Anson showed integral $n = 2$ values using thin-layer chronopotentiometry (*134*).

Wurster's Blue, the oxidized form of N,N,N',N'-tetramethyl-PPD (also frequently abbreviated in biochemical studies as TMPD) shows two well-defined 1e oxidations over most of the aqueous pH range. The first 1e transfer is reversible and uncomplicated. But a complex follow-up reaction that occurs at the second stage is still not understood completely (*135*). This reaction is of importance, since the compound is widely used in establishing P:O ratios, etc., in oxidative phosphorylation studies. In general, the EPR spectra of the Wurster-type cation radicals are well known, although that of Wurster's Red has defied complete interpretation, owing to its complexity (and possibly mixed EPR spectra in aqueous media).

The oxidation of N-phenyl-PPD has been discussed in Section 10-3.A under *p*-aminodiphenylamine. The hydrolysis rate of the quinoneimine is relatively slow and HMO calculations predict that the hydrolysis probably proceeds as

although the alternative pathway below is possible.

D. Benzidines

Much of the electrochemistry of benzidines has already been encountered earlier under anilines and triphenylamines. In general, benzidines undergo 2e oxidations to the fully oxidized diquinoids, which are really diimines with extended pi systems. The intermediate 1e cation radicals of the N,N' di- and tetra-substituted benzidines are moderately stable in both aqueous and nonaqueous media.

The 3,3'-dimethyl and 3,3'-dimethoxy derivatives are important compounds.

o-Tolidine (o-T)   o-Dianisidine (o-DIA)

Contrary to the potentiometric indications of cation radical stability by Oldfield and Bockris (*133*), no EPR signals were obtained in aqueous solutions for o-DIA, and only a single line with no hyperfine could be observed for o-T (*49*). This was surprising, for considerable stability would be predicted. Actually, o-DIA has been used for some time as a model example (in a strong acid solution) of a reversible 2e oxidation. However, an electrochemically observable 2e oxidation certainly does not preclude the existence of the semiquinone via dismutation (e.g., PPD oxidation). Kuwana and co-workers have apparently explained this difficulty recently by showing that the semiquinone of o-T is present as the dimer. This work was done using their optically transparent electrodes involving simultaneous transmission spectroscopy and electrochemistry

TABLE 10-17

Hydrolysis Rate of Quinonediimine of o-Tolidine

pH	k_h, sec^{-1} [a]
2.99	3.8×10^{-4}
3.20	4.4×10^{-4}
4.00	6.4×10^{-4}
4.55	15×10^{-4}

[a] At 25°C.

AROMATIC HYDROXY COMPOUNDS

(*139–140*). The rate of hydrolysis of the *o*-T diimine was also measured by this technique and (see Table 10-17) is quite slow at pH 2–5 (*140*).

$E_{1/2}$ data for various benzidines are contained in the original literature (*72,97–99,104*).

10-7. AROMATIC HYDROXY COMPOUNDS

A. Phenols

Aromatic and heterocyclic hydroxy compounds are very important in biogenic and metabolic systems. Unfortunately, little anodic electrochemistry has been done and the area deserves serious attention. Severe filming of electrode surfaces hinders mechanism studies.

The antioxidant activity of phenols has prompted several studies of $E_{1/2}$'s vs MO and Hammett function parameters (*31,141*). Early investigations were based on Feiser's critical oxidation potentials (*94*), but more recent ones have employed polarographic $E_{1/2}$'s (*19,78,142*). Little mechanistic information is provided by these studies. Both 1*e* and 2*e* processes have been indicated (*19,141*). An excellent summary of the chemical oxidation behavior of phenols is given by Scott (*143*). This includes the work of Waters and co-workers on EPR identification of phenoxy radicals formed in ceric oxidations. The monographs by Waters (*5*) and Stewart (*6*) contain additional data. The earlier preparative oxidations of phenols are treated in Fichter (*7*) and Allen (*8*). Further data on $E_{1/2}$-antioxidant behavior are summarized by Nash (*144*) and Panketh (*96*).

Opinions on the number of electrons transferred in the voltammetric oxidation of phenol in aqueous solution are divided between 1*e* (*145–147*) and 2*e* (*148*). Owing to the highly irreversible character of the phenol wave, the serious filming ordinarily present, and the questionable model compounds (usually hydroquinone) used for limiting-current comparisons, it is doubtful if any of these interpretations is totally reliable. However, the 1*e* oxidation appears most likely.

The most definitive studies of simple phenols have been done by Vermillion and Pearl (*149*). In acetonitrile, phenol is practically nonionized and they showed that the initial reaction probably corresponds to

$$\text{PhOH} \longrightarrow \text{PhO}^+ + 2e + H^+$$

To minimize polymeric film formation, Vermillion worked with 2,6-di-*t*-butyl-*p*-cresol (2,6-DTBC):

$$\text{[structure: phenol with OH, two R groups ortho, CH}_3\text{ para]} \quad (R = t\text{-butyl})$$

In controlled-potential electrolyses of 2,6-DTBC in acetonitrile, continually buffered by addition of tetraethylammonium hydroxide in methanol, they isolated 65% of 2,6-di-*t*-butyl-4-methyl-4-methoxy-cyclohexadienone:

$$\text{[structure: cyclohexadienone with R, R, CH}_3\text{, OCH}_3\text{]} \quad (R = t\text{-butyl})$$

This product can be rationalized by methanol addition to a mesomeric structure of the initial phenoxonium ion:

$$\text{[reaction scheme: phenol} \xrightarrow{-H^+, -2e} \text{phenoxonium mesomers]}$$

$$\text{[phenoxonium + CH}_3\text{OH} \longrightarrow \text{product]}$$

This and other isolated products were quantitatively identified with coulometric *n* values (*149*).

They also showed that if excess strong base were added to a phenol in acetonitrile, providing the phenoxide anion, the electrode reaction then shifted to an initial 1*e* oxidation. The vanillinate anion was used in this study:

$$\text{[structure: benzene with CHO, OCH}_3\text{, O}^-\text{]}$$

AROMATIC HYDROXY COMPOUNDS

From the above results Vermillion and Pearl suggested that the aqueous-solution behavior of nonionized phenols (low pH) should correspond to a 2e process, with a shift to the phenoxide anion and 1e behavior at high pH's. Brief studies in 50% aqueous–isopropanol buffers appeared to confirm this, but further testing should be done.

Bobbitt and co-workers electrolytically oxidized the *i*-quinoline alkaloid-corypalline to the corresponding dimer in aqueous bicarbonate solution (*150*). The reaction undoubtedly corresponds to

Corypalline → "Dimer" ($-2e, -H^+$)

They obtained the identical product by photolytic oxidation, which supports the exciting suggestion by Joschek and Miller that strong similarities exist in the photochemical and electrochemical oxidations of phenols (*151*). Steuber and Dimroth have reported the $E_{1/2}$'s of a series of aryl- and arylcyano-substituted phenols taken at a graphite electrode in acetonitrile–water solution. Stable radical ion products resulting from 1e oxidations have been noted in many cases (*185*).

Much work is needed on nonbenzenoid phenols. A few $E_{1/2}$'s are available for naphthols and related compounds (*72*). Deys has shown that the anodic oxidation of the phenolic group of morphine can be used for quantitative determinations. Serotonin and other 5-hydroxy indoles have been studied briefly (*152*). Other phenolic oxidations are reviewed by Weinberg and Weinberg (*82*).

B. Hydroquinones

In this day, when ligand field theory replaces learning the color of copper sulfate, even freshman students know the thermodynamic significance of the quinhydrone electrode. Despite this successful application to pH measurements, the hydroquinone–quinone (H_2Q–Q) redox system is remarkedly irreversible under many conditions at most solid electrodes.

The exact nature of the charge-transfer process remains in question (*153–156*). For a review of previous electrode kinetics studies the reader is referred to the thesis of Lingren (*157*). The irreversible nature of the system is readily seen in Fig. 10-10. These are representative but approximate (± 20 mV) half-peak potentials taken from the cyclic voltammetry

366 APPLICATIONS TO ORGANIC COMPOUNDS

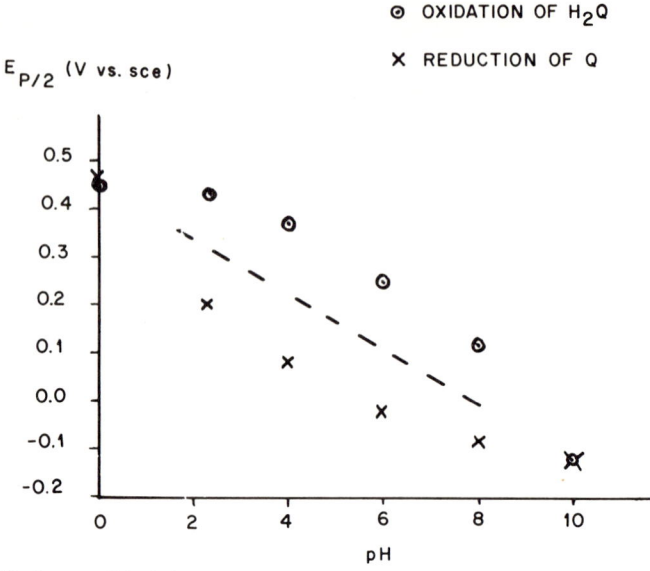

Fig. 10-10. Irreversible behavior of a hydroquinone–quinone system at carbon paste electrodes.

of H_2Q on a carbon paste (Nujol) electrode. The scan was always initiated in an anodic sense. Only at the pH extremes is the system fairly reversible. In the pH range 2–9 there is some 2- to 300-mV spread in the anodic and cathodic processes. The dashed line of Fig. 10-10 represents the theoretical 0.059 pH dependence of the system. The anodic and cathodic $E_{p/2}$'s only vaguely follow this line. Amusingly, the two extreme points fit very well! Somewhat similar results were obtained by Lingren on platinum using charge-integration techniques and pure solutions of H_2Q or Q for the anodic and cathodic segments, respectively (*157*).

There are many anomalies in the H_2Q–Q system at carbon paste electrodes. The $i_p/V^{1/2}C$ ratio for H_2Q oxidation is reasonably constant in the sweep range 1–16 V/min at any given acidity. However, this ratio varies almost linearly from a relative value of 70 at pH 2 to about 30 at pH 10. This behavior negates its use as a model system for *n*-value comparisons in peak voltammetry at carbon paste electrodes.

At the RDE (carbon paste) in this same range of acidities, the $i_L/\omega^{1/2}C$ ratio is $5.05 \pm 0.01 \times 10^3$ (from pH 2.3 to pH 6.0). The value for a known 2e, reversible system (*o*-dianisidine oxidation in 1 M $HClO_4$) is 5.06×10^3. In 1–2 M perchloric acid, the H_2Q value increases to 6.14×10^3 (but in

2 M H_2SO_4 is 4.66 × 10³) (*158*). Again, although at first glance the system appears to behave reasonably well at the RDE, there are unexplained discrepancies. Until a better understanding of the H_2Q–Q couple at platinum, gold, and carbon electrodes is available, it should be avoided as a *quantitative comparison system* in aqueous media.

Despite the difficulties, quantitative determinations of H_2Q are quite possible (*96,118,120,145,148*). Santhanam and Krishnan successfully analyzed H_2Q and methyl- and halo-substituted hydroquinones by controlled-potential coulometry (*163*). The complexity of H_2Q oxidation in 10–16 M sulfuric acid was unraveled by Mark and Atkin (*159*).

Substituted hydroquinones show $E_{1/2}$'s which are related to the usual substituent effects. Ryba and co-workers studied the $E_{1/2}$ variations of a large number of alkylated hydroquinones (and catechols). These were done at the dropping mercury electrode and converted to pH = 0 values (*160*).

Papouchado et al. have found what appears to be the first example, under voltammetric conditions, of an anodic hydroxylation of an aromatic system. This occurs with 2-substituted hydroquinones in acidic solutions from 6 M $HClO_4$ to pH 4.0. The 2-substituent must be a strong-electron-withdrawing group such as —NO_2, —CHO, or —COOH. In this case, the initial oxidation product, the 2-substituted-*p*-quinone, undergoes a 1,4 addition of H_2O to give the 3-hydroxy-substituted hydroquinone. As usual, this is an ECE reaction and the final product is the 2-(X)-3-hydroxybenzoquinone. The rates of the 1,4 addition reactions are readily measured (*161*).

Only a few studies exist on naphthalene and higher-hydrocarbon diols (*72,94*). Recently Chambers and Chambers studied in detail the oxidation of the biologically important 2-methyl-1,4-naphthoquinol-1-phosphate. The overall oxidation is a 2*e* process but proceeds via the semiquinone in alkaline solution. The potential relation between the electrochemical process and oxidative phosphorylation was suggested (*162*).

C. Catechols and Polyhydroxy Aromatics

The oxidation of catechol is a fairly reversible 2*e* process leading to the *o*-quinone. The interest in catechols centers on the high reactivity of this *o*-quinone, and it has provided an excellent substrate for the measurement of rapid 1,4 addition reactions. The first such reactions involved the important class of biogenic catecholamines, adrenaline, noradrenaline (also known as epinephrine and norepinephrine, respectively), dopamine, and related substances. Here the initial oxidation is to the *o*-quinone with a rapid intra-1,4 addition of the amino side chain. In the case of adrenaline, this follow-up reaction produces leucoadrenochrome, which is then further oxidized to adrenochrome. Details of the rate-constant measurements and their enzymatic significance are given in the original literature (126).

Detailed measurements of the 1,4 additions of substituted aromatic amines and amino acids to 4-methylcatechol have been made (122,127). The corresponding HCl additions were studied also (164). Ryba et al. (160) and Deys (152) have reported on $E_{1/2}$'s and analytical data for catechols.

Pyrogallol and other polyhydroxy aromatic systems have been reported briefly but no data are available on the mechanism of their oxidations (96,144).

D. Alkoxy-Substituted Aromatics

Hawley and Adams showed by cyclic voltammetry, reverse-current chronopotentiometry, and EPR spectra that the oxidation of *p*-methoxyphenol correspond to a 2e oxidation followed by rapid hydrolysis of the methoxy function to give *p*-benzoquinone:

The *p*-ethoxy derivative as well as 2-methoxy-4-methylphenol underwent similar EC reactions (122,165). A similar sort of reaction occurs with the aryloxy compound, 4,4'-oxydiphenol (122,135):

All these follow-up reactions are very rapid and their rates have not yet been measured.

Blackburn and Putman investigated this anomaly and found in anhydrous methanol that the reactive species could be assigned to the monoprotonated, methanolated crystal violet. Under similar circumstances, malachite green produced no tetramethylbenzidine in methanol (*178*). The hydration and protonation equilibria in the TPM dyes is very complex and obviously plays an important role in their overall electrooxidation.

B. Biochemically Important Molecules

The general porphyrin system is present in hemoproteins, cytochromes, oxidation–reduction enzymes, and, with slight modifications, in chlorophyll. Stanienda and Biebl examined the oxidation of various porphyrins at a platinum RDE in butyronitrile. The metal-free compounds were found to be more difficult to oxidize than the complexes. The oxidations corresponded to 1*e* reversible reactions in most cases (*179,180*). The RDE and cyclic voltammetry of chlorophyll a and b were also studied by Stanienda (*181*).

Davis and Orgeron used cyclic voltammetry to measure the rates of ligand substitution reactions in iron porphyrin complexes (*182*). Starting with a mixed cyanopyridine hemichrome they showed that the rate of substitution of pyridine for the cyano group was about 0.5 sec^{-1}. These are typical follow-up reactions after electron transfer and their rate measurements are of potential significance in biological redox processes. Recent work by Rollman and Iwamoto concerns the cyclic voltammetry, EPR, and optical spectra of phthalocyanines (*183*).

The oxidation of catecholamines and related biogenic amines has been mentioned previously. Serotonin, which is 5-hydroxytryptamine, has been found to be easily oxidized over the entire aqueous pH range (*52*).

$$HO-C_6H_4-CH_2-CH_2-NH_2$$

The relationships between the oxidation–reduction characteristics of the various phenylethylamine, amphetamine, and tryptamine derivatives of which various members are powerful hallucinogens deserve further study. It is interesting to note that a common property of the normally occurring serotonin, many tranquilizers such as the phenothiazines, and various psychotropic molecules is their ease of oxidation.

The biologically important oxidation of ascorbic acid, which is followed by a rapid hydration reaction, was recently studied by Perone and Kretlow. Excellent agreement on the follow-up rate were obtained via cyclic voltammetry and a potential step method. This rate is very rapid (ca. 10^3 sec^{-1})

and illustrates the utility of fast electrochemical techniques (*184*). This and other follow-up reactions were studied via pulse techniques by Jaenicke and Hoffmann (*228*).

Biochemically important redox systems are rich in amino, phenolic, and mercapto substituents. Especially with the latter two functions, if the problem of electrode filming could be overcome, many valuable investigations would be possible. The use of platinum electrodes for quantitative measurements with in vivo systems is always worrisome because of filming and adsorption. Much work remains to be done in this area with model test molecules.

10-10. ALIPHATIC HYDROCARBONS

Aliphatic hydrocarbons are difficult to electrooxidize. The reactions are ordinarily carried out at elevated temperatures in strong acid or base solution. In most instances the oxidation is complete—yielding carbon dioxide. This somewhat uninteresting final result is, of course, of utmost concern to those working with fuel cells. Extensive research on aliphatic hydrocarbon oxidations has been carried out in fuel-cell-development programs. The electrode systems and high current densities used makes it difficult to include these studies in the framework of voltammetry.

Adsorption phenomena and the nature of the electrode play a major role in the anodic oxidation, as indicated in some representative publications (*186–190*). Very definitive patterns of intermediates produced during oxidation of saturated hydrocarbons have been given by Brummer and Turner (*191*).

Anodic halogenations of aliphatics are well known (*7,8*). A recent chlorination of *n*-dodecane at porous carbon anodes illustrated the utility of voltammetry at small electrodes to examine the larger-scale flow-cell reaction (*192*).

10-11. ALIPHATIC ACIDS

The anodic oxidation of carboxylic acids (actually the carboxylate anions) is a very old and extensively studied reaction. The first such work was done by H. Kolbe in 1848 and uniformly the process is known as the Kolbe synthesis (or Kolbe reaction, electrolysis, etc.). As generally practiced, the carboxylate anion, $RCOO^-$, is electrolyzed in water, methanol, or other nonaqueous solvents. The desired product is the dimer R—R. The process is a useful synthetic tool.

Certain peculiarities of the Kolbe reaction have made it particularly difficult to study from the mechanistic viewpoint. It gives significant yields at very high anodic potentials, i.e., 2.1–2.2 V. This is considerably beyond the potential for oxygen evolution in aqueous media (ca. 1.7 V) or for methanol background oxidation. Yet, when the Kolbe reaction proceeds efficiently, very little O_2 is found among the products.

A great deal of study has been expended on this process, and the pertinent references can be found in a few reviews (8,193–195). Although several mechanisms of the Kolbe reaction have been favored over the years, one of the early ones now seems firmly established (193,194).

The initial charge transfer involves a 1e oxidation of the carboxylate anion:

$$RCOO^- \rightarrow RCOO\cdot + e$$

The resulting acyloxy radical rapidly decarboxylates:

$$RCOO\cdot \rightarrow CO_2 + R\cdot$$

(an estimated $t_{1/2}$ for acetoxyl radicals is ca. 10^{-9}–10^{-10} sec) (193). Coupling then occurs via two aliphatic free radicals to give the desired dimer:

$$2R\cdot \rightarrow R—R$$

Depending upon conditions, many side products are to be expected from cross reactions of R· and these are observed. In support of the above process, Geske (196) and Russell and Anson (197) showed discrete oxidation waves for acetate ion in acetonitrile, where background oxidation does not interfere. Geske's work showed i_L for acetate to be approximately proportional to a 1e process and Russell and Anson obtained typical Kolbe production of ethane under these conditions, but these points seem not to have been noted by the Kolbe types.

There are certain interesting limitations to the Kolbe synthesis and the most important is that simple aromatic acids (e.g., benzoic) do not undergo the usual reaction. If the aromatic nucleus is considerably removed from the —COOH function, the reaction is usually normal. Other substituents and unsaturated situations limit the usual dimer formation (194).

Eberson feels that one of the main side reactions (peraps in addition to reactions of R· with solvent, etc.) is further 1e oxidation of R· to the carbonium ion:

$$R\cdot \rightarrow R^+ + e$$

According to calculations, this process may be important where R· has ionization potentials < ca. 8 eV (*193*). Oxidations in which the reactive carbonium ion play a part have been demonstrated (*198*).

Since the Kolbe reaction is predominately of synthetic interest, the original literature should be consulted for further details. The relationship of anodic hydrocarbon substitutions to the Kolbe reaction was discussed in Section 10-2E. An interesting adjunct to this discussion is the case of the simplest aliphatic acid, HCOOH (*199–202*).

Ross et al. indicated a 1e oxidation of formate ion to the formyloxy radical in DMF as solvent. The formyloxy radical rapidly abstracts a proton from DMF:

$$HCOO^- \longrightarrow HCOO\cdot + e$$

$$HCOO\cdot + H-\overset{O}{\underset{\|}{C}}-N\overset{CH_3}{\underset{CH_3}{\diagdown}} \longrightarrow HCOOH + H-\overset{O}{\underset{\|}{C}}-N\overset{CH_2\cdot}{\underset{CH_3}{\diagdown}}$$

and the resulting radical presumably reacts with other HCOO· to give isolated products such as (*203*)

$$H-\overset{O}{\underset{\|}{C}}-N\overset{CH_2\cdot}{\underset{CH_3}{\diagdown}} + \cdot OOCH \longrightarrow H-\overset{O}{\underset{\|}{C}}-N\overset{CH_2O-\overset{O}{\underset{\|}{C}}-H}{\underset{CH_3}{\diagdown}}$$

In this case, the formyloxy radical is implicated in the formoxylation since HCOO⁻ is very easily oxidized. The corresponding acetoxylation of DMF was presumed to proceed similarly via acetoxyl radical (*203*). But Eberson and Nyberg (*61*) obtained the identical N-acetoxymethyl-N-methyl-formamide as did Ross et al., except at a controlled potential of +1.5 V— below that normally expected to oxidize acetate ion. Hence the acetoxylation proceeds via the Eberson-type substrate cation (in this case DMF⁺) and nucleophilic attack by acetate ion as discussed in anodic substitutions. Later work by Ross et al. on oxidations in DMF with nitrate ion present tends to indicate that nitrate ion rather than DMF is initially oxidized (*208*). The case of formate, with an oxidation potential less than that of the substrate, is difficult to study and the question remains unanswered (*61,208*). Cation formation in the oxidation of norborane carboxylic acids (*209*) and some other medium-ring carboxylates (*210*) has been proposed. These oxidations were carried out without potential control.

In an aqueous medium, oxalic acid is oxidized to CO_2 at platinum, but the reaction is inhibited by oxide formation on the electrode surface. A number of workers have studied this particular system (204–207). Bagotskii and co-workers have been concerned with effects of oxide-coated surfaces on the oxidation of various alcohols, aldehydes, and acids (211,212).

Although hardly carboxylic acids, some unusual acids which have been oxidized include *n*-butylboronic and uric acid. Although only preliminary results, the *n*-butylboronic acid appeared to give boric acid and unsaturated hydrocarbons somewhat analogous to a Kolbe process (213). The major products of the uric acid oxidation at graphite electrodes were alloxan and allantoin (214). Many other early studies of acid oxidations can be found in Fichter (7).

10-12. ALIPHATIC ALCOHOLS AND ALDEHYDES

The oxidation of the lower aliphatic alcohols has been studied for years. MacNevin and Sweet carried out some early analytical voltammetry of ethanol at platinum electrodes (215). The present interest, especially in methanol, as well as formaldehyde, stems from fuel-cell applications. For a background of this work the reader is referred especially to the work of Breiter et al. and the references contained therein (216–218). A succinct discussion of the behavior of some lower alcohols and aldehydes under ordinary voltammetric conditions is given by Liang and Franklin (219). Papova and co-workers investigated glycerol and other polyhydric alcohols (220).

Oxidation of aliphatic alcohols in nonaqueous media appears not to have been done, but Lund surveyed some 30 aromatic carbinols in acetonitrile. In the case of anisyl alcohol the corresponding anisaldehyde was formed. In many cases deformed waves and insulating films were obtained (221).

10-13. ALIPHATIC AMINES AND AMIDES

In spite of their basicity, aliphatic amines are too difficult to anodically oxidize in any quantitative fashion in aqueous solution. Dapo and Mann studied triethylamine oxidation in dimethylsulfoxide and found it to involve a 1*e* transfer with an irreversible follow-up reaction (222). Russell also oxidized triethylamine at platinum but in acetonitrile (223). Both groups postulated a reaction scheme

$$(C_2H_5)_3N: \rightarrow (C_2H_5)_3N\cdot^+ + e \qquad E$$
$$(C_2H_5)_3N\cdot^+ + S \rightarrow (C_2H_5)_3NH^+ + \cdot S \qquad C$$

where the triethylamine cation radical rapidly extracts a proton from the solvent S. The protonated amine is not reducible and electroinactive. Hence the process represents an EC reaction. The resulting radicals ·CH$_2$SOCH$_3$ and ·CH$_2$CN from dimethylsulfoxide and acetonitrile, respectively, presumably form coupled products, but these have not been isolated.

In a further study of some 19 aliphatic amines in acetonitrile, Mann showed that tertiary amines were most easily oxidized, followed by

TABLE 10-18

Compound	E_p, V
Tripropylamine	0.87
Dipropylamine	0.99
Propylamine	1.37

secondary and primary (224). For propylamines the approximate peak potentials in acetonitrile (vs. SCE) are as given in Table 10-18.

The oxidation mechanisms of primary aliphatic amines are much different and more complex than that of the tertiary compounds. Entirely different results obtain under voltammetric and controlled-potential electrolysis conditions. This is an excellent example of a reaction in which the greater time gate of a controlled-potential electrolysis allows interactions of various intermediates to occur which are not seen on the short time scale of a single potential sweep. Thus Barnes and Mann found that the current–time curve in the controlled-potential oxidation gave a peak response—indicating slow interactions producing electroactive species (225). A thorough and careful examination of products via UV spectrophotometry and gas chromatography allowed postulation of two general reaction schemes pertinent to the fast and slow electrolyses. In general the oxidation of the primary amines in acetonitrile led to identifiable quantities of aldehydes, ammonia, nitrogen, and aliphatic hydrocarbons. The original literature should be consulted for details on the individual compounds (225).

O'Donnell and Mann found that the oxidation of tertiary amides in acetonitrile was much like that of the corresponding amines; i.e., the product was essentially the protonated amide (226). The reaction scheme is

$$RCON(CH_2R')_2 \rightarrow RC\dot{O}N(CH_2R')_2 + e$$
$$RC\dot{O}N(CH_2R')_2 + CH_3CN \rightarrow RCONH^+(CH_2R')_2 + \cdot CH_2CN$$
$$2 \cdot CH_2CN \rightarrow (CH_2CN)_2$$

Reactions similar to the first two steps were discussed by Ross et al. in Kolbe reaction studies (55).

10-14. ALIPHATIC HALIDES

A recent study of Miller and Hoffmann showed that alkyl iodides were oxidized in acetonitrile to the alkyl carbonium ion, which then reacted with solvent probably as

$$RI \xrightarrow{-e} R^+ + 1/2 I_2$$
$$R^+ + CH_3CN \longrightarrow R-N=C^+-CH_3$$
$$R-N=C^+-CH_3 + H_2O \longrightarrow CH_3\overset{O}{\underset{\|}{C}}-NHR$$

Although the I_2 in acetonitrile undergoes further electrooxidations and the scheme is not necessarily this simple, the formation of the N-alkylacetamide is easily established by product identification.

In contrast, an aryl iodide does not involve carbon–iodine bond breaking. Instead, although its lifetime is too short for cyclic voltammetry detection, the initial product is presumably the iodoaryl cation radical. This then attacks an unoxidized molecule of aryliodide with final formation of a-diaryliodonium ion. It is interesting that it could be shown that the follow-up reaction does not consist of dimerization of two ArI·+ radicals by trapping experiments with benzene (227).

10-15. REDUCTION PROCESSES

A great number of the early reduction studies were at solid electrodes of lead and various amalgams (7,8) and it is not the intent to review them here. Platinum in nonaqueous media serves very well for a large number of reductions. Thus many of the anion radicals generated in DMF and acetonitrile for EPR studies were carried out at platinum. The voltammetry of such systems is well defined, although reversibility and heterogeneous rate constants, etc., may vary widely from those obtained at mercury surfaces. Carbon electrodes can be used, but carbon paste has very restricted utility in nonaqueous media.

The reduction of oxygen is a particularly important problem and of special significance with regard to in vivo measurements of oxygen tension. The Clark electrode, in which a platinum surface is covered with cellophane, has been used in a variety of modifications for physiological studies. This and other oxygen-measuring solid electrodes are reviewed in the excellent monograph *Electroanalytical Methods in Biochemistry* by Purdy (*229*).

REFERENCES

1. G. J. Hoijtink, *Rec. Trav. Chim.*, **76**, 885 (1957).
2. T. A. Gough and M. E. Peover, *Polarography—1964*, Macmillan, New York, 1966, p. 1017.
3. M. E. Peover and B. S. White, *J. Electroanal. Chem.*, **13**, 93 (1967).
4. M. E. Peover, in *Electroanalytical Chemistry*, Vol. 2 (A. J. Bard, ed.), Dekker, New York, 1967, Chap. 1.
5. W. A. Waters, *Mechanisms of Oxidation of Organic Compounds*, Wiley, New York, 1964.
6. R. Stewart, *Oxidation Mechanisms*, Benjamin, New York, 1964.
7. F. Fichter, *Organische Elektrochemie*, Theodore Steinkopf, Dresden and Leipzig, 1942.
8. M. J. Allen, *Organic Electrode Processes*, Chapman and Hall, London, 1958.
9. S. Wawzonek, *Anal. Chem.*, *1949–1964*, biannual reviews.
10. D. J. Pietrzyk, *Anal. Chem.*, **38**, 278R (1966).
11. S. Swann, *Univ. Ill. Bull.*, **45**(69) (1948); *Electrochem. Tech.*, **1**, 308 (1963); **5**, 53, 101, 393, 549 (1967).
12. W. M. Clark, *Oxidation-Reduction Potentials of Organic Systems*, Williams & Wilkins, Baltimore, 1960.
13. H. Lund, *Acta Chem. Scand.*, **11**, 1323 (1957).
14. G. J. Hoijtink, *Rec. Trav. Chim.*, **77**, 555 (1958).
15. E. S. Pysh and N. C. Yang, *J. Am. Chem. Soc.*, **85**, 2124 (1963).
16. J. W. Loveland and G. R. Dimeler, *Anal. Chem.*, **33**, 1196 (1961).
17. W. C. Neikam, G. R. Dimeler, and M. M. Desmond, *J. Electrochem. Soc.*, **111**, 1190 (1964).
18. W. C. Neikam and M. M. Desmond, *J. Am. Chem. Soc.*, **86**, 4811 (1964).
19. C. Parkanyi and R. Zahradnik, *Collection Czech. Chem. Commun.*, **30**, 4287 (1965).
20. A. Stanienda, *Z. Physik. Chem. (Frankfurt)*, **33**, 170 (1962).
21. L. Eberson and K. Nyberg, *J. Am. Chem. Soc.*, **88**, 1686 (1966).
22. R. E. Visco and E. A. Chandross, *J. Am. Chem. Soc.*, **86**, 5350 (1964).
23. R. E. Sioda and W. S. Koski, *J. Am. Chem. Soc.*, **87**, 5573 (1965).
24. P. A. Malachesky, L. S. Marcoux and R. N. Adams, *J. Phys., Chem.*, **70**, 2064 (1966).

REFERENCES

25. R. C. Larson, R. I. Iwamoto, and R. N. Adams, *Anal. Chim. Acta*, **25**, 371 (1961).
26. M. E. Peover, *Electrochim. Acta*, in press.
27. E. Gileadi, *J. Electroanal. Chem.*, **11**, 137 (1966).
28. J. Phelps, K. S. V. Santhanam, and A. J. Bard, *J. Am. Chem. Soc.*, **89**, 1752 (1967).
29. L. S. Marcoux, J. M. Fritsch, and R. N. Adams, *J. Am. Chem. Soc.*, **89**, 5766 (1967).
30. A. Streitweiser, *Molecular Orbital Theory for Organic Chemists*, Wiley, New York, 1961.
31. B. Pullman and A. Pullman, *Quantum Biochemistry*, Wiley (Interscience), New York, 1963.
32. K. Higasi, H. Baba, and A. Rembaum, *Quantum Organic Chemistry*, Wiley (Interscience), New York, 1965.
33. C. A. Coulson and A. Streitweiser, *Dictionary of Pi Electron Calculations*, Freeman, San Francisco, 1965.
34. L. S. Marcoux, Ph.D. thesis, Univ. Kansas, Lawrence, 1967.
35. A. Zweig, A. H. Maurer, and B. G. Roberts, *J. Org. Chem.*, **32**, 1322 (1967).
36. G. Cauquis, J. P. Billon, J. Raison, and Y. Thibaud, *Compt. Rend.*, **257**, 2128 (1963).
37. R. L. Hansen, P. E. Toren, and R. H. Young, *J. Phys. Chem.*, **70**, 1653, 1657 (1966).
38. A. Zweig, G. Metzler, A. Maurer, and B. G. Roberts, *J. Am. Chem. Soc.*, **88**, 2864 (1966); **89**, 4091 (1967).
39. L. S. Marcoux and R. N. Adams, unpublished data, 1967.
40. R. Zahradnik and J. Koutecky, *Advan. Heterocyclic Chem.*, **5**, Chap. 2 (1965).
41. C. Parkanyi and R. Zahradnik, *Bull. Soc. Chim. Belges*, **73**, 57 (1964).
42. J. Volke, *Talanta*, **12**, 1081 (1965).
43. J. Volke, *Phys. Methods Heterocyclic Chem.*, **1**, Chap. 6 (1965).
44. W. R. Turner and P. J. Elving, *Anal. Chem.*, **37**, 467 (1965).
45. R. F. Nelson, D. W. Leedy, E. T. Seo, and R. N. Adams, *Z. Anal. Chem.*, **224**, 184 (1967).
46. J. P. Billon, *Bull. Soc. Chim. France*, **1960**, 1784.
47. J. P. Billon, G. Cauquis, and J. Combrisson, *Bull. Soc. Chim. France*, **1960**, 2062.
48. P. Kabasakalian and J. McGlotten, *Anal. Chem.*, **31**, 431 (1959).
49. L. H. Piette, P. Ludwig, and R. N. Adams, *Anal. Chem.*, **34**, 916 (1962).
50. L. H. Piette and I. S. Forrest, *Biochim. Biophys. Acta*, **57**, 419 (1962).
51. M. J. Allen and V. J. Powell, *J. Electrochem. Soc.*, **105**, 541 (1958).
52. H. Sharp, D. Stump, and R. N. Adams, unpublished data, 1968.
53. A. P. Tomilov and M. Ya Fioshin, *Russ. Chem. Rev. (English Transl.)*, **32**, 30 (1963).
54. F. H. Merkle and C. A. Discher, *J. Pharm. Soc.*, **53**, 620 (1964); *Anal. Chem.*, **36**, 1639 (1964).
55. S. D. Ross, M. Finkelstein, and R. C. Petersen, *J. Am. Chem. Soc.*, **86**, 4139 (1964).
56. F. D. Mango and W. H. Bonner, *J. Org. Chem.*, **29**, 1367 (1964).
57. M. Leung, J. Herz and H. W. Salzberg, *J. Org. Chem.*, **30**, 310 (1965).
58. H. W. Salzberg and M. Leung, *J. Org. Chem.*, **30**, 2873 (1965).
59. L. Eberson, *Acta Chem. Scand.*, **17**, 2004 (1963).
60. L. Eberson and K. Nyberg, *Acta Chem. Scand.*, **18**, 1568 (1964).
61. L. Eberson and K. Nyberg, *J. Am. Chem. Soc.*, **88**, 1686 (1966); *Tetrahedron Letters*, **1966**, 2389.
62. L. Eberson, *J. Am. Chem. Soc.*, **89**, 4669 (1967).

63. K. Koyama, T. Susuki and S. Tsutsumi, *Tetrahedron Letters*, **1965**, 627.
64. V. D. Parker and B. E. Burgert, *Tetrahedron Letters*, **1965**, 4065.
65. T. Inoue, K. Koyama, T. Matsouka, K. Matsouka, and S. Tsutsumi, *Tetrahedron Letters*, **1963**, 1409.
66. T. Inoue, K. Koyama, and S. Tsutsumi, *Bull. Chem. Soc. Japan*, **37**, 1597 (1964).
67. T. Inoue and S. Tsutsumi, *Bull. Chem. Soc. Japan*, **38**, 661 (1965).
68. B. Belleau and N. L. Weinberg, *J. Am. Chem. Soc.*, **85**, 2525 (1963).
69. N. L. Weinberg and E. A. Brown, *J. Org. Chem.*, **31**, 4054, 4058 (1966).
70. K. E. Friend and W. E. Ohnesorge, *J. Org. Chem.*, **28**, 2435 (1963).
71. S. S. Lord and L. B. Rogers, *Anal. Chem.*, **26**, 284 (1954).
72. R. N. Adams, J. H. McClure, and J. B. Morris, *Anal. Chem.*, **30**, 471 (1958).
73. T. Kuwana and R. N. Adams, *Anal. Chim. Acta*, **20**, 51, 60 (1959).
74. D. M. Mohilner, R. N. Adams, and W. J. Argersinger, *J. Am. Chem. Soc.*, **84**, 3618 (1962).
75. D. W. Leedy, Ph.D. thesis, Univ. Kansas, Lawrence, 1968.
76. J. D. Voorhies and S. M. Davis, *J. Phys. Chem.*, **67**, 332 (1963).
77. I. Fox, R. W. Taft, and J. M. Schempf, *U.S. Dept. Comm. Rep.*, *280* (1959).
78. J. C. Suatoni, R. E. Snyder, and R. O. Clark, *Anal. Chem.*, **33**, 1894 (1961).
79. C. Olson, H. Y. Lee, and R. N. Adams, *J. Electroanal. Chem.*, **2**, 396 (1961).
80. J. Bacon and R. N. Adams, *J. Am. Chem. Soc.*, **90** (1968).
81. S. Wawzonek and T. W. McIntyre, *J. Electrochem. Soc.*, **114**, 1025 (1967).
82. N. L. Weinberg and H. R. Weinberg, *Chem. Rev.*, **68**, 449 (1968).
83. K. S. V. Santhanam and V. R. Krishnan, *Z. Anal. Chem.*, **206**, 33 (1964).
84. W. M. Fox and W. A. Waters, *J. Chem. Soc.*, **1964**, 6010.
85. D. Hawley and R. N. Adams, *J. Electroanal. Chem.*, **10**, 376 (1965).
86. W. P. Jencks, *Progr. Phys. Org. Chem.*, **2**, 63 (1964).
87. E. H. Cordes and W. P. Jencks, *J. Am. Chem. Soc.*, **84**, 832 (1962); **85**, 2843 (1963).
88. K. Koehler, W. Sandstrom, and E. H. Cordes, *J. Am. Chem. Soc.*, **86**, 2413 (1964).
89. E. Knobloch, *Proc. 1st Polarog. Congr., Prague, 1951*, p. 764.
90. P. A. Malachesky, Ph.D. thesis, Univ. Kansas, Lawrence, 1966.
91. C. R. Christensen and F. C. Anson, *Anal. Chem.*, **36**, 495 (1964).
92. P. A. Malachesky, K. B. Prater, G. Petrie, and R. N. Adams, *J. Electroanal. Chem.*, **16**, 41 (1968).
93. M. F. Marcus and M. D. Hawley, *J. Electroanal. Chem.*, **18**, 175 (1968).
94. L. F. Feiser, *J. Am. Chem. Soc.*, **52**, 4915, 5204 (1930).
95. Z. Galus and R. N. Adams, *J. Phys. Chem.*, **67**, 862 (1963).
96. G. E. Panketh, *J. Appl. Chem.*, **7**, 512 (1957).
97. F. T. Eggertsen and F. T. Weiss, *Anal. Chem.*, **28**, 1008 (1956).
98. R. F. Nelson, Ph.D. thesis, Univ. Kansas, Lawrence, 1966.
99. V. Dvorak, I. Nemec, and J. Zyka, *Microchem. J.*, **12**, 99, 324, 350 (1966).
100. K. E. Heusler and H. Schurig, *Z. Physik. Chem.*, (*Frankfurt*), **47**, 117 (1963).
101. T. Mizoguchi and R. N. Adams, *J. Am. Chem. Soc.*, **84**, 2058 (1962).
102. Z. Galus and R. N. Adams, *J. Am. Chem. Soc.*, **84**, 2061 (1962).
103. Z. Galus, R. M. White, F. S. Rowland, and R. N. Adams, *J. Am. Chem. Soc.*, **84**, 2065 (1962).
104. E. T. Seo, R. F. Nelson, J. M. Fritsch, L. S. Marcoux, D. W. Leedy, and R. N. Adams, *J. Am. Chem. Soc.*, **88**, 3498 (1966).
105. F. Fichter and E. Rothenberger, *Helv. Chim. Acta*, **5**, 166 (1922).

REFERENCES

381

106. C. A. H. Chambers and J. K. Lee, *J. Electroanal. Chem.*, **14,** 309 (1967).
107. J. E. Dubois, P. C. Lacaze, and A. Aranda, *Compt. Rend.*, **260,** 3383 (1965).
108. N. L. Weinberg and E. A. Brown, *J. Org. Chem.*, **31,** 4058 (1966).
109. N. L. Weinberg and T. B. Reddy, *J. Am. Chem. Soc.*, **90,** 91 (1968).
110. F. Fichter and P. Schönmann, *Helv. Chim. Acta*, **19,** 1411 (1936).
111. A. Zweig, J. E. Lancaster, M. T. Neglia, and W. H. Jura, *J. Am. Chem. Soc.*, **86,** 4130 (1964).
112. B. M. Latta and R. W. Taft, *J. Am. Chem. Soc.*, **89,** 5172 (1967).
113. R. F. Nelson, E. T. Seo and J. M. Fritsch, in press.
114. R. F. Nelson and R. N. Adams, *J. Am. Chem. Soc.*, **90,** 3925 (1968).
115. G. Petrie and R. N. Adams, unpublished data, 1968.
116. L. Michaelis, M. P. Schubert, and S. Granick, *J. Am. Chem. Soc.*, **61,** 1981 (1939).
117. R. E. Parker and R. N. Adams, *Anal. Chem.*, **28,** 828 (1956).
118. P. J. Elving and A. F. Krivis, *Anal. Chem.*, **30,** 1645, 1648 (1958).
119. H. Y. Lee and R. N. Adams, *Anal. Chem.*, **34,** 1587 (1962).
120. G. A. Ward, *Talanta*, **10,** 261 (1963).
121. H. B. Mark and F. C. Anson, *Anal. Chem.*, **35,** 722 (1963).
122. M. D. Hawley, Ph.D. thesis, Univ. Kansas, Lawrence, 1965.
123. L. K. J. Tong, *J. Phys. Chem.*, **58,** 1090 (1954).
124. L. K. J. Tong and M. C. Glesmann, *J. Am. Chem. Soc.*, **78,** 5827 (1956).
125. L. K. J. Tong, M. C. Glesmann, and R. L. Bent, *J. Am. Chem. Soc.*, **82,** 1988 (1960).
126. M. D. Hawley, S. V. Tatawawadi, S. Piekarski, and R. N. Adams, *J. Am. Chem. Soc.*, **89,** 447 (1967).
127. S. Piekarski and R. N. Adams, unpublished data, 1967.
128. M. D. Hawley and R. N. Adams, unpublished data, 1966.
129. M. T. Melchior and A. H. Maki, *J. Chem. Phys.*, **34,** 471 (1961).
130. D. B. Julian and W. R. Ruby, *J. Am. Chem. Soc.*, **72,** 4719 (1950).
131. R. L. Bent, J. C. Dessloch, F. C. Duennebier, D. W. Fassett, D. B. Glass, T. H. James, D. B. Julian, W. R. Ruby, J. M. Swell, J. H. Sterner, J. R. Thirtle, P. W. Vittum, and A. Weissberger, *J. Am. Chem. Soc.*, **73,** 3100 (1951).
132. T. Kuwana, Ph.D. thesis, Univ. Kansas, Lawrence, 1957.
133. L. F. Oldfield and J. O'M. Bockris, *J. Phys. Colloid Chem.*, **55,** 1255 (1951).
134. C. R. Christensen and F. C. Anson, *Anal. Chem.*, **36,** 495 (1964).
135. S. V. Tatawawadi, S. Piekarski, M. D. Hawley, and R. N. Adams, *Chem. Listy*, **61,** 624 (1967).
136. D. J. Macero and R. A. Janeiro, *Anal. Chim. Acta*, **27,** 585 (1962).
137. J. A. Friend and N. K. Roberts, *Australian J. Chem.*, **11,** 104 (1958).
138. J. D. Voorhies and J. S. Parsons, *Anal. Chem.*, **31,** 516 (1959).
139. J. W. Strojek, T. Kuwana, and S. W. Feldberg, *J. Am. Chem. Soc.*, **90,** 1353 (1968).
140. T. Kuwana and J. Strojek, *J. Electroanal. Chem.*, **16,** 471 (1968).
141. T. Fueno, T. Ree, and H. Eyring, *J. Am. Chem. Soc.*, **63,** 1940 (1959).
142. H. N. Simpson, C. K. Hancock, and E. A. Meyers, *J. Org. Chem.*, **30,** 2678 (1965).
143. A. I. Scott, *Quart. Rev. (London)*, **19,** 1 (1965).
144. R. A. Nash, Ph.D. thesis, University Connecticutt, Storrs, 1958.
145. V. I. Ginzburg, *Zh. Fiz. Khim.*, **33,** 1504 (1959).
146. N. E. Khomutov and V. I. Bystrov, *Zh. Fiz. Khim.*, **36,** 2246 (1962).
147. J. F. Hedenberg and H. Freiser, *Anal. Chem.*, **25,** 1355 (1953).
148. V. F. Gaylor, P. J. Elving, and A. L. Conrad, *Anal. Chem.*, **25,** 1078 (1953).

149. F. J. Vermillion and I. A. Pearl, *J. Electrochem. Soc.*, **111**, 1392 (1964).
150. J. M. Bobbitt, J. T. Stock, A. Marchand, and K. H. Weisgraber, *Chem. Ind. (London)*, **1966**, 2127.
151. H. I. Joschek and S. I. Miller, *J. Am. Chem. Soc.*, **88**, 3269, 3273 (1966).
152. H. P. Deys, *Pharm. Weekblad*, **99**, 737 (1964).
153. M. A. Loshkarev and B. I. Tomilov, *Zh. Fiz. Khim.*, **34**, 1753 (1960); **36**, 132 (1962).
154. B. I. Tomilov and M. A. Loshkarev, *Zh. Fiz. Khim.*, **36**, 1902 (1962).
155. Y. Lu-an, Y. B. Vasil'ev, and V. S. Bagotskii, *Zh. Fiz. Khim.*, **38**, 205 (1964).
156. V. S. Bagotsky and Y. B. Vasilyev, *Electrochim. Acta*, **9**, 869 (1964).
157. W. E. Lingren, Ph.D. thesis, Univ. Washington, Seattle, 1962.
158. W. A. Latham and R. N. Adams, unpublished results, 1967.
159. H. B. Mark and C. L. Atkin, *Anal. Chem.*, **36**, 514 (1964).
160. O. Ryba, J. Petranek, and J. Pospisil, *Collection Czech. Chem. Commun.*, **30**, 843, 2157 (1965).
161. L. Papouchado, G. Petrie, J. H. Sharp, and R. N. Adams, *J. Am. Chem. Soc.*, **90** (1968).
162. C. A. Chambers and J. Q. Chambers, *J. Am. Chem. Soc.*, **88**, 2922 (1966).
163. K. S. V. Santhanam and V. R. Krishnan, *Z. Physik. Chem.*, *(Frankfurt)*, **39**, 10 (1963).
164. R. N. Adams, M. D. Hawley, and S. W. Feldberg, *J. Phys. Chem.*, **71**, 851 (1967).
165. M. D. Hawley and R. N. Adams, *J. Electroanal. Chem.*, **8**, 163 (1964).
166. A. Zweig, W. G. Hodgson, and W. H. Jura, *J. Am. Chem. Soc.*, **86**, 4124 (1964).
167. J. J. O'Connor and I. A. Pearl, *J. Electrochem. Soc.*, **111**, 335 (1964).
168. D. G. Davis and E. Bianco, *J. Electroanal. Chem.*, **12**, 254 (1966).
169. M. M. Nicholson, *J. Am. Chem. Soc.*, **76**, 2539 (1954).
170. H. V. Drushel and J. F. Miller, *Anal. Chim. Acta*, **15**, 389 (1956); *Anal. Chem.*, **29**, 1456 (1957).
171. K. S. V. Santhanam and V. R. Krishnan, *Z. Physik. Chem.*, *(Frankfurt)*, **34**, 312 (1962).
172. A. Zweig and J. E. Lehnsen, *J. Am. Chem. Soc.*, **87**, 2647 (1965).
173. A. Zweig, W. G. Hodgson, W. H. Jura, and D. L. Maricle, *Tetrahedron Letters*, **26**, 1821 (1963).
174. J. D. Voorhies and R. N. Adams, *Anal. Chem.*, **30**, 346 (1958).
175. J. D. Voorhies and N. H. Furman, *Anal. Chem.*, **30**, 1656 (1958).
176. Z. Galus and R. N. Adams, *J. Am. Chem. Soc.*, **84**, 3207 (1962); **86**, 1666 (1964).
177. D. A. Hall, M. Sakuma, and P. J. Elving, *Electrochim. Acta*, **11**, 337 (1966).
178. T. R. Blackburn and R. C. Putnam, in press.
179. A. Stanienda, *Naturwiss.*, **52**, 105 (1965).
180. A. Stanienda and G. Biebl, *Z. Physik. Chem.*, *(Frankfurt)*, **52**, 254 (1967).
181. A. Stanienda, *Naturwiss.*, **24**, 731 (1963); *Z. Physik. Chem. (Leipzig)*, **229**, 257 (1965).
182. D. G. Davis and D. J. Orgeron, *Anal. Chem.*, **38**, 179 (1966).
183. L. D. Rollman and R. T. Iwamoto, *J. Am. Chem. Soc.*, **90**, 1455 (1968).
184. S. P. Perone and W. J. Kretlow, *Anal. Chem.*, **38**, 1760 (1966).
185. F. W. Steuber and K. Dimroth, *Ber.*, **99**, 258 (1965).
186. H. Dahms and J. O'M. Bockris, *J. Electrochem. Soc.*, **111**, 728 (1964).
187. J. W. Johnson, H. Wroblowa, and J. O'M. Bockris, *J. Electrochem. Soc.*, **111**, 863 (1964).
188. E. Gileadi, B. T. Rubin, and J. O'M. Bockris, *J. Phys. Chem.*, **69**, 3335 (1965).

189. E. Gileadi, G. Stoner, and J. O'M. Bockris, *J. Electrochem. Soc.*, **113**, 585 (1966).
190. M. J. Schlatter, *Advan. Chem. Ser.*, **47**, 292 (1965).
191. S. B. Brummer and M. J. Turner, *J. Phys. Chem.*, **71**, 2825 (1967).
192. F. N. Ruehlen, G. B. Wills, and H. M. Fox, *J. Electrochem. Soc.*, **111**, 1107 (1964).
193. L. Eberson, *Acta Chem. Scand.*, **17**, 2004 (1963).
194. B. E. Conway, *Theory and Principles of Electrode Processes*, Ronald Press, New York, 1965, Chap. 8.
195. G. E. Svadkovskaya and S. A. Voitkevich, *Russ. Chem. Rev. (English Transl.)*, **29**, 161 (1959).
196. D. H. Geske, *J. Electroanal. Chem.*, **1**, 502 (1960).
197. C. D. Russell and F. C. Anson, *Anal. Chem.*, **33**, 1282 (1961).
198. L. Eberson and K. Nyberg, *Acta Chem. Scand.*, **18**, 1567 (1964).
199. R. P. Buck and L. R. Griffin, *J. Electrochem. Soc.*, **109**, 1005 (1962).
200. J. Giner, *Electrochim. Acta*, **9**, 63 (1964).
201. R. A. Munson, *J. Electrochem. Soc.*, **111**, 372 (1964).
202. M. W. Breiter, *J. Electrochem. Soc.*, **111**, 1298 (1964).
203. S. D. Ross, M. Finkelstein, and R. C. Petersen, *J. Am. Chem. Soc.*, **86**, 2745 (1964).
204. J. Giner, *Electrochim. Acta*, **1**, 42 (1961).
205. J. J. Lingane, *J. Electroanal. Chem.*, **1**, 379 (1960).
206. F. C. Anson and F. A. Schultz, *Anal. Chem.*, **35**, 1114 (1963).
207. J. W. Johnson, H. Wroblowa, and J. O'M. Bockris, *Electrochim. Acta*, **9**, 639 (1964).
208. S. D. Ross, M. Finkelstein, and R. C. Petersen, *J. Am. Chem. Soc.*, **88**, 4657 (1966); **89**, 4088 (1967).
209. E. J. Corey, N. L. Bauld, R. T. LaLonde, J. Casanova, and E. T. Kaiser, *J. Am. Chem. Soc.*, **82**, 2645 (1960).
210. J. G. Traynham and J. S. Dehn, *J. Am. Chem. Soc.*, **89**, 2139, 2799 (1967).
211. O. A. Khazova, Yu. B. Vasilev, and V. S. Bagotskii, *Izv. Akad. Nauk SSSR*, **9**, 1531, 1787 (1965).
212. V. S. Bagotskii and Yu. B. Vasilyev, *Electrochim. Acta*, **9**, 869 (1964).
213. A. A. Humffray and L. F. G. Williams, *Chem. Commun.*, **1965**, 616.
214. W. A. Struck and P. J. Elving, *Biochemistry*, **4**, 1343 (1965).
215. W. M. MacNevin and T. R. Sweet, *Quart. J. Studies Alc.*, **12**, 46 (1951).
216. M. W. Breiter and S. Gilman, *J. Electrochem. Soc.*, **109**, 622 (1962).
217. S. Gilman and M. W. Breiter, *J. Electrochem. Soc.*, **109**, 1099 (1962).
218. M. W. Breiter, *Electrochim. Acta*, **9**, 827 (1964).
219. C. Liang and T. C. Franklin, *Electrochim. Acta*, **9**, 517 (1964).
220. T. I. Papova, N. A. Siminova, and V. S. Bagotskii, *Zh. Fiz. Khim.*, **38**, 2452 (1964).
221. H. Lund, *Acta Chem. Scand.*, **11**, 491 (1957).
222. R. F. Dapo and C. K. Mann, *Anal. Chem.*, **35**, 677 (1963).
223. C. D. Russell, *Anal. Chem.*, **35**, 1291 (1963).
224. C. K. Mann, *Anal. Chem.*, **36**, 2424 (1964).
225. K. K. Barnes and C. K. Mann, *J. Org. Chem.*, **32**, 1474 (1967).
226. J. F. O'Donnell and C. K. Mann, *J. Electroanal. Chem.*, **157**, 163 (1967).
227. L. L. Miller and A. K. Hoffmann, *J. Am. Chem. Soc.*, **89**, 593 (1967).
228. W. Jaenicke and H. Hoffmann, *Z. Elektrochem.*, **66**, 803, 814 (1962).
229. W. C. Purdy, *Electroanalytical Methods in Biochemistry*, McGraw-Hill, New York, 1965.

AUTHOR INDEX

Numbers in parentheses are reference numbers and indicate that an author's work is referred to although his name is not cited in the text. Numbers in italics give the page on which the complete reference is listed.

A

Abarbarchuk, I. L., 34, *42*
Ackerman, E., 220(23), 225(23), *262*
Adams, R. N., 26(8, 9), 27(10, 11), 29(14), 30(43), 31(38), 33(40, 41), 34, *41*, *42*, 52(8–10), *65*, 85(31, 35), 87(92), 88(39), 90(35), 91(31), 93(31), 94(42), 96(58), 98, 101(106), 102(42), *108–110*, *112–114*, 129(32), 130, 137(39), 144(45, 46), 155, 156(77), *161*, *162*, 164(7), 165(8–10), 169(16), 171(16), 177(16), 182(51), *185*, 198(20), 205 (56, 57), *209*, 221(51), 222(56–58), 227(51), 229(51, 57), 236–238, 244 (131), 252(142), 253, 254(119), 255 (123), 257(81), 259(137), 261(131, 137–140), *263–265*, 273(26), 283(77), 284(50, 79), 285, 291(65), 296(75, 76), *300*, *301*, 309(24, 25, 29), 311(29), 314(39), 315(29), 322(39, 45), 323 (49, 52), 326(29), 327, 329(74), 330(80), 331(74), 334(79), 335(80), 336, 339(85, 92), 340(85), 344(72), 345, 346(72), 351(101–104), 354(104, 114), 356(49, 114, 117, 119), 358 (126), 360(73, 117, 119, 128, 135), 361(73, 135), 362(49), 363(72, 104), 365(72), 367(72, 158, 161), 368(126, 127, 135, 164), 370(176), 371(52), *378–382*
Agar, J. N., 63, *65*, 71(4), 73(9), *107*

Alberts, G. S., 249, 253, *264*
Albery, W. J., 96, 98, 101, *109*, *110*, *112*, *113*
Alden, J. R., 144(46), *161*, 296(76), *301*
Allen, M. J., 273(21), *300*, 304, 323, 363, 372(8), 373(8), 377(8), *378*, *379*
Altman, S., 198, *208*
Anderson, E. W., 231(67), *263*
Anson, F. C., 61(42, 45, 46), *65*, *66*, 177, 183(53), *185*, *186*, 198, 199(22), 200, 201(28, 30, 33–35), 202(35), 203, 207, *209*, 245, 258(88), *264*, 339 (91), 356, 360(134), 361(134), 373, 375(206), *380*, *381*, *383*
Arakelyan, R. A., *111*
Aranda, A., 352(107), *381*
Argersinger, W. J., 257(81), *264*, 329(74), 331(74), *380*
Arvia, A. J., 78, 80, 104(112), 105, *109*, *110*, *113*, 223, *264*
Asada, K., 64(34), *65*
Atamanenko, N. N., 4(19), *17*
Aten, A. C., 244, *265*
Atkin, C. L., 367, *382*
Auerbach, C., 254, *265*
Ayabe, Y., 125, 126(31), 145(31), *161*
Aykazyan, E. A., 85(32), 90(32), *108*, *110*, *111*, 221(28, 54), 228(54), *262*, *263*
Azim, S., 83(25), 85, *108*, *112*, *113*, 284 (49), *301*

B

Baba, H., 310(32), 314(32), *379*
Bacon, J., 329, 335(80), *380*
Badoz-Lambling, J., 165, *185*
Bagotskii, V. S., *112*, *113*, 225, 262, 365(155, 156), 375(211, 212, 220), *382*, *383*
Bair, E. J., 296, *301*
Baker, B., 201(27), *209*
Bard, A. J., 33(47, 48), *42*, 55, *65*, 172, *173*, 174, 183, *185*, *186*, 222, 226, 227(104), 231, 253, 257, 258(82, 87), *262–264*, 309(28), 311(28), 315(28), *379*
Bardin, M. B., *111*, *112*
Barnes, K. K., 376(225), *383*
Batasova, L. V., *113*
Bauld, N. L., 374(209), *383*
Baumann, F., 23(4), *41*, 191, 194, 195, *208*, 272, *300*
Bazán, J. C., 78, 80(119), *109*, *110*, *113*, 223, *264*
Beacom, S. E., *112*
Beilby, A. L., 25, *41*, 62, *65*, 220(14), 224, *262*
Bell, R. P., 96, *109*, *112*
Belleau, B., 325(68), 326(68), *380*
Belyaeva, V. A., *111*, *112*
Belyanchikov, M. P., *111*
Bent, R. L., 358(125), 360(131), 361(125), *381*
Berezina, S. I., 90(44), *108*, *111*
Berger, R. L., 220(23), 225(23), *262*
Berkey, R., 221(100), *264*
Berndt, D., 203, *209*
Bersohn, R., 259(133), *265*
Berzins, T., 62, *65*, 138, *160*, 176, 177, 178(43), *185*, 272, *300*
Besezine, N. P., 101(103), *110*, *112*
Bianco, E., 369, *382*
Biebl, G., 371, *382*
Bierowski, M., *112*
Billon, J. P., 316(36), 322(46, 47), *379*
Bircumshaw, L. L., 71(7), *107*, *110*
Blackburn, T. R., 371, *382*
Blaedel, W. J., 79(81), *109*
Blaha, E. W., 221(100), *264*
Blurton, K. F., 85, *109*, *113*

Bobbitt, J. M., 365, *382*
Bockris, J. O'M., 13(24), *17*, 362, 372 (186–190), 375(207), *381*, *383*
Bodnar, L. H., 226, *263*
Bogotskaya, I. A., 103, *109*, *110*
Bogulavskii, L. I., 104(62), *109*, *110*, 224(9), *262*
Böld, W., 151, *161*, 203, *209*
Bonnemay, M., 202(41), *209*
Bonner, W. H., 323, *379*
Booman, G. L., 58, 62, *65*, 296, *301*
Borisova, T. I., 205(50), *209*
Boronenkov, V. N., *113*
Bowden, F. P., 191(10), *208*
Breiter, M. W., *111*, 151, 152, *161*, 199, 200, 201(29), 203, *209*, 221(26), 233(77), *262*, *263*, 374(202), 375, *383*
Bricker, C. E., 14(26), *17*, 198(20), 200 (25), 203(25), *208*, *209*
Brooks, W., 25(25), *41*
Brown, E. A., 325(69), 326(69), 352, 369(69), *380*, *381*
Bruckenstein, S., 98, 101(100, 101), *110*, *112–114*, 274, 286(51), *300*, *301*
Brummer, S. B., 372, *383*
Brunner, E., 3, *16*
Bublitz, D. E., 171(27, 28), *185*, 221(53), 228(53), *263*
Buck, R. P., 102(108), *110*, *112*, 130, 152, *161*, 374(199), *383*
Budevskii, E., *111*
Buob, K., 63(25), 64(25), *65*
Burgert, B. E., 325, 369(64), *380*
Busch, R. H., 198, *208*
Butler, J. A. V., 191(16), *208*
Byrne, J. T., 122(13, 14), *160*, 205(53), *209*
Bystrov, V. I., 363(146), *381*

C

Carman, R. L., 85(57), *109*, *113*
Carrington, A., 259(134), *265*
Carrozza, J. S. W., 80(118), 104(112), 105, *109*, *110*, *113*
Casanova, J., 374(209), *383*
Caspari, W. A., 3, *16*
Catherino, H. A., 78(84), *109*
Cauquis, G., 316, 322(47), *379*
Cavalier, B., *113*

AUTHOR INDEX

Chambers, C. A. H., 352, 367(162), *381*, *382*
Chambers, J. Q., 144(46), *161*, 296(76), *301*, 367(162), *382*
Chandross, E. A., 309(22), *378*
Chovnyk, N. G., *111*
Christensen, C. R., 258(88), *264*, 339(91), 360(134), 361(134), *380*, *381*
Christie, J. H., 61(42, 43), *65*, 253, *265*
Chuang, L., 26(37), *42*
Clark, R. O., 333(78), 334(78), 363(78), *380*
Clark, W. M., 304, 337, *378*
Coetzee, J. F., 29(15), *41*
Cohen, S. H., 122, *160*
Combrisson, J., 322(47), *379*
Conrad, A. L., 37(5), 38(5), *41*, 278(41, 42), 279(41), 291(42), *300*, *301*, 363 (148), 367(148), *381*
Conway, B. E., *113*, 373(194), *383*
Cooke, W. D., 15, *17*, 139, *160*, 291, *301*
Cordes, E. H., 337, 340(88), 342(88), *380*
Corey, E. J., 374(209), *383*
Coulson, C. A., 310(33), *379*
Cozzi, D., 287, *301*
Crittenden, A. L., 58(14), 61(49), 62(14), *65*, *66*, 118, *160*, 220(14), 224, 225(16), *262*
Cyranski, R., 143(44), *161*

D

Daguenet, M., *114*
Dahms, H., 372(186), *382*
Dale, J. M., 138(68), *161*
Dapo, R. F., 375, *383*
Darlington, R. K., 31(38), *42*, 257(108), *264*
Davies, P. W., 286(52), *301*
Davis, D. G., 171(25), *185*, 202(39), *209*, 225, 226, *262*, *263*, 369, 371, *382*
Davis, R. E., *113*
Davis, S. M., 333, 339, *380*
Day, R. J., 226, *263*
DeFord, D. D., 296, *301*
Dehn, J. S., 374(210), *383*
Delahay, P., 13(25), 16, *17*, 37(32), *42*, 45(1), 50(4), 58(13), 61(15), 62(15), *64*, *65*, 76(14), 94(14), *108*, 119, 120(3), 123(17, 18), 124, 126(3), 128, 134, 137(22), 138, 145(3), *160*, 168(12, 13), 175–177, 178(12, 43), 183, *185*, *186*, 232, 233(77), 235(78), 241, *263–265*, 267, 272, 295, *300*, *301*
Delimarskii, Y. K., 4, *17*, 34, *42*, *111*, *112*, *114*
DeMars, R. D., 139, *160*
Denton, E. B., 74(13), *108*
Desideri, P. G., 287, *301*
Desmond, M. M., 309(17, 18), 326(17, 18), *378*
Dessloch, J. C., 360(131), *381*
De Vries, W. T., 137, *161*
Deys, H. P., 365, 368, *382*
Dezael, C., *113*
Dezider'ev, G. P., 90(44), *108*, *111*
Dimeler, G. R., 30(18), *41*, 309(16, 17), 322, 326(16, 17), *378*
Dimroth, K., 365, *382*
Dirscherl, W., 278, *300*
Discher, C. A., 323, *379*
Dogonadze, R. R., 95, *109*, *111*
Douglass, D. C., 231(67), *263*
Dracka, O., 181(48), *185*
Drushel, H. V., 33, *42*, 231, *263*, 369, *382*
Dubois, J. E., 352(107), *381*
Duennebier, F. C., 360(131), *381*
Durdin, J. V., *113*
Dvorak, V., 346, 347, 352(99), 353, 360(99), 363(99), *380*

E

Eberson, L., 309(21), 324(60), 325(61, 62), 326(61), 369(61, 62), 373(193), 374 (61, 198), *378*, *379*, *383*
Edelman, I. S., 226(40), *262*
Eggertsen, F. T., 272(11), *300*, 346, 360 (97), 363(97), *380*
Eisenberg, M., 63(23), *65*, 74(12), 105, *108*, *110*, 220(11), 223, *262*
Elving, P. J., 25(7), 26(37), *41*, *42*, 102 (111), *110*, *113*, 278(41), 279(41), 280, *300*, *301*, 322, 356(118), 360, 363(148), 367(118, 148), 370(177), 375(214), *379*, *381–383*
El Waakad, S. E. S., 191(13, 14), *208*

Emara, S. H., 191(13), *208*
Enke, C. G., 200(25), 203(25), *209*, 296(73), *301*
Erlinger, D. B., 122(11), *160*
Esin, O. A., *113*
Evans, D. H., 171, *185*
Evans, N. T. S., 286(53), *301*
Evans, V. R., 205(54), *209*
Everett, G. W., 168(15), 169(15), 170(15), 171(22), *185*
Everhart, M. E., 225, *262*
Ewald, A. H., 287, *301*
Eye, J. D., 286(55), *301*
Eyring, H., 363(141), *381*

F

Fassett, D. W., 360(131), *381*
Fedorova, A. I., 85(32), 90(32), 104(61, 62), *108–110*, 221(28), 224(9), *262*
Feiser, L. F., 337, 339(94), 356, 358, 363, 367, *380*
Feldberg, S. W., 200, 203, *209*, 252(141), 254, 255(123), *264*, *265*, 363(139), 368(164), *381*, *382*
Feldman, G. A., 98, 101, *110*, *113*
Felinovskii, V. Y., *114*
Feoktistov, L. G., *114*
Ferrett, D. J., 16, *17*, 105, *109*, 273, 275(27), *300*
Fichter, F., 304, 309, 331(7), 351, 353, 363, 372(7), 375, 377(7), *378*, *380*, *381*
Finkelstein, M., 323(55), 374(203, 208), 377(55), *379*, *383*
Fioshin, M. Ya., 322(53), *379*
Fisher, D. J., 296(70, 71), *301*
Fishman, E., 231, *263*
Forrest, I. S., 323(50), *379*
Fouad, M. G., 64(38, 39), *65*
Fox, H. M., 372(192), *383*
Fox, I., 333, 334(77), *380*
Fox, W. M., 336, *380*
Franklin, T. C., 375, *383*
Freiser, H., 272(10), *300*, 363(147), *381*
French, E. J., 64(35), *65*
French, W. G., 202, *209*
Fried, I., 26(37), *42*, 102(111), *110*, *113*

Friend, J. A., 360(137), *381*
Friend, K. E., 227, *263*, 309(70), 326, *380*
Fritsch, J. M., *114*, 261(138), *265*, 309(29), 311(29), 315(29), 326(29), 351(104), 353(113), 354(104), 363(104), *379–381*
Frumkin, A. N., 4, *17*, 84, 94, 98, 99(51), 100, 101(51), *108*, *110*, *111*, *113*
Fueno, T., 363(141), *381*
Fujinaga, T., 34, *42*
Fulinski, A., *113*
Furlani, C., 181(45), *185*
Furman, N. H., 14, *17*, 164(7), 171(23), *185*, 227, *263*, 273(26), 291(62), *300*, *301*, 370, *382*
Furness, W., 292, *301*

G

Galus, Z., 85(31, 35), 90(35), 91(31), 93(31), 94(42), 102(42), *108*, *112*, 144(45), *161*, 222(58), 236, 237, 253, 261(140), *263–265*, 284(50), 296(75), *301*, 345, 351(102, 103), 370(176), *380*, *382*
Gardiner, K. W., 139(25), *160*
Gaskill, H. S., 74(13), *108*
Gaylor, V. F., 37(5), 38(5), *41*, 278, 279(41), 291, *300*, *301*, 363(148), 367(148), *381*
Geissler, W., *111–113*, 221(33), 226(33), *262*
Geske, D. H., 29, 30(16), 39, *41*, *42*, 179, *185*, 257(79), 258(82), *263*, *264*, 373, *383*
Gierst, L., 163, 168(5), 175, 176(34), *185*, 233, *263*
Gileadi, E., 152, *162*, 318(27), 372(188, 189), *379*, *383*
Gilman, S., 152, *161*, 201, *209*, 375(216, 217), *383*
Giner, J., 200, 201(31), *209*, 374(200), 375(204), *383*
Ginzburg, V. I., 363(145), 367(145), *381*
Girina, G. P., *114*
Glass, D. B., 360(131), *381*
Glasstone, S., 4, *17*, 71(5), *107*, 163(1), *184*
Glesmann, M. C., 358(124, 125), 361(124, 125), *381*

AUTHOR INDEX

Gokhshtein, A. Y., 130, 137, *161*
Gokhshtein, Y. P., 130, 137, *161*
Goncharenko, V. P., *112*
Goodrich, R. B., 4(17), *17*, 189(2), *208*
Goodwin, S., 31(38), *42*
Gorbachev, S. V., 79, *109*, *111*, *112*
Gorodetski, Y. S., *113*
Gorodyskii, A. V., 4(21), *17*
Gouda, T., 64(39), *65*
Gough, T. A., 304(2), 309(2), 315(2), 327(2), 335. 360(2), *378*
Grabowski, Z. R., 143(43), *161*
Granick, S., 356(116), *381*
Gregory, D. P., 83(23, 24), 87, 89, 90(23), 91, *108*, *110*, *111*, 220(30), 224, *262*
Greiser, H., 205(55), *209*
Griess, J. C., 122(11–13), *160*, 205, *209*
Griffith, L. R., 152, *161*, 374(199), *383*
Gurinov, Yu. S., 79, *109*

H

Haber, F., 3, *17*
Haberland, D., *113*, *114*, 253, *264*
Hagihara, B., 286(54), *301*
Haissinsky, M., 122, *160*
Hale, J. M., 102, *110*, *112*, 255, *265*
Hall, D. A., 370, *382*
Hancock, C. K., 363(142), *381*
Hanselman, R. B., 30, *41*
Hansen, R. L., 314, *379*
Hansen, W. N., 257(109, 110), *264*
Harras, J. E., 225(16), *262*
Harris, E. D., 107(74), *109*, 277, *300*
Hawley, M. D., 88(39), *108*, 155, *162*, 182(51), *185*, 221(61), 222(56, 61), 228(61), 245, 251(61), 252(141, 142), *263*, *265*, 336, 339(85), 340(85), 342, 356(122), 357(122), 358(126), 359 (122), 360(128, 135), 361(135), 368 (122, 126, 135, 164), *380–382*
Hedenberg, J. F., 205(55), *209*, 272(10), *300*, 363(147), 381
Hendel, J., *111*, 226(93), *264*
Herman, H. B., 183, *186*, 253, *264*
Herz, J., 324(57), *379*
Heusler, K. E., 102, *110*, *112*, 114, 346, *380*

Heyndrickx, A., 39(34), *42*, 273(23), *300*
Heyrovsky, J., 120(5), *160*, 267(2), *300*
Hickling, A., 71(5), *107*, 163(1), *184*, 191(11, 12, 15), *208*
Higasi, K., 310(32), 314(32), *379*
Himmelblau, D. M., 226, *263*
Hine, F., 64(34), *65*
Hintermann, H. E., *113*
Hodgson, W. G., 369(166), 370(173), *382*
Hoffman, K., *111*, 221(26), *262*
Hoffmann, A. K., 377, *383*
Hoffmann, H., 372, *383*
Hogge, E. A., 83, 87, 88, *108*, *110*, 220(13), 223, 224, *262*
Hoh, G., 171(27–29), *185*, 221(53), 228(53), *263*
Hoijtink, G. J., 221(102), 244, *264*, *265*, 304, 309(14), 326(14), *378*
Holleck, L., 94, *109*, *112*
Hollyer, R. N., *112*
Horanyi, Gy., *112*
Horsfield, A., 259, *265*
Horvath, G. L., *113*
Hsueh, L., 92(115, 116), *110*, *113*
Humffray, A. A., 375(213), *383*
Hurwitz, H., 233, *263*

I

Ibl, N., 63, 64(25, 38), *65*, 73(10), 74, 102(65), *107*, *109*
Ilkovic, D., 120(5), *160*
Ingram, D. J. E., 259(132), *265*
Inoue, T., 325(65–67), *380*
Interrante, L. V., 202(38), *209*
Ivanov, Yu. B., 98–101(51), *108*, *110*, *111*
Iwamoto, R. T., 31, *41*, 122(16), *160*, 184, *185*, 291(65), *301*, 309(25), 371, *379*, *382*
Izutsu, K., 34(45), *42*

J

Jacobs, E. S., 139, *161*, 281, *301*
Jacq, J., *113*
Jaenicke, W., 372, *383*
Jahn, D., 84, 91(27), 94–96, *108*, *111*, *112*, 224(35), *262*
James, T. H., 360(131), *381*

Janeiro, R. A., 360(136), *381*
Janz, G. L., 291, *301*
Javick, R. A., 77(16), 78(82, 83), 106(71), *108*, *109*
Jencks, W. P., 337(86), *380*
Jiminez, L. R., 28, *41*
Johns, R. H., 168-170(15), 171(22), *185*
Johnson, D. C., *112-114*
Johnson, G. R., 85(34), *108*, *112*
Johnson, J. W., 372(187), 375(207), *382*, *383*
Jones, H. C., 296(70, 71), *301*
Jordan, J., 19(1), 23(1), 28, 39(33-35), *41*, *42*, 71(3), 77(16), 78, 106, *107-109*, 190(3, 4), *208*, 220(23), 224, 225, 241, 242(127), *262*, *265*, 272, 273(22, 23), 275(33), *300*
Joschek, H. I., 365, *382*
Julian, D. B., 16, *17*, 120, *160*, 190(3, 8), *208*, 272(9), *300*, 360(130, 131), *381*
Juliard, A. L., 144, *161*, 163, 168(5), 176, *185*
Jura, W. H., 353(111), 369(166), 370(173), *381*, *382*

K

Kabanov, B. N., 88, *108*, *110*, *111*, *113*
Kabanova, O. L., *111*, *112*
Kabasakalian, P., 322(48), *379*
Kacherova, S. A., 4(18), *17*
Kaiser, E. T., 374(209), *383*
Kalinowski, M. K., 143(43), *161*
Kambara, T., 16(29), *17*, 104(67), *109*, 118(1), *160*, 190(3, 5), *208*, 273(25), *300*
Kamiyama, F., 34(45), *42*
Karaoglanoff, Z., 4, *17*, 163(4), 168(4), *184*
Karmazin, V. I., 287, *301*
Karp, S., 226, 259(90), *263*, *264*
Kastening, B., 94(94), *109*, *112*
Keller, H. E., 102(108), *110*, *112*
Kelly, M. T., 296, *301*
Kemula, W., 143, *161*, 256, *263*
Keshaven, K., 286(55), *301*
Kevlegan, G. H., 63(24), *65*
Khazova, O. A., 375(211), *383*
Kholpanov, L. P., 90(45), *108*, *111*, *112*, *114*

Khomutov, N. E., 363(146), *381*
Kijowska, K., *113*
Kimla, A., 80, *109*
King, D. M., 207, *209*
King, R. M., 181, *185*
Klatt, L. N., 79(81), *109*
Kleinberg, J., 122(16), *160*, 171(29), *185*
Kleinerman, M., 233(77), *263*
Knobloch, E., 337, *380*
Knorr, C. A., 151, *161*, 203, *209*
Koehler, K., 337(88), 340, 342(88), *380*
Kolthoff, I. M., 4, 13(23), *17*, 19(1), 23(1), 29(15), 32, 39, *41*, *42*, 50(4), 52, 53(5, 6), 62, 63, *65*, 71(3), 104, *107*, *109*, 120(7), 121, 123, *160*, 189, 190(3, 4), *192*, 193, 202(40), *208*, *209*, 220(3), 221(59), 224, 225, 228, *262*, *263*, 267(1), 272, 273(22-24), 275(20, 33), 292, *300*
Koski, W. S., 309(23), *378*
Koutecky, J., 94, 95(54), *109*, *111*, 253, *264*, 319(40), *379*
Koyama, K., 325(65, 66), 369(63), *380*
Kraichman, M. B., 83, 87, 88, *108*, *110*, 220(13), 223, 224, *262*
Kraus, D. P., 122(11), *160*
Kresse, K., *112*
Kretlow, W. J., 139, *161*, 371, *382*
Krishnan, V. R., 335, 367, 369, *380*, *382*
Krivis, A. F., 280, *301*, 356(118), 360, 367(118), *381*
Krushcheva, E. I., *113*
Kublik, Z., 143(40, 41, 44), 152, *161*
Kulyavik, Y. Y., *113*
Kuwana, T., 28, 29(14), *41*, 171(27, 28), 183, *185*, *186*, 202, 205, *209*, 221(48, 53), 227(48), 228(48), 257(109, 110), *263*, *264*, 327, 360(73), 361(73, 132), 362, *380*, *381*

L

Lacaze, P. C., 352(107), *381*
Laitinen, H. A., 4, *17*, 52, 53(5), 62, 63, *65*, 104, *109*, 121, *160*, 199, 200, *209*, 220(3), 221(101), *262*, *264*, 273, 275(20), *300*

AUTHOR INDEX 391

LaLonde, R. T., 374(209), *383*
Lamb, B., 222(57), 229(57), *263*
Lancaster, J. E., 353(111), *381*
Landerl, J. H., 37(5), 38(5), *41*, 278(42), 291(42), *301*
Landsberg, R., *111–114*, 221(33), 226, 253, *262*, *264*
Langer, A., 287, *301*
Larson, R. C., 31, *41*, 291(65), *301*, 309 (25), *379*
Latham, W. A., 367(158), *382*
Latimer, W. M., 193, *208*, *209*
Latta, B. M., 353, *381*
Lauer, G., 61(42, 43, 45, 46), *65*, *66*
Lawrence, G. L., 25(25), *41*
Layoff, J., 244(131), 261(131), *265*
LeBlanc, M., 3, *16*
Lee, H. Y., 85(35), 90(35), *108*, *112*, 137(39), 144(45), 152, *161*, 221(55), 222(58), 228(55), *263*, 296(75), *301*, 334(79), 356(119), 360(119), *380*, *381*
Lee, J. K., 196, 198, *208*, 221(51), 222 (57), 227(51), 229(51, 57), *263*, 352, *381*
Leedy, D. W., 156(77), *162*, 257(108), 261(138, 139), *264*, *265*, 322(45), 329(75), 344(75), 350, 351(104), 354(104), 363(104), *379*, *380*
Lehnsen, J. E., 369, *382*
Leung, M., 324(57, 58), 333, *379*
Levich, V. G., 63, *65*, 71(6), 73(8), 76(6, 8), 80, 81(8), 83, 88(38), 89(8), 90, 91(41), 92, 94(43), 95(54, 55), 96, 98–101(51), 102(59, 64), 103, 104(63), *107–111*, 220(5, 24), 221(28, 54), 228(54), 253, *262*, *264*
Lewis, G. P., 85(30), *108*, *112*, 226, *262*
Liang, C., 375, *383*
Liang, K., *113*, 253(117), *264*
Lim, S. C., 287, *301*
Lin, C. S., 74(13), *108*
Lindsey, A. J., 107(74), *109*, 277, *300*
Lingane, J. J., 4, 13(23), *17*, 50(4), 55, *64–66*, 120(6, 7), 123, *160*, 170, *185*, 198, 199(22), 200, 201(32, 33), 202(37), *209*, 224, 225, 228, 253, 257(83), *262–265*, 267(1, 6), 273(20), 275(20), 292, *300*, 375(205), *383*

Lingren, W. E., 365, 366, *382*
Lohman, F., *114*
Lord, S. S., 19(2), 34, 37(2), *41*, 139, *160*, 190(3, 6), 205(6), *208*, 271, 272(8), 278, *300*, 327, 333, *380*
Loshkarev, M. A., 365(153, 154), *382*
Loveland, J. W., 30(18), *41*, 309(16), 322, 326(16), *378*
Lu-an, Y., *112*, 365(155), *382*
Ludwig, P., 323(49), 356(49), 362(49), *379*
Lund, H., 30(19), *41*, 227, *263*, 278, *300*, 309(13), 318, 319, 326, 333, 375, *378*, *383*
Lyalikov, Y. S., *111*, 287, *301*

M

McCall, D. W., 231, *263*
McCawley, F. X., *113*
McClure, J. H., 169(16), 171(16), 177(16), *185*, 205(57), *209*, 259, *264*, 327(72), 344–346(72), 363(72), 365(72), 367 (72), *380*
McDuffie, B. J., 14(26), *17*
Macero, D. J., 54, *65*, 224, *262*, 360(136), *381*
McEwen, W. E., 171(29), *185*
McGlotten, J., 322(48), *379*
McIntyre, T. W., 333, 335, *380*
McLachlan, A. D., 259(134), *265*
MacNevin, W. M., 201(27), *209*, 375, *383*
Maki, A. H., 30, *41*, 360(129), *381*
Malachesky, P. A., 33(41), *42*, 96, 101 (106), *109*, *110*, *113*, *114*, 242(118), 244(118, 131), 254(119), 261(118, 131), *264*, 265, 309(24), 337, 339(92), 340, *378*, *380*
Malev, V. V., *113*
Malmstadt, H. V., 296, *301*
Mamantov, G., 138, *161*, 168(13), 183(56), *185*, *186*
Mango, F. D., 323, *379*
Mann, C. K., 375, 376(225), *383*
Manning, D. L., 138(68), *161*
Marchand, A., 365(150), *382*
Marchiano, S. L., 80(117–119), 104(112), *110*, *113*

AUTHOR INDEX

Marcoux, L. S., 30(43), 33(41), *42*, 96(58), *109*, *113*, *114*, 218(99), 227(122), 254(119), 255(123), 261(138), *264*, *265*, 283(77), 284, *301*, 309(24), 310(34), 311(29), 314(39), 315(29), 316(34), 319(34), 322(39), 326(29, 34), 327(34), 351(104), 354(104), 363(104), *378–380*
Marcus, M. F., 342, *380*
Maricle, D. L., 259, *264*, 370(173), *382*
Mark, H. B., 177, *185*, 245, 257, *264*, 356, 367, *381*, *382*
Marple, T. L., 15, *17*, 139(26), *160*
Martin, K. J., 217, *262*
Matsouka, K., 325(65), *380*
Matsouka, T., 325(65), *380*
Matsuda, H., 125, 126(31), 145(31), 146, *161*
Mattax, C. C., 183(55), *186*
Mattson, J. S., 257(112), *264*
Maurer, A. H., 315(38), 316(35), 322(38), *379*
Mayell, J. S., 258(84), *264*
Mayrath, B., 31(38), *42*
Meek, J. S., 107(75), *109*, 277, *300*
Mehl, W., *114*
Meites, L., 226, 259, *263*, *264*, 267(4), *300*
Melchior, M. T., 360(129), *381*
Merkle, F. H., 323, *379*
Merriam, E. S., 3, *17*, 71(2), 104, *107*
Metzler, G., 315(38), 322(38), *379*
Meyer, R. E., *113*
Meyers, E. A., 363(142), *381*
Michaelis, L., 356, *381*
Miller, B., *114*
Miller, C. S., 225, *262*
Miller, F. J., 25, *41*
Miller, H. H., 4(17), *17*, 189(2), *208*
Miller, J. F., 33, *42*, 231, *263*, 369, *382*
Miller, L. L., 377, *383*
Miller, S. I., 365, *382*
Miller, T. A., 221(51), 222(57), 227(51), 229(51, 57), 244(131), 261(131), *263*, *265*
Milner, G. W. C., 48(3), *65*, 120, 123(8), *160*, 267(5), *300*
Minc, S., *113*

Mizoguchi, T., 351(101), *380*
Mohilner, D. M., 257(81), *264*, 329, 331(74), *380*
Molch, D., *112*
Morgan, E., 58(14), 62(14), *65*, 225, *262*
Moros, S. A., 259(89), *264*
Morpurgo, G., 181(45), *185*
Morris, J. B., 24(6), *41*, 53, *65*, 169(16), 171(16), 177(16), *185*, 205(57), *209*, 280, *301*, 327(72), 344–346(72), 363(72), 365(72), 367(72), *380*
Mueller, L., 101(104, 105), *110*, *112*
Mueller, T. R., 27(10, 11), *41*, 52, *65*, 129(32), 130, *161*, 220(6), 221(6), 238, *262*, *263*
Müller, O. H., 80, *109*, 190(3, 7), *208*, 267(3), 292, *300*
Muller, R. H., 64, *65*
Müller, S., *111*, *114*, 221(33), 226(33), *262*
Müller, W., 226(93), *264*
Munson, R. A., 226, *264*, 374(201), *383*
Murray, R. W., 178(42), *185*
Mussini, T., *112*

N

Nagai, T., 274, *300*
Nagy, F., *112*
Napp, D. T., *113*, *114*
Nash, R. A., 363, 368(144), *381*
Naylor, P. F. D., 286(53), *301*
Neglia, M. T., 353(111), *381*
Neikam, W. C., 309(17, 18), 326(17, 18), *378*
Nekrasov, L. N., 98(51), 99(51), 100(48, 51), 101(51), *108*, *110–112*
Nelson, R. F., 33(40), *42*, 156(77), *162*, 231(143), 261(138, 139), *265*, 322(45), 346, 351(104), 352(98), 353, 354(104), 356(114), 363(98, 104), *379–381*
Nemec, I., 346(99), 347(99), 352(99), 353(99), 360(99), 363(99), *380*
Nernst, W., 3, *17*, 67, 71(2), 104, *107*
Newman, J., 90, 92(114–116), *110*, *113*, *114*
Newson, J. D., 85(33), 90(33), *108*, *111*, 220(31), 224, *262*

AUTHOR INDEX

Nicholson, M. M., 34, *41*, 62, *65*, 128, 137, 139, *160*, 272, *300*, 369, *382*
Nicholson, R. S., 126(61), 130, 131, 134, 136(61), 137, 140(61, 62, 66), 141 (61), 142(61), 146, 147–149(73), 153, 155, 158, *161*, *162*, 248, *264*
Nightengale, E. R., 105, *109*, 202(40), *209*, 274(28), *300*
Nikelly, J. G., 139, *160*
Nilsson, O., 76, 104(77), *109*
Nitzsche, R., *112*, *113*
Noack, D., 278, *300*
Noack, J., 29, *41*
Nyberg, K., 309(21), 324(60), 325(61), 326(61), 369(61), 374(61, 198), *378*, *379*, *383*

O

O'Brien, R. N., 64(47, 48), *66*, *114*
O'Connor, J. J., 369, *382*
O'Donnell, J. F., 376, *383*
Oehme, F., 278, *300*
Ohnesorge, W. E., 227, *263*, 309(70), 326, *380*
Okada, S., 64(34), *65*
Oldfield, L. F., 362, *381*
Olmstead, M. L., 155, *161*
Olson, C. L., 26(9), 27(10), *41*, 52(8), *65*, 79(80), 85(35), 90(35), *108*, *109*, *112*, 222(58), *263*, 334, *380*
O'Neil, R. C., 139(24), *160*
Orgeron, D. J., 371, *382*
Orlemann, E. F., 221(59), 226, 228, *262*, *263*
Oshe, A. I., *113*
Osteryoung, R. A., 61, *65*, *66*, 257(109, 110), *264*
Otto, K., 278, *300*

P

Palke, W. E., 183(53), *186*
Panchenko, I. D., *111*, *112*, 287(58), *301*
Panketh, G. E., 345, 346(96), 360(96), 363, 367(96), 368(96), *380*
Papouchado, L., 218(99), *264*, 367, *382*
Papova, T. I., *113*, 375, *383*
Parkanyi, C., 309(19), 318, 319(41), 326(19), 344(19), 345, 363(19), *378*, *379*
Parker, R. E., 165(8), *185*, 205(56), *209*, 356(117), 360(117), *381*
Parker, V. D., 325, 369(64), *380*
Parsons, J. S., 171(24), *185*, 360, 369, *381*
Paunovic, M., 183, *186*
Pavlopoulos, T., 64, *65*, 223, *262*, 272, *300*
Pearl, I. A., 363, 364(149), 369, *382*
Peover, M. E., 227(103), *264*, 304, 309 (2–4), 315(2, 3), 317, 326(3), 327, 335, 360(2), *378*, *379*
Perone, S. P., 137, 139, *161*, *162*, 371, *382*
Peters, D. G., 170, *185*, 228, *263*
Petersen, R. C., 323(55), 374(203, 208), 377(55), *379*, *383*
Peterson, J. M., 183, *186*
Petranek, J., 367(160), 368(160), *382*
Petrie, G., 101(106), *110*, *114*, 339(92), 344, 367(161), *380–382*
Phelps, J., 33(47), *42*, 227(104), *264*, 309, 311, 315(28), *379*
Phillips, C. S. G., 16, *17*, 105, *109*, 275(27), *300*
Piekarski, S., 358(126), 360(135), 361 (135), 368(126, 127, 135), *381*
Pietrzyk, D. J., 304, *378*
Piette, L. H., 323(49, 50), 356(49), 362(49), *379*
Pilgram, M., 52–54(7), 57(7), *65*, 216(1), 219(1), 220(1), 221(32), 226 (32), *262*
Pleskov, Yu. K., 85(29), 90(29), *108*, *111*, 221(54), 228(54), *263*
Podesta, J. J., 80(118), *110*, *113*
Polcyn, D. S., 137, *161*
Polestra, F. M., 223, 229, *262*, *263*
Pominov, V. G., *111*
Pons, B. S., 257(111, 112), *264*
Pool, K. H., 61, *66*
Popov, A. I., 29, 30(16), *41*
Pospisil, J., 367(160), 368(160), *382*
Pourbaix, M., 20(3), *41*
Powell, V. J., 273(21), *300*, 323, *379*
Prager, S., *113*
Prater, B. G., 30(43), *42*, 221(51), 227 (51), 229(51), *263*, 283(77), *301*

AUTHOR INDEX

Prater, K. B., 30(43), *42*, 87(92), 98, 101(106), *109*, *110*, *113*, *114*, 222 (57), 229(57), *263*, 283(77), 285, *301*, 339(92), *380*
Price, J. E., 171, *185*
Pullman, A., 310(31), 326(31) 363(31) *379*
Pullman, B., 310(31), 326(31), 363(31), *379*
Purdy, W. C., 378, *383*
Putnam, G. L., 74(13), *108*
Putnam, R. C., 371, *382*
Pysh, E. S., 309(15), 326(15), 344(15), *378*

R

Raison, J., 316(36), *379*
Randles, J. E. B., 124, *160*, 244, *265*
Ranz, W. E., 77(16), *108*
Reddy, T. B., 32, *42*, 352, *381*
Ree, T., 363(141), *381*
Reilley, C. N., 164(7), 168(15), 169(15), 170, 171, 178(42), 181, *185*, 276(26), 291(62), *300*, *301*
Reinmuth, W. H., 124, 128, 130, 140(35), *161*, 164, 176, 177, 178(35, 40, 41, 50), 181(47), 182(47), 183, *185*, *186*, 231, 241, 253, *263*, *265*
Reishakrit, L. S., *113*
Rembaum, A., 310(32), 314(32), *379*
Remick, A. E., 182, *185*
Reuter, L. H., 286(55), *301*
Reynolds, G. D., 4, *17*
Riddiford, A. C., 71(7), 81(91), 83, 85(33), 87, 89, 90(23, 33), 91, 102 (91), *107–113*, 220(30, 31), 224, *262*, 284(49), *301*
Robert, J., *114*
Roberts, B. G., 315(38), 316(35), 322(38), *379*
Roberts, E. R., 107, *109*, 277, *300*
Roberts, N. K., 360(137), *381*
Robinson, C. V., 226(40), *262*
Roffia, S., 287, *301*
Rogers, G. T., *112*
Rogers, L. B., 4, 15, *17*, 19(2), 34, 37(2), *41*, 122, 139(24), *160*, 189, 190(3, 6), 205(6, 53), *208*, *209*, 271, 272(8), 278, *300*, 327, 333, *380*
Rollman, L. D., 371, *382*
Rosano, H. L., 287(61), *301*
Rosenfield, C., 64(47), *66*
Rosner, D. E., *113*
Ross, J. W., 195, *208*
Ross, S. D., 323, 374(203), 377, *379*, *383*
Ross, T. K., 79, *109*
Rothenberger, E., 351, *380*
Rouse, T. O., 183(52), *186*, 220(7), *262*
Rowland, F. S., 261(140), *265*, 351(103), *380*
Rozental, K. I., 204, 205(51), *209*
Rubin, B. T., 372(188), *382*
Ruby, W. R., *113*, 120, *160*, 190(3, 8), *208*, 253(117), *264*, 267, 272(9), *300*, *301*, 360(130, 131), *381*
Ruehlen, F. N., 372(192), *383*
Ruetschi, P., 85(30), *108*, *112*, 226, *262*
Rulfs, C. L., 224, *262*
Runner, M. E., 221(100), *264*
Russ, R., 3, *17*
Russell, C. D., 30, *42*, 183(53), *186*, 373, 375, *383*
Ryabokon, V. D., 287(58), *301*
Ryba, O., 367, 368, *382*

S

Sakuma, M., 370(177), *382*
Salomen, E., 3, *16*
Salzberg, H. W., 323, 333, *379*
Sanborn, R. H., 226, *262*
Sand, H. J. S., 4, *17*, 163(3), 168(3), *184*
Sandstrom, W., 337(88), 340(88), 342 (88), *380*
Santhanam, K. S. V., 33(47), *42*, 227 (104), *264*, 309(28), 311(28), 315(28), 335, 367, 369, *379*, *380*, *382*
Sauerwein, W., *113*
Saveant, J. M., 140(63–65), *161*, *162*
Sawyer, D. T., 202(38), *209*, 226, *263*
Schempf, J. M., 24(6), *41*, 53, *65*, 280, *301*, 333(77), 334(77), *380*
Schlatter, M. J., 372(190), *383*
Schmidt, H., 29, *41*
Schönmann, P., 353, *381*

AUTHOR INDEX

Schubert, M. P., 356(116), *381*
Schultz, F. A., 201(30), 203(30), *209*, 375(206), *383*
Schurdak, E. J., 33, *42*, 227, *263*
Schurig, H., 102, *110*, *112*, *114*, 346, *380*
Schwarz, W. M., 140(57), 158, *161*, *162*, 253, *265*
Schwarzer, O., *113*
Scott, A. I., 363, *381*
Seo, E. T., 156(77), *162*, 261(138, 139), *265*, 322(45), 351(104), 353(113), 354, 363(104), *379–381*
Sevcik, A., 123(19), 124, 143, *160*
Shain, I., 23(4), *41*, 118, 126(61), 130, 131, 134, 136(61), 137, 140(61, 66), 141(61), 142(61), 146, 153, 158, *160–162*, 191, *194*, *195*, *208*, 217, 248, 249, 253, *262*, *264*, *265*, 272, *300*
Shalit, H., 144, *161*
Shams El Din, 191(14), *208*
Sharma, L. R., 79(80), *109*
Sharp, H., 323(52), 371(52), *379*
Sharp, J. H., 367(161), *382*
Sheps, S. Q., 287(61), *301*
Shibata, S., 202(42, 43), *209*
Shilina, G. V., *111*, *112*, *114*
Shumilova, N. A., *113*
Shurygin, P. M., *113*
Silverman, H. P., 275(34), *300*
Siminova, N. A., 375(220), *383*
Simnad, M. T., 205(54), *209*
Simpson, H. N., 363(142), *381*
Sioda, R., 256, *263*, 309(23), *378*
Siver, Yu. G., 88(36), 102(110), *108*, *110*, *111*
Skobets, E. M., 4(20), *17*, 287, *301*
Smit, W. M., 220(10), 224, *262*
Smith, D. E., 140(58), *161*
Smith, D. L., 25(7), *41*, 280, *301*
Smith, J. G., 61(49), *66*
Snead, W. K., 182, *185*
Snyder, R. E., 333(78), 334(78), 363(78), *380*
Sobkowski, J., *113*
Solon, E., 231, 258(85, 87), *263*, *264*
Soos, Z. G., 55, *65*
Srinivasan, S., 152, *162*

Stanienda, A., *112*, *113*, 309, 326(20), 371, *378*, *382*
Stehney, A. F., 4(17), *17*, 122(10), *160*, 189(2), *208*
Sterner, J. H., 360(131), *381*
Steuber, F. W., 365, *382*
Stewart, R., 304, 363, *378*
Stiehl, G. L., 123(18), *160*, 295, *301*
Stock, J. T., 365(150), *382*
Stoner, G., 372(189), *383*
Štráfelda, F., 80, *109*
Streitweiser, A., 309(30), 310(30, 33), 326(30), *379*
Streuli, C. A., 30, *41*
Strickland, J. D. H., 64, *65*, 223, *262*, 272, *300*
Strojek, J. W., 363(139, 140), *381*
Struck, W. A., 375(214), *383*
Stump, D., 323(52), 371(52), *379*
Suatoni, J. C., 333, 334(78), 363(78), *380*
Sundheim, B. R., *113*
Susuki, T., 325(63), 369(63), *380*
Suter, E., *113*
Svadkovskaya, G. E., 373(195), *383*
Swann, S., 304, *378*
Sweet, T. R., '375, *383*
Swell, J. M., 360(131), *381*
Swofford, H. S., 85(57), *109*, *113*

T

Tachi, I., 16(29), *17*, 104(67), *109*, 118(1), *160*, 190(3, 5), *208*, 273(25), *300*
Taft, R. W., 333(77), 334(77), 353, *380*, *381*
Tamamushi, R., 244, *265*
Tanaka, N., 189, 190 (3), *190*, *192*, 193, *208*, 244, *265*, 273(24), *300*
Taniguchi, H., 291, *301*
Tarasevich, M. R., *113*
Tatawawadi, S. V., 231(69), *263*, 358(126), 360(135), 361(135), 368(126, 135), *381*
Taylor, K. J., *112*
Tedoradse, G., 84, 94, *108*, *111*
Temianko, V. S., *111*

Testa, A. C., 178(41, 50), 181(47), 182(47), *185*, 231, 253, *263*
Thibaud, Y., 316(36), *379*
Thiele, R., *112*, *114*
Thirtle, J. R., 360(131), *381*
Tobias, C. W., 63(23), 64(35), *65*, 74 (11, 12), 105(11), *108*, *110*, *113*, 220(11), 223(11), *262*
Tomilov, A. P. 322, *379*
Tomilov, B. I., 365(153, 154), *382*
Tong, L. K. J., *113*, 253, *264*, 358, 361 (123–125), *381*
Toome, V., 52–54(7), 57(7), *65*, 216(1), 219(1), 220(1), *262*
Toren, E. C., 296(73), *301*
Toren, P. E., 314(37), *379*
Toshev, S., *111*
Traynham, J. G., 374(210), *383*
Tremmel, C. G., 267, *301*
Trümpler, G., 63(25), 64(25), *65*, 76, *108*
Tsukamoto, T., 16, *17*, 104, *109*, 118, *160*, 190(3, 5), *208*, 273(25), *300*
Tsutsumi, S., 325(63, 65–67), 369(63), *380*
Turner, D. R., 85(34), *108*, *112*
Turner, M. J., 372, *383*
Turner, W. R., 322, *379*

V

Van Dalen, E., 137, *161*
Van Schooten, J., 221(102), *264*
Vashchenko, V. V., *111*
Vasil'ev, Yu. B., *112*, *113*, 365(155, 156), 375(211, 212), *382*, *383*
Vassiliades, T., 231, *263*
Vermillion, F. J., 363, 364(149), *382*
Vertes, Gy., *112*
Veselovskii, V. I., 204, *209*
Vetter, K. J., 203, *209*, 241, *265*
Vianello, E., 140(63–65), *161*, *162*, 287, *301*
Vidovich, G. L., 104(61, 62), *109*, *110*, 224(9), *262*
Vielstich, W., 84, 91(27), 94–96, *108*, *111*, *112*, 152, *162* 224(35), *262*
Visco, R. E., *114*, 309(22), *378*
Vittum, P. V., 360(131), *381*

Vogel, V., 152, *162*
Vogt, H., 94(94), *109*, *112*
Voitkevich, S. A., 373(195), *383*
Volke, J., 319(42, 43), *379*
von Stackelberg, M., 52, 53(7), 54, 57, *65*, *111*, 216, 219(1), 220(1), 221(32), 226(32), *262*
Voorhies, J. D., 33, 34, *41*, *42*, 165(9, 10), 171(23, 24), *185*, 227, *263*, 333, 339, 360, 369, 370, *380–382*

W

Wagner, C., 63(22), *65*
Wang, J. H., 223, 224, 226, 229, *262*, *263*
Ward, G. A., 171, *185*, 228(52), *263*, 280, *301*, 356(120), 360(120), 367(120), *381*
Waters, W. A., 304, 336, 363, *378*, *380*
Wawzonek, S., 221(100, 101), *264*, 304, 333, 335, *378*, *380*
Weber, H. F., 3, *17*, 163(2), 168(2), *184*
Weinberg, H. R., 304, 345, 365, *380*
Weinberg, N. L., 304, 325(68, 69), 326, 345, 352, 365, 369(69), *380*, *381*
Weininger, J. L., 199, 200, *209*
Weisgraber, K. H., 365(150), *382*
Weiss, F. T., 272(11), *300*, 346, 360(97), 363(97), *380*
Weissberger, A., 360(131), *381*
White, B. S., 227(103), *264*, 304(3), 309 (3), 315(3), 326(3), 327(3), *378*
White, R. M., 261(140), *265*, 351(103), *380*
Wijnen, M. D., 220(10), 224, *262*
Wilke, C. R., 63(23), *65*, 74(11, 12), 105 (11), *108*, *110*, 220(11), 223(11), *262*
Will, F. G., 151, *161*, 203, *209*
Williams, L. F. G., 375(213), *383*
Wills, G. B., 372(192), *383*
Wilson, R. E., 4, *17*
Wilson, W. H., 191(12), *208*
Winstram, L. O., 257(112), *264*
Wojtowicz, J., *113*
Woolf, L. A., 221(25), 225, *262*
Wragg, A. A., 79(79), *109*
Wranglén, G., 64(33), *65*, 76, 104(77), *109*

AUTHOR INDEX

Wroblowa, H., 372(187), 375(207), *382, 383*
Wyche, C., *113*

Y

Yang, N. C., 309(15), 326(15), 344(15), *378*
Yoshizawa, S., 64(34), *65*
Young, R. H., 314(37), *379*
Youtz, M. A., 4, *17*
Yukhtanova, V. D., 104(62), *109–112*, 224(9), *262*

Z

Zahradnik, R., 309(19), 318, 319(40, 41), 326(19), 344(19), 345, 363(19), *378, 379*
Zeller, H., 76, *108*
Zembura, Z., *111–113*
Zimmerman, J. F., 53, 57, *65*, 88(39), 91(47), *108*, 128, 130(33), *161*, 177(38), 184(38), *185*, 216, 220(4), 222(56), 240(4), *262, 263*
Zittel, H. E., 25, *41*
Zlotowski, I., 4, *17*
Zweig, A., 315(38), 316, 322, 353, 369, *379, 381, 382*
Zyka, J., 346(99), 347(99), 352(99), 353(99), 360(99), 363(99), *380*

SUBJECT INDEX

A

Acetonitrile, as solvent, 29–31
  purification of, 30
Acids, aliphatic, 372
Adsorbed hydrogen films, 189–191
Alcohols, 375
Aldehydes, 375
Amides, aliphatic, 375
Amines, aliphatic, 375
  aromatic, 327–356
Anodic stripping, 139
Aromatic compounds, basic oxidation patterns, 305–308
  miscellaneous oxidations, 370
Auxiliary electrodes, 12

B

Benzonitrile, as solvent, 31–32
Biochemically important molecules, oxidation of, 371–372
Boron carbide electrodes, 28
Boundary layers, diffusion, 70, 91
  hydrodynamic, 68, 81

C

Carbon paste electrodes, construction of, 280–283
  nonaqueous, 30
  potential limits of, 26–27
Carbon rod electrodes, construction of, 278–280
  potential limits of, 24–25
Charge integration, 58, 62
Chronopotentiometry, 105–172
  adsorption and oxide effects in, 174
  at cylindrical electrodes, 170–171
  equipment for, 166, 297–299
  electrode processes, studied via current programs, 177–183
  electrode processes, studied via $i\tau^{1/2}$ variations, 172–177
  precursor reactions in, 176–177
Conical microelectrodes, 77–78
Controlled potential coulometry, 257–259
Convection, 44
  forced, 67–107
  in unstirred solution, 63–64
Current function, 130–132
  tables of, 131
Current potential curves, 10
Current sweep voltammetry, 164–165
Current–voltage curves, 5–10
Cyclic voltammetry, 143–159
  charge transfer rates by, 147–149
  peak height measurements of, 152–158
  product identification by, 149–152

D

Diamines, aromatic, 356
Diffusion current, 10
Diffusion coefficients, evaluation of, 214–231
  tables of, 219–220
  tracer determination of, 227, 229
Diffusion processes, cylindrical, 61–62
  linear, 45–61
  to spherical electrodes, 62–63
Digital simulation techniques, 254
Dimethylsulfoxide, as solvent, 32
DME, *see* Dropping mercury electrode
Double layer, charging of, 37, 174

400 SUBJECT INDEX

Dropping mercury electrode (DME), relation to solid electrodes, 14–16

E

ECC reactions, rate measurements of, 251
  working curves for, 254
ECE reactions, rate measurements of, 247–251
  rotated disk treatment of, 253–255
  working curves for, 250, 254
Electroanalytical techniques, correlations of, 231–240
Electrode area, determinations, electrochemical, 56
  geometric, 57
  measured by $it^{1/2}$, 50, 56–58
  measured by chronopotentiometry, 214
Electrode sensitivity, 38–40
Electrode systems, classification of, 11–13
Electrolytic cells, 267–270
Electron paramagnetic resonance (EPR), applications to electrochemistry, 259–261
Electron transfer coefficient, 126, 135
EPR, see Electron paramagnetic resonance
Error function, in diffusion processes, 47–49
  table of, 48

F

Fick's first law, 46
Fick's second law, 47
Foil electrodes, construction of, 276–277
  chronopotentiometry at, 166, 171
  convection at, 63
Follow-up reactions, 247–255, 305

G

Gold electrodes, adsorbed hydrogen films on, 189–191
  oxidation of, 191–205
  potential limits of, 25

H

Halides, aliphatic, 377
Heat transfer, related to mass transfer, 72

Heterogeneous rate constants, 230
  determination of, relaxation methods, 240–241
  determination of, steady state methods, 92–94, 240–244
  determination of, via cyclic voltammetry, 147–149
Homogeneous chemical reactions, 244–255
Hydrocarbons, oxidation of, aliphatic, 372
  aromatic, 308–327
Hydroxy compounds, aromatic, 363

I

Instantaneous current equations, cylindrical electrode, 61
  nonshielded electrode, 55
  semiinfinite linear diffusion control, 50
  spherical electrode, 62
$it^{1/2}$ (chronoamperometric) measurements, 50–61
  for electrochemical areas, 56–58
  linear diffusion, 52–61
  working curves for ECE reactions, 250

K

Kinematic viscosity, 75
Kinetic layer, 95

L

Limiting current equations, conical microelectrode, 77
  general, 70, 106
  plate electrode, 76
  rotated disk electrode (RDE), 83
  rotated wire electrode, 105
  tubular electrode, 79
Linear diffusion electrode, design, 51

M

Mass transfer, coefficient, 106, 241
  by forced convection, 67–110
  in quiet solutions, 43–64
  with turbulent flow, 102–104
Mercury chloride film anode, 28

SUBJECT INDEX

Mercury plated electrodes, 28
Metal deposition, voltammetry of, 138
Methylene chloride, as solvent, 33
Migration processes, 44

N

Nernst diffusion layer, 67–71
Nernst equation, 5–10, 116
Nitro compounds, as solvents, 33
Nonaqueous solvents, potential ranges of, 29–34
    purity criteria for, 34–36

O

Order of electrochemical reactions, 94
Oxides on electrode surfaces, 191–205

P

Peak polarograms, general introduction to, 124–127
    for metal depositions, 138–139
    reproducibility of, 128
    reversibility of, 135–137
    theory of, 130–134
Physicochemical studies of electrode reactions, 255–261
Plate electrodes, 76–77
Platinum electrodes, adsorbed hydrogen films on, 189–191
    oxidation of, 191–205
    potential limits of, 22
    pretreatment of, 206–208
Polarization, polarized electrodes, 13–14
Potential limits, aqueous solutions, 19–29
    nonaqueous solutions, 29–34
    various electrodes, *see* specific electrodes (platinum, gold, etc.)
Precursor reactions, chronopotentiometric treatment of, 176–177
    rotated disk electrode, treatment of, 94–96
Propylene carbonate, as solvent, 32
Pyrolytic graphite electrodes, 25–26

R

Randles-Sevcik constant, evaluation of, 128–129

Randles-Sevcik equation, 124
Rapid voltage sweep methods, 122–124
Reduction processes, 337
Reference half-cells, 12, 288–291
Residual currents, 36–37
Reverse-current chronopotentiometry (RCC), 178–183
    follow-up reactions via, 181–182
Rotated disk electrode (RDE), 80–92
    application to electrode mechanisms, 92–102
    bibliography on, 110–114
    construction of, 283–286
    ECE reactions, treatment of, 253–255
    equations of (Levich equation), 83
    Gregory and Riddiford correction, 89
    order of reaction from RDE, 94
    performance tests for, 88
    precursor reactions, treatment of, 94–96
    rotation rates, 88, 106
    turbulence at, 103
    working curves for mechanism studies, 254
Rotated ring disk electrode, 96–98
    applications to mechanism studies, 99–102
    collection efficiency of, 99–102
    construction of, 96, 285

S

Sand equation, 169
Single sweep voltammetry, 124–140
Specific limiting current (SLC), 39–40

T

Transition times (in chronopotentiometry), 168, 172
    measurement of, 183–184
Tubular electrodes, 78–79
Tungsten electrodes, 28

U

Unusual electrode systems, 28

V

Vibrating wire electrodes, 107

Voltammetry, conventions of, 11
  cyclic, 143–159
  instrumentation for, 291–297
  single sweep, 124–140

W

Wire electrodes, construction of, 270–276
  diffusion currents at, 56
  rotated wires, 104–107
  vibrated wires, 107
Working curves, for potentiostatic ECE, 250
  for RDE applications, 254
Working electrodes, 11–12, 270–288